Status and Management
of Tropical Coastal Fisheries in Asia

Edited by

Geronimo Silvestre
and
Daniel Pauly

1997

ASIAN DEVELOPMENT BANK
6 ADB Avenue, Mandaluyong City, Philippines

INTERNATIONAL CENTER FOR LIVING AQUATIC RESOURCES MANAGEMENT
MCPO Box 2631, 0718 Makati City, Philippines

Status and Management of Tropical Coastal Fisheries in Asia

Edited by

Geronimo Silvestre
and
Daniel Pauly

1997

Printed in Manila, Philippines.

Published by the Asian Development Bank (ADB), 6 ADB Avenue, Mandaluyong City,
and the International Center for Living Aquatic Resources Management, MCPO Box 2631,
0718 Makati City, Philippines.

Silvestre, G.T. and D. Pauly, Editors. 1997. Status and management
    of tropical coastal fisheries in Asia. ICLARM Conf. Proc. 53, 208 p.

Technical editorial support: Len R. Garces and Rowena Andrea V. Santos
Copyediting: Danny B. Abacahin, Rita Kapadia and Marie Sol M. Sadorra
Proofreading: Kristine F. Santos, Rita Kapadia and Len R. Garces
Layout: Albert B. Contemprate
Cover concept: Geronimo Silvestre, Len R. Garces and Daniel Pauly
Cover design: Alan Siegfrid C. Esquillon
Artwork: Roberto N. Cada, Albert B. Contemprate and Alan Siegfrid C. Esquillon
Indexing: Kristine F. Santos, Francisco Torres, Jr. and Ma. Graciela R. Balleras

Cover: Design emphasizes the duality of coastal fisheries in South and Southeast Asia.
The need for management planning through participatory consultation among stakeholders
is also emphasized in resolving the conflict between small- and large-scale fisheries and
attaining sustainable exploitation of fisheries resources.

ISSN 0115-4435
ISBN 971-8709-02-9

ICLARM Contribution No. 1398

Contents

Toward a Generic Trawl Survey Database
Management System

Ecological Community Structure Analysis:
Applications in Fisheries Management

Appendices

Foreword

The fisheries sector employs more than 10 million families in the Asia Pacific region, and earns about $8 billion in foreign exchange for the Bank's developing member countries (DMCs) each year. At a per capita consumption of about 12 kg per annum, fish supplies 30 percent of the total animal protein consumption in the region. The region's fisheries contributed 38 million t or 45 percent of the world catch in 1993. Keeping in view the contribution of the fisheries sector to the economy and the well-being of the people of the region, the Bank has always attached a high priority to the fisheries sector. Commencing with its first fisheries loan in 1969, the Bank has made loans of over US$1 055 milion to a total of 51 fisheries projects in 17 DMCs. The Bank has also provided technical assistance grants amounting to about US$28 million for 82 country-specific project preparation and advisory services and 15 regional programs.

The fisheries sector has undergone many changes in emphasis and direction over time. The Bank has attempted to respond to these changes by reorienting its own policies. One of the fundamental developments in the sector during the last few decades was that marine capture fisheries came under heavy pressure, with marine fisheries resources, particularly in the coastal areas, being excessively exploited in many coastal nations in Asia. There is an urgent need for sound fisheries resources management measures to arrest the continued deterioration of coastal fisheries resources. To adapt its operations to these changes, the Bank has redirected its lending focus from marine capture fisheries development in 1970s to aquaculture development and coastal fisheries resources management and related environment conservation in the 1980s and 1990s.

Recognizing the problems of degradation of coastal fisheries resources and the resultant adverse impact on fishing communities in the coastal areas, most DMC governments have made efforts to improve coastal fisheries resources management and initiated various programs to improve the social and economic conditions of coastal communities. However, effective fisheries resources management strategies have not been put into place in many countries partly due to a lack of reliable fisheries resource information and databases essential for this exercise.

In support of national government initiatives in improving the management of coastal fish stocks, the Bank approved a technical assistance (TA) for the Regional Study and Workshop on Sustainable Exploitation of Coastal Fish Stocks in Asia. As part of the TA, a Workshop was organized which brought together researchers and resource managers to review the status of coastal fish stocks and existing resource databases, and propose appropriate follow-up actions and activities. The Workshop also provided an opportunity for participants to share and exchange experiences in fisheries resource assessment and management relevant to the development of appropriate strategies for coastal fisheries management.

The Workshop has identified constraints and issues related to this sector and recommended appropriate management directions and follow-up actions. The prototype database presented in the Workshop has been well received by the participants and should be further modified, improved and adapted by the participating countries for their coastal fisheries management programs. The Workshop is the first step towards assessing the prevailing situation and elaborating directions for the rehabilitation and improved management of coastal fisheries in the region.

The Workshop participants also reached a consensus on the need for regional efforts in addressing the problems at hand. ICLARM should play a lead role in coordinating the regional efforts and assisting the participating governments.

The success of the Workshop should ultimately be measured by the implementation of its recommendations. Therefore, we must find the means to realize the follow-up activities. The Bank will continue its support to regional efforts for coastal fisheries management and sustainable utilization of coastal fish stocks in Asia. We hope that the country participants will convey the workshop recommendations to their Governments to seek the necessary support for carrying out follow-up actions and activities.

Muhammad A. Mannan
Manager
Forestry and Natural
Resources Division
Asian Development Bank

Foreword

This Workshop, made possible by a grant from the Asian Development Bank (ADB) to the International Center for Living Aquatic Resources Management (ICLARM), brought together resource researchers and managers to examine the management of coastal fish stocks and existing resource databases in South and Southeast Asia.

The vision of ICLARM is to enhance the well-being of present and future generations of poor people in the developing world through improved production, management and conservation of living aquatic resources. The coastal fish stocks of Asia are among the most important in the world for the protein they provide, the livelihoods and communities they support and marine environments which sustain them.

This Workshop on 'Sustainable Exploitation of Tropical Coastal Fish Stocks in Asia' was one of the activities undertaken by the Center and its partners in Asia.

The results of the Workshop, documented in this volume, highlight the severe problems related to the management of coastal fish stocks throughout the region. All countries recognize that it is time to remedy these problems and that solutions require multiple action. During the Workshop it became clear that effective and broad-based action plans must be developed and adopted by the participating countries to deal with the problems identified. National capabilities in all aspects of coastal resources management should be strengthened to provide a base for formulating national policies, strategies and programs of action for the management and sustainable utilization of coastal fish stocks. Information on the state of the resources and the environment are at the core of management plannning

Following the Workshop, we anticipate the development of fish resource databases as a solid foundation for appropriate strategies and action plans at the regional and national levels. We will also look towards the strengthening of personnel and technical capabilities for coastal fisheries assessment and management in the region.

Meryl J. Williams
Director General
ICLARM

Preface

Coastal fisheries are an important component of the agricultural sector and rural economy of developing countries of South and Southeast Asia. In 1994, coastal fisheries employed roughly 8 million fishers, and contributed the bulk of the 13.3 million t of marine landings and US$9 billion of fishery exports from these countries. Fish comprises an important part of the diet, contributing as much as 70% of the animal protein intake in many coastal areas. Many of the coastal fisheries, however, are impacted by a number of issues, such as overfishing and habitat degradation, with adverse consequences on fishing incomes, fish supply to consumers and living conditions in coastal communities. Moreover, competition among small and large-scale fishers in shallow fishing grounds has intensified resource depletion and increased social conflicts. Several major fishing grounds in the region, such as Manila Bay, Gulf of Thailand, Java Sea, are heavily fished, with the abundance of fish stocks as low as 10-15% of their unexploited levels.

In view of these problems and their impact on coastal communities, many developing countries in the region are placing a greater emphasis on improving the management of their coastal fisheries. A number of policies and action programs for dealing with these issues have been established. The question is whether their magnitude and comprehensiveness are sufficient to reverse the numerous interrelated problems, and whether they have considered the available resources and related information. Many surveys of coastal fishery resources were undertaken in these areas from the 1960s to the 1980s to help guide fisheries development and management efforts. Much of the information in these surveys remains underutilized, having been only superficially analyzed. No account was taken of the diversity of fish stocks and its effects on the yields which the resources can sustain. These surveys can potentially provide an independent statistical baseline of resource conditions, as well as directions and strategies for improved management.

The Workshop on 'Sustainable Exploitation of Tropical Coastal Fish Stocks in Asia' was organized by ICLARM, with funding from the Asian Development Bank (ADB) to help assess the prevailing situation and identify follow-up actions for improved management of coastal fisheries in South and Southeast Asia. The Workshop was conducted in Manila, Philippines, on 2-5 July 1996. It aimed to:

1. examine the status of coastal fish stocks and fisheries in developing South and Southeast Asian countries and directions for their improved management;
2. review previous fish stock surveys and their potential for retrospective analyses in support of management efforts;
3. elaborate relevant database guidelines in support of such retrospective analyses; and
4. explore follow-up action and support activities for regional collaboration in coastal fisheries management.

These proceedings document the papers presented at the Workshop and include a synopsis of the Workshop's main results and recommendations.

The Workshop program (see Appendix I) was divided into (four) sessions designed to include inputs to five interrelated working group discussions: (1) Management Issues, (2) Management Strategies/Actions, (3) Scope for Regional Collaborative Efforts, (4) Trawl Database Guidelines/Prototype, and (5) Indicative Planning for Follow-up Regional Collaboration. A total of 27 resource persons (see Appendix II) actively participated in the workshop sessions and working group discussions. The overview paper by the editors and the seven country papers (Khan et al. - Bangladesh; Priyono and Sumiono - Indonesia; Abu Talib and Alias - Malaysia; Barut et al. - Philippines; Maldeniya - Sri Lanka; Eiamsa-ard and Amornchairojkul - Thailand; and Thuoc and Long - Vietnam) give situational updates and provided key inputs for the first two working group discussions on management issues and strategies/actions. The contributions by Martosubroto and by Mansor and Mohd-Taupek provided important perspectives of issues and strategies/actions by outlining the activities of two important international fisheries organizations, FAO and SEAFDEC, respectively, involved in this region. All the papers outlined ongoing efforts and activities and, hence, provided a basis for discussions on regional collaboration efforts in support of coastal fisheries management.

The paper by Gayanilo et al. outlines a trawl survey database prototype – a key element required to assemble, standardize and analyze the large number of extant but underutilized surveys. The contribution by McManus on community structure analyses illustrates how potential analyses can be conducted on existing trawl surveys, once these are entered in a database, to provide insights and options for improved management. These two papers, together with the information on trawl surveys summarized in Appendix III, were the main inputs for the working group discussion on trawl database guidelines/prototype. The paper by Gayanilo et al. incorporates the revisions and suggestions resulting from the deliberations of the working group. It also includes the insights gained by Mr. Tøre Strømme, the leader of the Norwegian project whose vessel, the *Dr. Fridtjoft Nansen*, has surveyed the waters of the majority of countries included in this Workshop and many more in Africa and south and central America.

A synopsis of the main results and recommendations of the Workshop is given by Silvestre and Pauly. It includes the outcome of the working group discussion on indicative planning for follow-up regional collaboration. The Workshop points to the need for action on a broad range of issues at the local, national and international levels. Consensus was achieved on a number of interventions which can assist these countries in the management of coastal fisheries. Of these, catalytic elements (consisting of training, research and planning activities) have been packaged into a proposal for expanded regional technical assistance for possible funding by ADB.

The Workshop and the proceedings presented here would not have been possible without the assistance of various agencies and colleagues. We particularly wish to acknowledge the funding support provided by ADB. The assistance of BFAR and PCAMRD during the conduct of the Workshop is also gratefully acknowledged. Dr. M.J. Williams, Dr. M.V. Gupta, Dr. M. Mannan, Mr. W.D. Zhou and Dr. T. Gloerfelt-Tarp gave valuable advice and assistance during the preparation and conduct of the Workshop and the completion of the proceedings. Mr. P.V. Manese, Ms. R. Funk and Ms. L. Arenas provided administrative support. Mr. L. Garces, Ms. R.A.V. Santos, Mr. R. Pabiling, Mr. D. Bonga and Ms. K.F. Santos provided technical support during the Workshop and the preparation of the proceedings. Thanks are also due, for editorial assistance, to Ms. M.S.M. Sadorra, Ms. R. Kapadia, Mr. D. Abacahin, Mr. D. Satumba, Mr. A.B. Contemprate, Mr. R.N. Cada and Mr. A.S. Esquillon.

The Editors

Synopsis and Recommendations of the ADB/ICLARM Workshop on Tropical Coastal Fish Stocks in Asia[1]

G. SILVESTRE
International Center for Living Aquatic Resources Management
MCPO Box 2631, Makati City 0718, Philippines

D. PAULY
International Center for Living Aquatic Resources Management
MCPO Box 2631, Makati City 0718, Philippines
and
Fisheries Centre, University of British Columbia
2204 Main Mall, Vancouver, BC, Canada V6T 1Z4

SILVESTRE, G. and D. PAULY. 1997. Synopsis and recommendations of the ADB/ICLARM Workshop on tropical coastal fish stocks in Asia, p. 1-7. *In* G. Silvestre and D. Pauly (eds.) Status and management of tropical coastal fisheries in Asia. ICLARM Conf. Proc. 53, 208 p.

Abstract

A summary is presented of the results of a Workshop held in Manila on 2-5 July 1996 devoted to the 'Sustainable Exploitation of Tropical Coastal Fish Stocks in Asia'. Participants came from Bangladesh, Indonesia, Malaysia, the Philippines, Sri Lanka, Thailand and Vietnam, and from FAO, SEAFDEC, ADB and ICLARM. The regional overview and country reports document the prevailing coastal fisheries situation and the host of issues impacting the sustainability of, and improvement of benefits derived from, coastal fisheries in South and Southeast Asia. Management directions and interventions requiring increased attention include: limited entry and effort rationalization schemes; gear, area and temporal restrictions on fishing; improvement of marketing and post-harvest facilities; enhancement of stakeholders awareness and participation; reduction of coastal environmental impacts; institutional upgrading and strengthening; and enhancement of fisheries research and information.

Consensus was achieved on the usefulness of compiling and analyzing past trawl surveys to establish benchmarks for stock rehabilitation, supplement existing statistical baselines and improve management directions and strategies. A prototype database and analytic tool for this purpose (i.e., TrawlBase) was presented and evaluated during the Workshop using data from available surveys conducted in South and Southeast Asia. The potential scope for regional collaborative efforts in the area of coastal fisheries research and management was also evaluated given the background of (planned and ongoing) activities at the local, national and regional levels. Consensus was achieved on selected elements which can act as catalysts in assisting these countries in research and management of their coastal fisheries. An indicative framework for a regional project incorporating these elements is briefly outlined.

Introduction

The 'Workshop on Sustainable Exploitation of Tropical Coastal Fish Stocks in Asia' was conducted on 2-5 July 1996 at the Hyatt Regency Hotel, Manila, Philippines. The Workshop was organized by ICLARM with funding support from ADB. It aimed, as an initial step, to bring together fishery scientists and managers to evaluate the fisheries management situation and the relevant resource databases in the South and South-east Asian region, and to propose directions for follow-up actions toward improved management of coastal fisheries. The specific objectives of the workshop were to:

- examine the status of coastal fish stocks and fisheries in the region and directions for their improved management;
- review previous fish stock surveys and their

---

[1] ICLARM Contribution No. 1386.

potential for retrospective analyses in support of management efforts;

- elaborate database guidelines for such retrospective analyses; and
- explore follow-up action and support activities for regional collaboration in coastal fisheries management.

The Workshop program and list of participants is given in Appendix I and II, respectively. A total of 27 participants attended the Workshop. Of these, 12 were representatives from seven of the ADB's developing member countries, namely: Bangladesh, Indonesia, Malaysia, the Philippines, Sri Lanka, Thailand and Vietnam. Three of the participants were from international fisheries organizations with major involvement in the region (FAO and SEAFDEC). The other participants were staff members of ADB and ICLARM, and observers from various fisheries institutions.

To realize the objectives of the Workshop, the program was divided into four sessions: (1) status of coastal fisheries in tropical Asia; (2) coastal fisheries management issues, strategies and actions; (3) retrospective analysis of trawl surveys; and (4) workshop summary and recommendations. The papers presented and the discussions during the first three sessions were intended to provide inputs to five interrelated working group discussions on: (1) management issues, (2) management strategies and actions, (3) scope for regional collaborative efforts, (4) trawl database guidelines/prototype, and (5) indicative planning for follow-up regional collaboration.

Management Issues and Interventions

The overview paper by Silvestre and Pauly (this vol.) and the seven country papers in this volume (i.e., Khan et al. - Bangladesh; Priyono and Sumiono - Indonesia; Abu Talib and Alias - Malaysia; Barut et al. - Philippines; Maldeniya - Sri Lanka; Eiamsa-ard and Amornchairojkul - Thailand; and Thuoc and Long - Vietnam) provided regional and country-specific updates of the coastal fisheries situation. These were supplemented by perspectives from representatives of two important international fisheries organizations (i.e., Martosubroto - FAO; Mansor and Mohd-Taupek - SEAFDEC) involved in the region. These contributions provided the key inputs to the first two working group discussions on management issues, strategies and actions.

The working group discussions on management issues covered a multiplicity of issues impacting coastal fisheries in South and Southeast Asia (Table 1). The consensus was that the coastal fish stocks and fisheries in the region sorely need improved management to sustain and/or improve the benefits derived from them. The main issues requiring attention are:

- overfishing, i.e., excessive fishing effort;
- inappropriate exploitation patterns;
- post-harvest losses;
- large vs. small-scale fisheries conflicts;
- habitat degradation;
- information and research inadequacies; and
- institutional weaknesses and constraints.

The working group discussions on management strategies and actions covered a number of interventions to resolve or mitigate these issues. A range of management directions and interventions were discussed (Table 2). It was acknowledged that some policies and programs of action in line with the interventions listed in Table 2 do exist in the developing countries of South and Southeast Asia. The concern, however, is about the adequacy of the scale and scope of these to reverse or mitigate the problems which have been identified.

There was consensus on the need for increased attention in the following areas:

i. limited entry and effort rationalization schemes;
ii. gear, area and temporal restrictions;
iii. improved marketing and post-harvest facilities;
iv. enhancement of stakeholders awareness and participation;
v. reduction of coastal environmental impacts;
vi. institutional upgrading and strengthening; and
vii. enhancement of fisheries information and research.

Several of the contributions to this volume discussed these issues. The developing member countries of the ADB and the relevant international organizations were encouraged to support and strengthen their programs in these areas.

The host of management interventions requiring increased attention entails concerted action at various levels of the institutional hierarchy. Consensus was achieved that enhancement of research and information inputs into the policy and decision-making process, however, deserves higher priority to facilitate progress in the other intervention areas. For instance, the failure to properly assess and/or communicate the gravity of the resource situation in many areas has held up the adoption of effective entry and effort rationalization schemes, and relevant gear, area and temporal restrictions on fishing. Coastal environmental impacts which aggravate the fisheries resource situation are subsequently viewed with less urgency -- given limited appreciation of resource conditions. Lack of information on the distribution (both spatial and temporal) of resources and landings, as well as

Table 1. Summary of major issues affecting coastal fisheries in selected countries in tropical Asia.

Key issues/constraints	Bangladesh	Indonesia	Malaysia	Philippines	Sri Lanka	Thailand	Vietnam	Others
Overfishing	✓	✓	✓	✓	✓	✓		✓
Inappropriate exploitation patterns		✓	✓		✓	✓	✓	
Use of destructive fishing practices	✓	✓	✓	✓	✓		✓	✓
Small- and large-scale fisheries conflicts	✓		✓	✓	✓		✓	✓
Post-harvest losses	✓			✓	✓			✓
Siltation/sedimentation	✓	✓	✓	✓	✓		✓	✓
Coastal soil erosion	✓	✓	✓		✓			✓
Flooding	✓	✓		✓		✓	✓	
Habitat degradation/destruction	✓	✓	✓	✓	✓	✓	✓	✓
Mangrove conversion	✓	✓	✓	✓		✓		✓
Reduced biodiversity	✓	✓	✓	✓	✓	✓	✓	✓
Industrial pollution	✓	✓		✓	✓	✓		✓
Toxic mine tailings	✓	✓		✓	✓			✓
Agrochemical loading	✓	✓	✓	✓	✓	✓	✓	✓
Domestic/sewage pollution	✓	✓	✓	✓	✓		✓	✓
Oil spills	✓	✓	✓	✓	✓	✓	✓	✓
Inadequate information/research support for management				✓				✓
Inadequate policy and legal framework								✓
Limited personnel and technical capabilities								✓
Limited resources/funding							✓	✓
Lack of institutional coordination/collaboration				✓	✓			✓
Insufficient/ineffective law enforcement								✓

4

Table 2. List of key management interventions.

Limited entry and effort reduction
- zoning or establishment of marine protected areas
- restructuring of relevant policy and regulatory frameworks
- redirection of systems of subsidies/support
- direct exit interventions
- enhancement of alternative livelihood and occupational mobility

Gear, area and temporal restrictions
- technological control/limitations (i.e., trawl ban)
- gear selectivity
- spatial restrictions (i.e., marine sanctuaries)
- temporal restrictions (i.e., seasonal closures)

Improvement of marketing and post-harvest facilities
- strategic location of salt, ice and cold storage facilities
- enhanced private sector participation
- improved processing and handling techniques
- strategic rural road infrastructure
- extension, training and credit support

Enhancement of awareness and participation of stakeholders
- stakeholder participation in the decision-making process
- enhancement of fishers organizations
- education/awareness programs
- devolution/decentralization of management authority

Reduction of environmental impacts
- adoption of ICZM and integrated coastal fisheries management approaches
- adoption of multiple use zonation schemes
- coastal habitat restoration/rehabilitation
- ban on destructive fishing methods
- establishment of EIA systems and precautionary approach
- enforcement of regulations and penalties/incentive systems

Institutional strengthening/upgrading
- technical, personnel and facilities upgrading
- improvement of financial capability and mandates of organizations
- enhancement of organizational coordination/collaboration
- improvement of transparency, accountability and participation in decision-making process

Enhancement of research and information
- appropriate size and siting of fish sanctuaries or marine protected areas
- resource enhancement and habitat rehabilitation techniques
- selective fishing
- appropriate fisheries management reference points
- ecosystem modeling
- policy and institutional studies

---

the extent of post harvest losses, has impaired progress in improving marketing and post-harvest facilities. These in turn have affected the urgency of making the recquisite institutional reforms and fostering stakeholders participation in the management process. Obviously, problems must first be assessed and perceived as one before solutions to them may be prescribed. Information and research are indispensable in this process of 'problem-solution' identification and the empirical debate that fisheries management entails.

Substantive discussions were generated relative to the limited resources and support given to research and information activities in many South and Southeast Asian countries. Moreover, it was noted that full utilization of available information and research results for improved management requires emphasis. Particular attention in this regard was drawn to the numerous trawl surveys conducted in South and Southeast Asia from the 1960's to the 1980's. Much of the data generated during these surveys have been analyzed only superficially and much of their information content remain underutilized. While enhanced support and resources for information and research remain largely valid, full utilization of information for which resources have already been expended deserve

Table 3. Main activity areas for expanded regional collaborative efforts in coastal fisheries research and management.

- Research/information activities
  - a. Trawl survey database development
    - i. national workshops
    - ii. database programming/improvements
    - iii. data inputting

  - b. Trawl survey and related data analyses
    - i. population/stock analyses
    - ii. community analyses
    - iii. biosocioeconomic analyses

- Management policy/planning
  - a. National strategies and action plans
    - i. strategic coastal fisheries management reviews
    - ii. strategic policy and management planning
    - iii. action planning and indicative investment programs

  - b. Regional strategies for resources rehabilitation
    - i. resource baselines comparative analyses
    - ii. issues/interventions reviews
    - iii. problem/opportunity structures reviews
    - iv. action planning and indicative investment programs

- Training and networking activities
  - a. Training/networking
    - i. stock assessments (particularly methods standardization)
    - ii. community analyses
    - iii. biosocioeconomic analyses
    - iv. fisheries policy analyses and planning

  - b. Workshops
    - i. regional (comparisons, lessons, synthesis)
    - ii. national (database development/planning)

Trawl Database Prototype

The contribution by Gayanilo et al. (this vol.) and McManus (this vol.) provided key inputs for the working group discussion on database guidelines/prototype. The former presented a prototype for a trawl survey database (TrawlBase) - an important tool to assemble and analyze the numerous surveys conducted in the region which remain largely underutilized. The latter illustrated the utility of ecological community structure analysis (which can be conducted using extant surveys) in developing improved management strategies/actions. Trawl surveys conducted in the region (Appendix III) and representative data sets from these surveys were examined for both resource trends and improvement of the database prototype. The resource trends that we identified were incorporated into the country-specific contributions to this volume and the suggested database revisions were incorporated into the TrawlBase prototype of Gayanilo et al. (this vol.).

The consensus was that the retrospective analysis of these surveys and their collation into TrawlBase (in terms of stock size, community structure, bio-socioeconomic and management implications) hold great potential for improvement in the management of coastal fisheries. Moreover, TrawlBase can serve as a resource database, independent of existing fisheries statistical systems, to provide a baseline for sustainability and help improve information inputs into the management process. Ensuring the regional coverage of TrawlBase will strongly supplement national and regional (e.g., FAO, SEAFDEC) initiatives to strengthen the fisheries statistical systems of developing countries in South and Southeast Asia.

Scope for Regional Collaboration

Substantive inputs were elicited during the working group discussions on the potential scope for regional collaborative efforts. The primary emphasis was given to initiatives which can act as catalysts in the efforts of the developing countries of South and Southeast Asia to address the seven management issues listed above. These initiatives were considered in the light of efforts and activities (both planned and ongoing) at the local, national and regional levels, so as to avoid duplication and assure complementarity. A preliminary list of activities for regional collaboration was assembled and refined through several stages of discussion. Following the working group discussion on database guidelines/prototype, the list was finalized and the results are summarized in Table 3. The potential activities for regional collaboration include elements for research and information support, management policy and planning, and training and networking.

increased attention. Consensus was achieved that assembling and analyzing extant trawl surveys in the region can help elaborate, among others, the following:

- resource condition, potential and exploitation level through time essential for limited entry and effort rationalization schemes;
- spatio-temporal trends for operationalizing gear, area and temporal restrictions on fishing;
- resource distribution for strategic location of marketing and post-harvest facilities;
- combined effect of fishing and environmental impacts on resource condition;
- historical (i.e., sustainability) baselines as benchmark for resource rehabilitation; and
- improved agenda for resource research and monitoring.

These elements, in turn, can help foster attention on the requisite institutional intervention areas.

Table 4. Indicative framework for regional collaborative project in the area of coastal fisheries research and management.

Design summary	Target
1. Sector/area goal	
• Help improve management and sustainable utilization of coastal fishery resources in developing countries of South and Southeast Asia.	• Enhance personnel capabilities and information at the national level and develop action plans for improved coastal fisheries management.
2. Objectives	
• Enhance resource databases and management information consistent with resource management needs of developing South and Southeast Asian countries.	• Develop a trawl survey database (TrawlBase) and elaborate national and (South and Southeast Asia) regional resource situation.
• Assist selected countries in South and Southeast Asia in developing appropriate strategies and action plans for improved management of coastal fisheries.	• Elaborate national/regional fisheries situation and development of strategies and action plans (including indicative investment program) for improved coastal fisheries management.
• Strengthen national capabilities in coastal fisheries assessment and management.	• Training of national staff in TrawlBase use and relevant analytic tools, and provision of requisite hardware/software support and technical advice.
3. Components/activities	
3.1. Research, information and training	

• Trawl database development
  - TrawlBase design, programming, revisions and distribution
  - national data inputting
  - regional database consolidation
	• Development of TrawlBase in support of resource assessment and management efforts.

• Training
  - training of national staff in the use/elaboration of TrawlBase
  - national staff training in stock assessment, ecological community analyses, biosocioeconomic analyses and CRM/fisheries policy and planning
	• Upgrading of technical/personnel capabilities in resource assessment and management.

• Data analyses/assessments
  - analysis of resource situation using TrawlBase and related information
  - stock, community structure, biosocioeconomic and policy analysis, and implications for fisheries management
  - national/regional trends, issues and interventions
	• Detail and update assessment of coastal resource and fisheries situation in support of management planning.

3.2. Management policy and planning	

• National and regional workshops
  - national strategic action planning and consultative workshops
  - regional strategic action planning and lessons/synthesis workshops
	• Facilitate consultation and adoption of management strategies and action plans (including indicative investment program) at the national and regional levels.

• National strategies and action planning
  - strategic national fisheries management reviews incorporating research results
  - strategies and problem-opportunity structures
  - action planning and indicative investment programming
	• Draft national management strategies and action plans.

• Regional strategies and action planning
  - strategic regional fisheries management reviews incorporating research results
  - regional strategies and problem-opportunity structures
  - regional action planning and indicative investment programming
	• Draft regional management strategies and action plans.

4. Inputs/indicative cost	
• US$3.3 million	• Contributions by national governments and international donors/organizations.

The working group discussion on indicative planning for follow-up regional collaboration elicited substantive inputs on the nature, scope and scale of a future regional project for coastal fisheries management. Building upon the results of the four other working group discussions, the elements of a project framework emerged after several iterations. The results of the final iteration are summarized in Table 4. They outline the goal, objectives, components/activities and corresponding targets (as well as funding input) for a collaborative regional project.

The project requires collaborative work among multidisciplinary teams of scientists from select developing member countries of ADB and from ICLARM, to be performed in close coordination with managers at the national level and the staff of concerned international organizations. The scope of the proposed regional project includes the following main activities:

- training in the use of TrawlBase;
- development of resource databases principally on extant surveys and their consolidation into a single TrawlBase with regional coverage;
- regional training courses (involving national scientists and/or managers) in the fields of stock assessment, assemblage/community, and biosocioeconomic analyses, and CRM/fisheries policy analysis and planning;
- review and analysis (including stock, community, and biosocioeconomic) of the resource base and related information in terms of their management implications;
- national workshops for data consolidation/generation and consultative planning;
- strategic review of the fisheries management situation and programs at the national and regional levels (including resource/management trends and opportunities);
- regional workshops to consolidate results of data analyses and to elaborate on regional trends, strategies and action programs; and

- development of strategies, action plans and indicative investment programs at the national and regional levels, based on the reviews and assessments conducted during the course of the project.

The consensus was that the proposed regional project will be invaluable in assisting the developing countries in South and Southeast Asia in formulating resource databases and action plans for the improved management of their coastal fisheries.

ICLARM was requested by the Workshop participants to build upon these elements and package a proposal for funding by ADB and other interested donors[2].

Conclusion

The results outlined above were discussed and revised during the last plenary session of the Workshop. The participants acknowledged that improved coastal fisheries management in South and Southeast Asia rests largely on the support that will be made available for the purpose in the context of other equally pressing social and developmental needs in these countries. The general consensus was that focusing available resources on the management intervention areas outlined above should pay substantive dividends for the region. Moreover, support for the regional collaborative efforts will assist these countries in strengthening institutional capabilities and developing policy interventions for the improved management of, and sustained benefits from, their coastal fisheries.

Acknowledgements

Thanks are due Mr. Len Garces and Ms. Bing Santos for assistance in compiling the information given in Table 1.

---

[2] A detailed project proposal to this effect has been prepared by ICLARM and submitted to ADB for possible funding.

Management of Tropical Coastal Fisheries in Asia:
An Overview of Key Challenges and Opportunities[1]

G. SILVESTRE
International Center for Living Aquatic Resources Management
MCPO Box 2631, Makati City 0718, Philippines

D. PAULY
International Center for Living Aquatic Resources Management
MCPO Box 2631, Makati City 0718, Philippines
and
Fisheries Centre, University of British Columbia
2204 Main Mall, Vancouver, BC, Canada V6T 1Z4

SILVESTRE, G. and D. PAULY. 1997. Management of tropical coastal fisheries in Asia: an overview of key challenges and opportunities, p. 8-25. *In* G. Silvestre and D. Pauly (eds.) Status and management of tropical coastal fisheries in Asia. ICLARM Conf. Proc. 53, 208 p.

Abstract

Coastal fisheries are an important component of the fisheries sector and rural economy of tropical developing countries in Asia — generating food, employment and foreign exchange. In 1994, marine landings of these countries were about 13.3 million t (roughly 16% of world marine landings), most originating from coastal areas. The coastal fishery resources consist dominantly of species with relatively high growth, natural mortality and turnover rates; and exhibit maximum abundance in shallow depths (less than 50 m). Fishers use a multiplicity of gears, with heavy concentration in nearshore areas where abundance, catch rates and shrimp availability are highest. The management of these coastal fisheries attempts to promote three main objectives: (1) productivity/efficiency, (2) distributional equity and (3) environmental integrity. Efficient institutional/administrative arrangements are sought to attain these objectives and to maintain a balance among them.

Coastal fisheries operate in a spectrum ranging from light fishing, essentially single sector (i.e., fisheries) situations to intense fishing and multisector use of the coastal area (and its adjacent terrestrial and marine zones). Issues impacting coastal fisheries multiply through this range, requiring increasingly comprehensive and integrated analytic frames and scope of action to sustain fisheries benefits. The key issues impacting coastal fisheries in the region include: (1) overfishing, (2) inappropriate exploitation patterns, (3) post harvest losses, (4) conflicts between large and small-scale fisheries, (5) habitat degradation, (6) inadequacy of management information and research and (7) institutional weaknesses and constraints. Appropriate management strategies and actions on a broad front are necessary, and success is largely premised on institutional capabilities and resources mobilization. Moreover, the ultimate mitigation of these factors rests on effectively addressing poverty and promoting overall economic development.

Introduction

Coastal fisheries are important components of the fisheries sector and rural economy of tropical developing countries in Asia. These fisheries provide food and employment to a significant portion of the population, as well as valuable foreign exchange to the economy (Hotta 1996). In 1994, marine landings of these developing countries were about 13.3 million t (i.e., 16% of world marine landings and 12% of world fisheries production). Roughly 8 million fishers were involved in marine fisheries and aggregate fishery exports were about $9 billion per year. Most of the marine landings originated from fishing operations in coastal shelves (between the shoreline and 200 m depth) especially on their shallower parts (from 0 to 50 m). However, these fisheries are adversely affected by a number of problems and constraints, with serious

---

[1] ICLARM Contribution No. 1379.

consequences for the income of fishers, the supply of fish to consumers and poverty in rural communities.

This paper attempts to provide an overview of the main issues confronting coastal fisheries in tropical developing Asian countries as well as the corresponding management directions to help resolve or mitigate them. Numerous works provide detailed reviews of the overall situation through time and represent a substantive background and source of materials for this synopsis. Among others, the work of Aoyama (1973), Shindo (1973), Marr (1976, 1981) and Pauly (1979) and the contributions in Tiews (1973), Pauly and Murphy (1982), and Pauly and Martosubroto (1996) elaborate the situation in the 1970s. For the 1980s, reviews include Soysa et al. (1982), Sivasubramaniam (1985), IPFC (1987 a and b), APO (1988), Pauly and Chua (1988), and Pauly (1989). More recently, FAO (1992, 1995a and b), Yanagawa and Wongsanga (1993) and Hotta (1996) provide detailed situational updates.

We have avoided the detailed conventional review approach for this synopsis. The works cited above and the country specific contributions to this volume provide sufficiently detailed treatments. We have concentrated instead on drawing from the available literature the commonalities in the main issues and opportunities occurring across the countries and logically structuring them into generic categories. Many of the problems have been building up for some time and now lead to inescapable conclusions. In many respects the substance of the required solutions remains the same, though the debate over implementation strategies to effectively resolve the problems continues vigorously.

We first provide, by way of background, some basic features of coastal fisheries in tropical developing countries in Asia. A synopsis of the main fisheries management objectives pursued in these countries is then presented in generic categories based on the multiplicity of detailed objectives sought by management. Consideration of the objectives is a logical necessity for evaluating the existing situation versus the desired state. The main management issues are presented, using selected site-specific assessments for illustration. The key management interventions currently being emphasized to address these issues are then briefly discussed. Lastly, the structure of the objectives, issues and interventions is summarized and trends affecting the feasibility of management success are briefly discussed.

Sectoral Background

The scope of this study includes fisheries in coastal areas, from the shoreline to 200 m depth, situated within the area bounded by 60°E longitude in the west, 135°E longitude in the east, 10°S latitude in the south, and 20°N latitude or the coast of mainland Asia in the north (Fig. 1). This geographical delineation includes the fisheries of 13 developing coastal states (excluding Singapore given its level of economic development and limited shelf area). Table 1 provides selected statistics pertaining to these countries, 5 in South Asia and 8 in Southeast Asia. They had a combined population of about 1.7 billion in 1996, the highest being in India and the lowest in the Maldives and Brunei Darussalam. Gross national product (GNP) per capita varied between US$215 per annum (Cambodia) and US$20 400 (Brunei Darussalam). It is generally low, with only 3 countries having a per capita GNP above $2 000 (Brunei Darussalam, Malaysia and Thailand). High population growth, low incomes and underdevelopment characterize many of these countries, though accelerated economic growth is improving these conditions, particularly in Southeast Asia (ADB 1996).

The marine jurisdictorial area of the countries covered here is extensive, spanning an aggregate of about 13 million km^2. This is roughly 1.5 times the extent of their combined land area, totaling 8.5 million km^2. The extent of the declared exclusive economic zones (EEZ) is highest for Indonesia, India and the Philippines and is lowest for Brunei Darussalam, Cambodia and Bangladesh (WRI 1995). Despite the large marine area, however, only 35% (4.6 million km^2) of the aggregate EEZ consists of shallow, productive continental shelves. The most extensive shelves are found off Indonesia, India, Malaysia and Vietnam. The highest shelf to EEZ ratios are found in Malaysia, Bangladesh and Indonesia where over 50% of the EEZ consists of shelves. Longhurst and Pauly (1987) provide a review of the biophysical characteristics and ecology of the tropical waters discussed here and point to the significance of coastal shelves to fisheries productivity. Moreover, mangroves, coral reefs and seagrasses line the coastal fringes of these shelves and enhance their productivity particularly in Southeast Asia where the peak in biodiversity of these habitats occur (McManus 1988; Fortes 1988, 1995). These coastal habitats are coming under increased stress from various human activities due to expanding populations and economies (Gomez et al. 1990; Sen Gupta et al. 1990; Chou 1994; Holmgren 1994; and Wilkinson et al. 1994).

Table 2 summarizes selected fisheries statistics of these countries for 1994. Annual fisheries production range from 6 000 t (Brunei Darussalam) to 4 540 000 t (India), with over half of the countries producing over 1 million t each. Overall fisheries production, including inland fishery and aquaculture, was about 20 million t, or a little over 18% of global fisheries production. Exports of fish and fishery products was about $8.8 billion, representing a significant source of foreign exchange for these economies. Over $1 billion

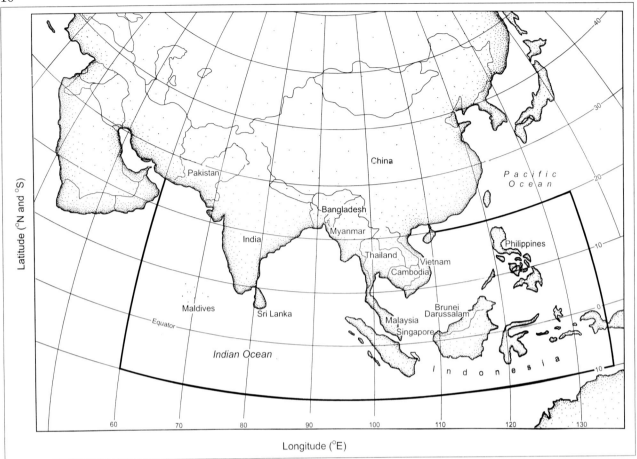

Fig. 1. Map illustrating geographical scope of this study and the location of the developing countries covered.

Table 1. Selected statistics for tropical developing countries in Asia. (Sources: ADB 1995, 1996; WRI 1995).

Country	Population (x 10⁶; 1996)	Per Capita GNP (US$; 1995)	Land Area (x 10³ km²)	Exclusive Economic Zone (EEZ) (x 10³ km²)	Continental Shelf (0-200 m depth) Area (x10³ km²)	As % of EEZ
Bangladesh	124.0	283	144.0	76.8	55	72
Brunei Darussalam	0.3	20 400	5.8	38.6	9	22
Cambodia	10.2	215	181.0	55.6	15	27
India	943.7	335	3 287.6	2 014.9	452	22
Indonesia	197.6	940	1 904.6	5 408.6	2 777	51
Malaysia	20.6	3 930	329.8	475.6	374	79
Maldives	0.3	900	0.3	959.1	-	-
Myanmar	47.7	890	676.6	509.5	230	45
Pakistan	133.2	465	796.1	318.5	59	18
Philippines	69.3	1 130	300.0	1 786.0	178	10
Sri Lanka	18.2	660	65.6	517.4	27	5
Thailand	61.4	2 680	513.1	257.6	86	33
Vietnam	76.3	250	331.7	722.1	328	45
Total	1 702.8	-	8 536.2	13 140.3	4 588	35

in fishery exports was registered by three countries, Thailand, Indonesia and India. Hotta (1996) estimates employment in fisheries (inland and marine fisheries, as well as aquaculture) to be about 11 million. Fish has traditionally been an important part of the diet of the population, particularly in Southeast Asia. Per capita fish consumption is highest in the Maldives, followed by the Philippines, Malaysia, Thailand and Brunei Darussalam, with annual consumption exceeding 25 kg. The lowest per capita consumption is in the three South Asian countries, namely: Pakistan, India and Bangladesh. These statistics indicate fisheries to be an important source of food, employment and foreign exchange.

Table 2. Selected 1994 fisheries statistics for tropical developing countries in Asia. (Sources: FAO 1994; Hotta 1996).

Country	Total fisheries production (x 10^3 t year^{-1})	Marine fisheries production (x 10^3 t year^{-1})	Fishery exports (US$ x 10^6 year^{-1})	Per capita fish consumption (kg·year^{-1})	Number of fishers (x 10^3)
Bangladesh	1 091	251	240	8.2	55
Brunei Darussalam	6	6	-	21.9	2
Cambodia	103	30	14	12.0	75
India	4 540	2 420	1 125	4.0	3 837
Indonesia	4 060	2 970	1 583	15.5	1 523
Malaysia	1 173	1 053	325	29.5	100
Maldives	104	104	37	126.0	22
Myanmar	824	599	103	15.5	696
Pakistan	552	418	153	2.2	308
Philippines	2 657	1 666	533	36.1	733
Sri Lanka	224	211	32	16.3	98
Thailand	3 432	2 798	4 190	25.3	61
Vietnam	1 155	817	452	13.4	266
Total	19 921	13 343	8 787	8.7	7 777

Table 2 also summarizes marine fisheries catches. Aggregate marine fisheries catches were about 13.3 million t (representing roughly 16% of world marine landings), which constitutes 67% of the total fisheries production for these countries. Hence, marine fisheries contributes the bulk of fisheries production. Marine fisheries production varied between 6 000 t (Brunei Darussalam) and about 3 million t (Indonesia). Five countries, Indonesia, Thailand, India, the Philippines and Malaysia, registered marine fisheries landings exceeding 1 million t, which is indicative of extensive coastal fisheries. It is estimated that about 7.8 million fishers are working in marine fisheries in the 13 countries covered here. The number of full-time and part-time fishers varies between 1 600 in Brunei Darussalam and about 3.8 million in India, and millions more are involved part time, including women and children (Pauly 1997). The bulk of marine fisheries yields and employment originates from fishing operations in shallow, coastal shelves, indicating that coastal fisheries account for a substantial part of the food and employment generated by the fishing sector and contributes significantly to foreign exchange earnings via export of shrimps, small pelagics and demersals.

The coastal fishery resources consist of highly diverse, multispecies complexes (Pauly 1979, Longhurst and Pauly 1987). These are dominantly species with relatively high growth, natural mortality and turn-over rates (Raja 1980; Ingles and Pauly 1984; Sivasubramaniam 1985; Chullasorn and Martosubroto 1986; Dwiponggo et al. 1986; and data in FishBase, Froese and Pauly 1996). A common feature of these resources is that they frequently exhibit maximum abundance in nearshore, shallow areas. Fig. 2 illustrates the depth distribution of resource abundance off Brunei Darussalam. Note that catch rates observed through time consistently show peak abundance in waters less than 50 m. Such a distribution of resource abundance is widespread across the South and Southeast Asian area. This is very different from the situation prevailing in the North Atlantic (which provided the early models for fisheries development and industrialization in South and Southeast Asia), where commercially viable fish abundance occurs down to depths of one kilometer and more.

Another feature of these coastal fishery resources is that many of the species exhibit increasing size with depth. Fig. 3 illustrates the size range of fishes in shallow (less than 15 m depth) versus deeper waters off Brunei Darussalam (Silvestre and Matdanan 1992). This highlights the significance of nearshore areas as nursery grounds and the serious implications of concentrated small and large-scale fishing in these areas. The abundance of very valuable shrimps only in nearshore waters and the favorable concentration of finfishes in areas less than 50 m depth has encouraged the concentration of fishing effort and incursion of trawlers in shallow grounds.

The abundance and diversity of coastal fishery resources has supported vibrant, small-scale fisheries for centuries in these countries (Butcher 1996). The period between the two world wars saw various attempts to 'modernize' these fisheries. These efforts were generally unsuccessful for a variety of technical and social reasons, not least of which includes the lack of dynamism of late colonial societies (Butcher 1994). The period immediately following the Second World War was different. Starting in the Philippines, a wave of technology and investments occurred which rapidly developed the demersal and, later, the pelagic fisheries in Southeast Asia (Pauly and Chua 1989). Mechanization of coastal fisheries also occurred in South Asia, although it appears to have been more diffuse.

12

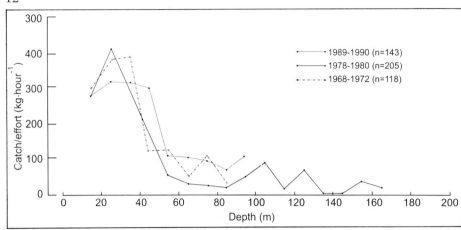

Fig. 2. Fish abundance off Brunei Darussalam (based on trawl surveys conducted around 1970, 1980 and 1990) typical of variation in resource abundance with depth observed in South and Southeast Asia. (Source: Silvestre and Matdanan 1992).

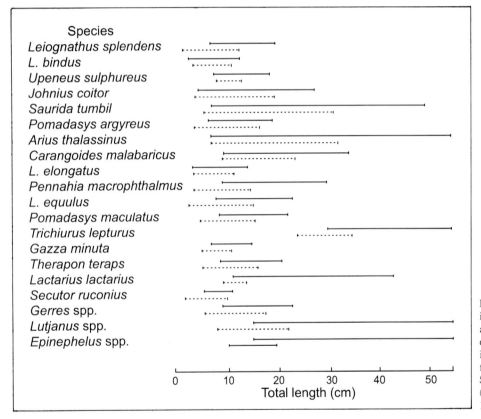

Fig. 3. Size range of fishes caught in areas less than (dashed line) and more than (solid line) 15 m depth off Brunei Darussalam illustrating trend of increasing fish sizes with depth observed in South and Southeast Asia. (Source: Silvestre and Matdanan 1992).

A multiplicity of gears are currently used to exploit the multispecies resources. These vary from relatively simple, inexpensive gears, like handlines and gillnets, using no water craft or dug-outs, to large trawls and purse seines using boats with powerful inboard engines. Sequential (and overlapping) deployment of these gears and small-/large-scale duality of coastal fisheries are common features. Fig. 4 illustrates these features in the case of Brunei Darussalam. The mix of gears used are concentrated in shallow grounds where abundance, catch rates and shrimp availability is highest. Many of the species are fished sequentially by different gears as they grow and move to deeper, offshore areas. Varied technological and biological interactions characterize the coastal fisheries exploitation regimes, making assessment and management rather difficult (FAO 1978; Pauly 1979; and Pauly and Murphy 1982).

The situation in Brunei Darussalam is unique in that the levels of exploitation are so low that major management problems have not occurred so far (Silvestre and Matdanan 1992). In the other countries, however, a heavy concentration of small and large-scale gears in many shallow coastal waters has led to overfishing, gear conflicts and dissipation of economic rent. Recent assessments have noted the increasing trend of overfishing of coastal fish stocks and habitat degradation (FAO 1995a; APFIC 1996). This has serious implications for fish supply as well as other benefits derived from coastal fisheries. In these

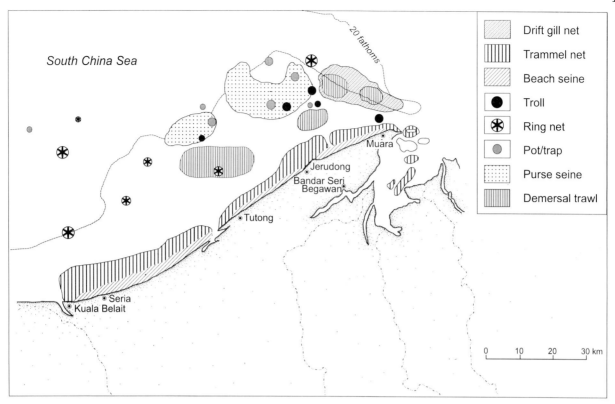

Fig. 4. Fishing area by gear type off Brunei Darussalam typical of those observed in other coastal areas in South and Southeast Asia. (Source: Khoo et al. 1987).

countries, food fish consumption is projected to grow from an aggregate of 14.2 million t in 1992 to 20 million t by 2010 (Hotta 1996).

Overview of Main Fisheries Management Objectives

Fisheries management may be viewed as a dynamic resource allocation process where the ecological, economic and institutional resources of a fisheries exploitation system are distributed with value to society (in the broad sense) as the overall goal. Some recent works covering the status of fisheries management science and related concepts are Anderson (1987), Caddy and Mahon (1995), Olver et al. (1995), Stephenson and Lane (1995), Williams (1996) and Caddy (in press). The fisheries management process includes the resolution of normative and empirical debates to determine the direction of resource allocation decisions. What constitutes value to society is ultimately determined in the political field, and highly influenced by existing needs (or perceptions of such needs), available knowledge and information (or access to them), and religious and cultural values or norms in society.

The coastal fisheries discussed here are set in a variety of natural and human conditions. There is, therefore, a wide diversity of specific objectives being pursued in their management. These objectives may be gleaned from national legislations, development plans and fisheries project documents. Some objectives are implicitly rather than explicitly stated, and many have been noted to be conflicting or incompatible when pursued simultaneously (Lawson 1978; Lilburn 1987). From the available literature we will summarize these diverse objectives into generic categories of objectives and management directions.

Fig. 5 gives a schematic representation of the conventional 'fishing system' framework in fisheries management. The arrows indicate the interactions between and among components of fishery resources and the fisheries relying on these. The framework emphasizes the essential dependence of fisheries on available resources for continued viability and a sustained flow of goods and benefits. It is a widely recognized principle of management in these countries that fisheries management systems must set up fishing regimes that appropriately match the productive capacity of the resource base.

Another feature of coastal fisheries management is the widening scope of 'fisheries management' itself.

14

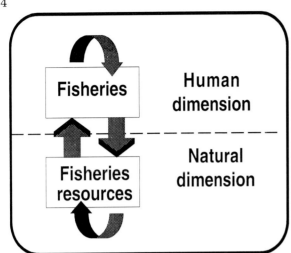

Fig. 5. Schematic representation of the conventional 'fishing system' framework in fisheries management.

Given the increasing multiplicity of issues impacting many coastal fisheries, fisheries management concerns (and objectives) have taken on a wider framework and scope of action in many areas. Fig. 6 illustrates an example of the scope of multidisciplinary work conducted in San Miguel Bay, Philippines (Silvestre 1996).

This encompasses: (1) fishery resources and the habitats (e.g., coral reefs, mangroves) and habitat characteristics (e.g., water quality) which sustains them; (2) other activities (e.g., forestry) which impact fisheries, the fishery resources and the natural environment; and (3) the socioeconomic development and policy framework within which fisheries and other economic activities operate. Similar to the situation in San Miguel Bay, coastal fisheries management in the South and Southeast Asian region increasingly entails the implementation of a wide range of measures within the confines of the traditional fisheries sector, as well as interventions requiring coordination with other sectoral agencies (e.g., forestry, agriculture) at various levels of the institutional hierarchy.

Within this frame of reference, Fig. 7 gives the typical hierarchy of objectives sought in the management of coastal fisheries in these countries. Consistent with sustainable coastal fisheries development as the overall goal, management entities attempt to: (1) optimize productivity/efficiency of the fisheries exploitation regime; (2) ensure that the benefits of production or improved productivity are distributed equitably; and (3) ensure that the productivity generated results in minimum damage to the resource

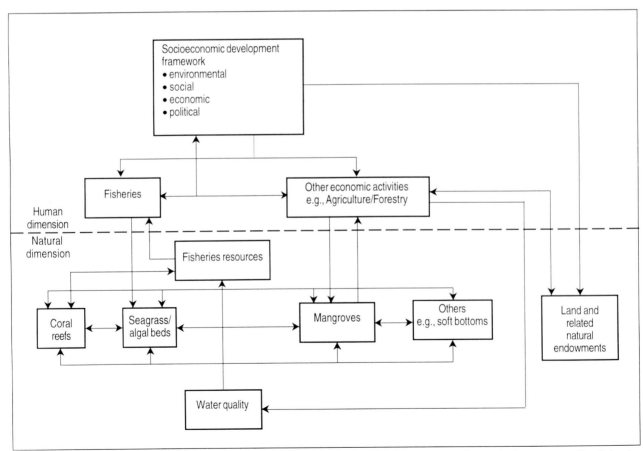

Fig. 6. Schematic representation of an expanded framework for fisheries management. Interrelations among the fisheries resources, the fisheries exploiting them, and other components of the human and natural dimensions are illustrated by arrows. (Source: Silvestre 1996).

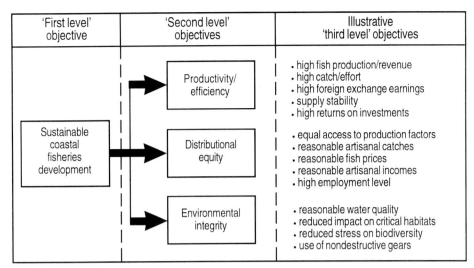

'First level' objective	'Second level' objectives	Illustrative 'third level' objectives
Sustainable coastal fisheries development	Productivity/ efficiency	• high fish production/revenue • high catch/effort • high foreign exchange earnings • supply stability • high returns on investments
	Distributional equity	• equal access to production factors • reasonable artisanal catches • reasonable fish prices • reasonable artisanal incomes • high employment level
	Environmental integrity	• reasonable water quality • reduced impact on critical habitats • reduced stress on biodiversity • use of nondestructive gears

Fig. 7. General goal and objectives in fisheries management.

base and the supporting natural environment. Environmental integrity also encompasses the intergenerational equity concerns embodied in the sustainable development concept of the Brundtland Commission report (WCED 1987). These three objectives are not always mutually compatible and the optimal balance among the three is highly dependent on situational realities and have been noted to vary temporally and spatially within individual countries. Apart from the three generic ('ends') objectives above, appropriate management systems/regimes are sought to effectively attain a balance among these objectives. Hence, institutional effectiveness is a fourth generic category of ('means') objective sought in coastal fisheries management in South and Southeast Asia.

Fig. 7 also gives typical 'third-level' objectives commonly encountered. These are translated into a number of policy instruments and management measures taking the form of regulatory instruments, market-based incentives, institutional measures, research agendas and/or government support investments. For example, the licensing scheme in many countries has productivity as the main rationale. The Indonesian trawl ban (Sardjono 1980) and the 15-km exclusive municipal fishing zones in the Philippines had equity as their primary consideration. The ban on the use of poisons and explosives in fishing in many countries has environmental integrity as the main driving force.

The Challenges: Overview of Key Management Issues

Coastal fisheries in the tropical developing countries of South and Southeast Asia operate in a spectrum ranging from light fishing, essentially single sector (i.e., fisheries) situations, to intense fishing and multisector use of the coastal area (and its adjacent terrestrial and marine zones). The number of negative factors impacting coastal fisheries multiply through this range, requiring increasingly comprehensive approaches and wider scope of action to sustain fisheries benefits. Many coastal fisheries are in (or moving into) the more industrialized, intensive stages of the fishing and coastal use spectrum, necessitating improved management efforts. We briefly outline below the main issues which require increased management attention.

Excessive Fishing Effort

High levels of fishing effort on coastal fish stocks, particularly in nearshore traditional fishing grounds, is a common management concern (Yanagawa and Wongsanga 1993; FAO 1995a and b; APFIC 1996; Hotta 1996). High fish demand (due to increasing population and incomes), burgeoning fishing populations combined with a lack of livelihood opportunities in rural areas, advances in fishing technology and accelerated industrial fisheries development has led to excessive fishing pressure and overfishing in many coastal areas. This has resulted in a leveling-off (if not decline) in landings; reduced catch rates, incomes and resource rents, and; intense competition and conflict among fishers. Fig. 8 illustrates the gravity of the issue of excessive fishing effort evident in some areas. In the case of the demersal and small pelagic fisheries in the Philippines (which are concentrated in very shallow waters), by the mid-1980s the level of effort exceeded what was required to harvest maximum economic yield by 150%-300% and maximum sustainable yield by 30%-130%. This implies dissipation of resource rents of about $450 million annually for the demersal and small pelagic fisheries combined. The developing countries of South and Southeast Asia can ill afford the economic losses resulting from overfishing. Although there are coastal areas which remain lightly fished

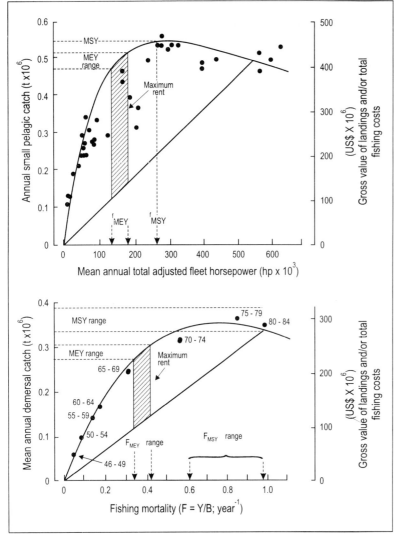

Fig. 8. Surplus production models of the Philippine small pelagic and demersal fisheries. (Sources: Silvestre and Pauly 1986; Dalzell et al. 1987).

(e.g., sparsely populated eastern Indonesia, parts of East Malaysia), the general consensus is that few coastal fish stocks can accommodate an expansion in fishing effort, and that many coastal fisheries in nearshore areas (particularly in the Gulf of Thailand, the Philippines, Bay of Bengal and western Indonesia) require significant reductions in fishing effort (Pauly and Chua 1988; FAO 1995a; APFIC 1996; Hotta 1996).

Inappropriate Exploitation Patterns

Inappropriate patterns of exploitation have led to suboptimal benefits from the exploitation of coastal fishery resources. This stems from the species and size selectivity of the mix of fishing gears used, i.e., their technological characteristics and spatio-temporal deployment in coastal fishing grounds. The selectivity of fishing gears and techniques for their assess-

ment are well documented in the literature (Hamley 1975; Pope et al. 1975; Sainsbury 1984; Silvestre et al. 1991). The theory of fishing illustrates the utility of influencing selectivity to maximize fish yields and related benefits (Beverton and Holt 1957; Ricker 1975; Gulland 1983). Armstrong et al. (1990) provides an update on the importance of selectivity to the conservation of fish stocks.

The concentration of fishing effort in shallow, coastal shelves is a problem across many areas in South and Southeast Asia. The use of explosives and poisons in fishing is also rampant in certain places. The use of fine-meshed nets by artisanal fishers in nearshore areas to catch fish (as well as milkfish and shrimp seeds for aquaculture) is a serious concern. The use of small-meshed nets by trawlers is leading to substantial losses. Fig. 9 illustrates the results of multispecies yield and value per recruit assessment of the trawl fishery operating in the Lingayen Gulf, Philippines. Note that the use of small-meshed (i.e., 2 cm) trawl codends is leading to losses of up to 20% and 35% of potential yield and value, respectively.

Post Harvest Losses

The magnitude of post harvest losses is another major concern. Alverson et al. (1994) estimates the extent of discards for the fishing areas discussed here to be over 5 million t. This is broken down as follows: western central Pacific - 2.8 million t; eastern Indian Ocean - 0.8 million t; and western Indian Ocean - 1.5 million t. This level of discards is high at roughly 40% of marine landings of the 13 developing countries covered here. There are doubts about the accuracy of these estimates, based as they are on limited observations with small spatio-temporal coverage, and better figures will become available in the coming years. However, we believe the level to be significant (see for example Khan and Alamgir, this vol.) for countries with substantial trawl fleets and a limited market for low-value marine fishes. Apart from discards, the extent of physical losses due to spoilage of landings should be limited given the possibility of conversion to fish sauce and related products (Pauly 1996a). Value loss of harvests due to reduced quality is a common concern.

Large and Small-scale Fisheries Conflicts

The question of who should have access and, thus, benefit directly from the use of coastal fishery resources is a primary consideration in the management of fisheries. Increased competition and conflict between the small and large-scale fishing sectors is characteristic of many coastal fisheries (Thomson 1988). Table 3 illustrates the uneven competition between the small-scale (i.e., municipal) and large-scale (i.e., trawl) fisheries in San Miguel Bay, Philippines. The trawlers, consisting of 89 units and belonging to only 40 households, obtain 85% of pure profit, 42% of catch value and 31% of the total catch in the San Miguel Bay fishery. The rest goes to 2 300 small-scale fishing units owned by 3 500 households and employing about 5 100 fishers. Social equity and relative factor endowments (i.e., abundant labor and limited capital) in these countries often require the resolution of these conflicts in favor of the small-scale sector, as occurred in Indonesia, the Philippines, Malaysia and Bangladesh. Competition and conflict persists due to the economic and political power of the industrial sector and requires increased management and enforcement efforts.

Trawling in coastal areas damages patch reefs as well as seagrass and soft-bottom communities (Longhurst and Pauly 1987).

Localized pollution, particularly in semi-enclosed coastal waters, is increasing in frequency due to pollutants from domestic, industrial, agricultural and mining sources (Gomez et al. 1990; Sen Gupta et al. 1990; Holmgren 1994; APFIC 1996; Hotta 1996). Deforestation is leading to increased flooding and alteration of hydrological regimes in coastal areas. The degradation of coastal habitats (e.g., coral reefs, mangroves, and seagrass/algal beds) is apparent in many areas due to the combined effects of siltation, pollution, alteration of hydrological regimes, habitat conversion and extractive activities like coral/sand mining and mangrove forestry (Fortes 1988; Chou 1994; Holmgren 1994; Wilkinson et al. 1994; Koe and Aziz 1995). Moreover, the threat of potential oil spills is increasing given increased oil tanker traffic and marine transport in the area. All these impacts have repercussions on coastal biodiversity and on the productivity of coastal fishery resources. For instance, the biomass decline associated with high effort in the surplus production models given in Fig. 8 may be aggravated by the degradation of coastal habitats in the Philippines (Barut et al., this vol).

Table 3. Summary of data on the duality of the fisheries in San Miguel Bay, Philippines. (Source: Smith et al. 1983).

Parameter	Medium + small trawlers	Small-scale fishery
Number of fishing units	89	2 300
Total horsepower	13 200	5 600
Number of owners	40	2 030
Number of households	40	3 500
Crew income/month (P)[a]	339-810	164-342
Number of fishers	500	5 100
% of total catch	31	69
% of total catch value	42	58
% of total rent	85	15

[a] P = Philippine peso; then US$1= P10

Habitat Degradation

Coastal fish stocks and the coastal environment which sustains them are coming under increased stress from fishing and other economic activities. On an onshore-offshore axis, Table 4 summarizes ongoing economic activities in coastal and adjacent terrestrial and marine zones. The table also provides a summary of the main impacts of these activities on the coastal environment. The use of explosives and poisons in fishing occurs in many coastal fishing grounds, leading to degradation of coral reefs (Gomez 1988; Pauly and Chua 1988; Silvestre 1990; Chou 1994).

Inadequacy of Information and Research

The inadequacy of information and research inputs into the complex decision-making process that constitutes coastal fisheries management is a commonly raised issue. The appropriateness of the scope, elements, timeliness and accuracy of the available statistical information has often been questioned. Many countries require improvements in fisheries statistics and databases to make real-time management of coastal fisheries feasible (see for example FAO/SEAFDEC/SIFR 1994). Fig. 10 illustrates the patchiness of information for conducting site-specific

Table 4. Generic coastal transect summarizing main activities and issues relevant to coastal fisheries and integrated coastal zone management in South and Southeast Asia.

Major zones	Terrestrial			Coastal		Marine	
	Upland (>18% slope)	Midland (8-8% slope)	Lowland (0<8% slope)	Interface (1 km inland from HHWL-30 m depth)	Nearshore (30m-200m depth)	Offshore (>200m depth -EEZ)	Deep sea (beyond EEZ)
Main resource uses/activities	Logging Mining Agriculture	Mining Agriculture	Urban development Industries Agriculture Tourism	Mining (coral/sand) Mangrove forestry Aquaculture Fisheries Tourism Industries Urban development	Artisanal fisheries Commercial fishing Marine transport Oil drilling	Marine transport Industrial fishing Offshore development	Marine transport Industrial fishing
Main environmental issues/impacts on the coastal zone	Siltation Flooding Toxic mine tailings	Agrochemical loading Erosion Siltation Flooding	Siltation Domestic pollution	Reduced biodiversity Habitat degradation and destruction Overfishing Industrial pollution Domestic pollution	Reduced biodiversity Overfishing Oil spills	Overfishing Oil spills	Oil spills

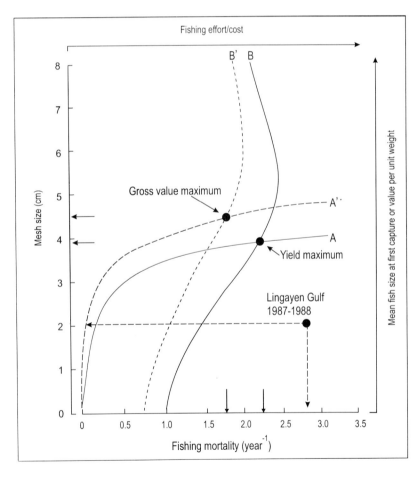

Fig. 9. Multispecies yield and value per recruit assessment of the trawl fishery in Lingayen Gulf, Philippines (Source: Silvestre 1990). Intersections of the 'eumetric' and 'cacometric' lines for yield (A and B) and gross value (A' and B') indicate mesh size and fishing mortality combinations where yield and value are maximized. Note excessively high effort and low mesh sizes in Lingayen Gulf which lead to losses of up to 20% and 35% relative to maxima in yield and value, respectively.

assessment in these countries. In this example from the Lingayen Gulf (Philippines), the spatial scope of available catch statistics does not meet assessment needs and effort information is not available. Assessment of the status of fisheries in the area is, therefore, possible only based on the results of independent trawl surveys and population censuses conducted in the past. The published results of these surveys and censuses allowed Silvestre (1990) to show that resource biomass was down to about 13% of its original level in the late 1940s, precluding further expansion of the fisheries.

The inadequacy of fisheries research in support of fisheries management efforts is also commonly cited. Much of the fisheries research is criticized for being too academic and peripheral to the management questions at hand, and for failing to take the extra step to elaborate requisite management options and measures. Many research results also remain unpublished leading to what Pauly (1995) refers to as the "shifting baseline syndrome" in fisheries. The short history of quantitative fisheries research, limitations in the available statistical baseline and limited research resources requires that past studies be documented, analyzed and made available for fisheries management purposes. For example, trawl surveys conducted

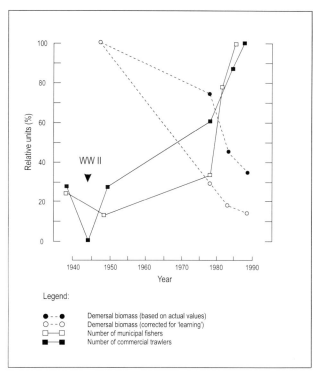

Fig. 10. Relative indices of demersal biomass, number of fishers, and number of commercial trawlers in Lingayen Gulf, Philippines from the 1930s to the 1980s. (Source: Silvestre 1990).

in many countries remain underutilized and potentially offer many insights for fisheries management (Silvestre et al. 1986; Pauly and Martosubroto 1996). Other areas commonly lacking research and information include: site-specific fisheries assessments; selectivity research; research on location and size of underfished stocks; marine protected areas; fish processing and marketing; socioeconomic research; and policy and institutional studies (IPFC 1987b; Yanagawa and Wongsanga 1993; FAO/SEAFDEC/SIFR 1994; APFIC 1996; Hotta 1996).

Institutional Weaknesses and Constraints

All these issues and concerns arise and persists due to the inability of existing institutions to deal with the changing realities of coastal fisheries. Problems and constraints commonly cited include: inadequacies in the policy and legal framework; limited personnel and technical capabilities; shortage of resources/funding; inadequate or overlapping mandates and functions; and a lack of institutional collaboration/coordination (IPFC 1987a and b; APFIC 1996; Hotta 1996). An increased emphasis on the participation of stakeholders and devolution of management authority to local levels are notable trends in many of the countries included in this study.

The Opportunities: Overview of Key Management Interventions

Given the multiplicity of issues impacting coastal fisheries, a variety of management interventions are prescribed in the available literature for their resolution or mitigation (Yanagawa and Wongsanga 1993; FAO 1995a and b; APFIC 1996; and Hotta 1996). We briefly outline below seven main categories of management interventions which we believe to be appropriate, given the status of coastal fisheries in these countries. Though many of these are in place, there is a common concern about the comprehensiveness and scale of the existing mix of measures to sufficiently reverse or mitigate the multiplicity of impacts on, and sustain the benefits derived from, coastal fisheries. Successful fisheries management will require effective implementation of a wide range of measures as well as fundamental shifts in management perspectives (Anderson 1987; Hilborn and Walters 1992; Pauly 1994, 1996b; Olver et al. 1995; Stephenson and Lane 1995; Caddy, in press).

Limited Entry and Effort Reduction

The establishment of viable systems of rights and access to limit entry into coastal fisheries is sorely lacking. Licensing schemes in many countries are still viewed as statistical and revenue generating exercises, rather than as effective management handles to limit entry and control fishing effort. In overfished coastal areas, the obvious need is for a reduction of fishing effort, particularly in nearshore, traditional fishing grounds. The requisite effort reduction in some areas is quite substantial as in the example for Philippine demersal and small pelagic fisheries shown in Fig. 8. In this case the reduction required is about half of prevailing effort levels. This kind of situation requires direct exit interventions, enhancement of alternative livelihood prospects and occupational mobility of fishers, restructuring of relevant policy and regulatory frameworks, and the redirection of subsidies and support towards improved rural/community development. Other measures outlined below are also directly relevant to requisite effort reduction schemes in overfished coastal fisheries.

Gear, Area and Temporal Restrictions

Measures influencing the species and size, and to a certain extent the sex and maturity stage, composition of catches include: (1) technological controls or limitations, e.g., gear restrictions such as mesh regulations, hook size control, trawl bans; (2) spatial restrictions, e.g., marine sanctuaries, area closures;

and (3) temporal restrictions, e.g., seasonal closures. Regulatory instruments include various forms of species and size restrictions on landings, as well as prohibitions on landing of gravid females. Table 5 uses selected regulations in effect in the Philippines to illustrate some of the forms that these selectivity measures may take. It should be noted that a creative use of other measures, such as incentives/disincentives, can be made to influence selectivity and the resulting exploitation patterns/levels of coastal fisheries.

While much of the theoretical and methodological aspects of gear selectivity are covered in the literature, there is a considerable scope for *in situ* information on selectivity to set up measures for site-specific management. Considerable opportunities exist for a more creative use of gear restrictions, zonation schemes, marine sanctuaries or protected areas (Bohnsack 1994), and seasonal closures to influence the selectivity of coastal fisheries (Silvestre 1995). The design and operation of measures to improve selectivity will vary depending on the number of species and fishing gears used. The complexity of the selectivity problem increases from single species, single gear situations to multigear, multispecies situations (Pauly 1979; Gulland 1983). This has hindered the more creative use of gear, area and temporal restrictions.

McManus (this vol.) points to faunal assemblages associated with spatial elements which can be tapped by managers in designing area restrictions, sanctuaries or zonation schemes (see also McManus 1986, 1989, 1996). The opening and closing of the fishing season for shrimps in Australia illustrates the potential for temporal restrictions, given similarities in the dynamics of exploited shrimp species (Rothlisberg et al. 1988; Staples 1991). Attention is also required in developing and dispersing appropriate hatchery techniques for cultured species, e.g., milkfish, shrimps, groupers. The restriction of gears with small-meshed nets in nearshore areas can succeed only if aquaculture dependence on wild seeds is curtailed.

Improvement of Marketing and Post Harvest Facilities

The level of discards and (value) loss in catches require increased management intervention (Alverson et al. 1994). Post harvest facilities (i.e., salt, ice and cold storage) are lacking in strategic locations in many areas. Private-sector participation in providing these facilities needs to be enhanced given the noted inefficiency of the public sector in maintaining such facilities. Development and dissemination of appropriate processing (e.g., *surimi*) and handling techniques also require attention, as does the development and maintenance of rural road infrastructures. Improved se-

lectivity of coastal fisheries is also important in reducing the magnitude of discards.

Enhancement of Awareness and Participation of Stakeholders

Enhancing the awareness and participation of stakeholders is necessary for better and more cost-effective management of coastal fisheries. Improved transparency and institutionalized participation of stakeholders in the management decision-making process is desirable. Other measures that can be implemented include: enhancement of fishers' organizations and other NGOs; education/awareness programs; devolution/decentralization of management authority; and appropriate extension, training and credit support for nonfishing activities.

Reduction of Environmental Impacts

The need for a reduction of the impacts of fishing and other economic activities on the coastal environment that sustains fisheries is evident in many countries. Efforts toward integrated coastal zone management (Chua and Pauly 1989; Clark 1992) and the adoption of integrated coastal fisheries management approaches (Silvestre 1996), will be necessary for the reduction of undesirable impacts on the coastal environment. Other areas requiring intervention include: wider adoption of multiple-use zonation schemes; restoration/rehabilitation of coastal habitats; curtailment of destructive fishing methods; adoption of appropriate environmental impact assessment systems; and improvement and enforcement of penalties/incentives systems. Progress in the wider use of the precautionary approach and (development of mechanisms for) 'internalization' of environmental costs is highly relevant to reducing coastal environmental impacts.

Institutional Strengthening/Upgrading

Concern about the issues above persists due to the inability of existing institutions to elaborate and effect the requisite management interventions. Strengthening of the policy, regulatory and organizational frameworks relevant to fisheries is urgently required. The areas identified as needing attention include: technical, personnel and facilities upgrading; improvement of financial capability and strengthening of mandates of organizations; enhancement of organizational coordination/collaboration; increased transparency, accountability and participation in the management decision-making process, and; the development of effective and cost-efficient monitoring,

Table 5. Illustrative examples of regulatory instruments affecting the selectivity of fishing operations in the Philippines. (Source: Silvestre 1995).

Regulatory instrument	Law/ordinance[a]	Specifications
1. Technological controls		
- mesh regulation	PD 704 (1975)	Prohibition of use of nets with mesh sizes less than 3 cm when stretched (nationwide).
	FAO 155 (1986)	Regulating the use of fine-meshed nets in fishing (nationwide).
- 'gear' ban	PD 704 (1975)	Prohibition of commercial trawling (less than 3 GT) in waters 7 fathoms deep or less (nationwide).
	FAO 163 (1986)	Prohibition on the operation of *muro-ami* and *kayakas* in all Philippine waters (nationwide).
	FAO 188 (1993)	Regulations governing the operation of commercial fishing boats in Philippine waters using tuna purse seine nets (nationwide).
	FAO 190 (1994)	Regulations governing *pa-aling* fishing operation in Philippine waters (nationwide).
	PD 704 (1975)	Prohibition on the use of explosives and poisons in fishing (nationwide).
2. Spatial restrictions		
- area closure	PD 704 (1975)	Prohibition of commercial fishing (with the use of boats more than 3 GT) in waters less than 7 fathoms (nationwide).
	LOI 1328 (1983)	Extended the ban on commercial trawls and purse seines within 7 km of the coastline in all provinces (nationwide).
	RA 7160 (1992)	Extended boundaries of municipal waters from 3 nautical miles (5.5 km) to 15 km from the shoreline (nationwide).
3. Temporal restrictions	FAO 9 (1950)	Regulation governing the conservation of the *ipon* goby fisheries of the Ilocos provinces; open season from November to January; closed season in September, October and February (area specific i.e., Ilocos Norte).
	FAO 136 (1982)	Closed season of five years for the operation of commercial fishing boats in San Miguel Bay (area specific).
4. Others	FAO 129 (1980)	Ban on the taking or catching, selling, possession, and transportation of *sabalo* (full grown *bangus* or milkfish) (nationwide).
	FAO 148 (1984)	Regulation for gathering, catching, taking or removing of marine tropical aquarium fish (nationwide).

[a] FAO = Fishery Administrative Order; PD = Presidential Decree; LOI = Letter of Instruction.

control and surveillance (MCS) systems (Flewelling 1995). The costs of improved management are substantial and exploration of appropriate cost-sharing schemes with industry (as the ultimate beneficiary) needs to be developed.

Enhancement of Research and Information

Management systems have to be supported by research and information. There is need for research in: appropriate size and siting of sanctuaries or protected areas; resource enhancement and habitat rehabilitation techniques; selective fishing; appropriate fisheries management reference points; ecosystem modeling (Christensen and Pauly 1995, 1996); and policy and institutional support. Documentation and retrospective analysis of existing information and past studies (e.g., trawl surveys) is important for purposes of comparison and for the potential insights they provide for the management of coastal fisheries. Establishment of statistical baseline information should be consistent with the MCS and management reference points appropriate to the situational realities obtaining in the individual countries. There should be more research collaboration and exchange of research and experiences between the countries given similarities in their resource base and development context.

Conclusion

In the 13 developing South and Southeast Asian countries covered in this study coastal fisheries generate food, employment and foreign exchange. Many factors impact the magnitude and sustainability of these benefits. Fig. 11 shows a logical structuring of the main objectives, issues and interventions relevant to coastal fisheries management in these countries and also provides a summary of the main points covered in this paper. There are three generic categories of ('ends') objectives for the management of coastal fisheries, viz., productivity/efficiency, distributional equity and environmental integrity. A fourth generic ('means') category, institutional effectiveness/efficiency, is often considered necessary for success in attainment of the main ('ends') objectives. Seven key issues affect the attainment of these objectives and the benefits derived from coastal fisheries. Seven key management interventions for the resolution or mitigation of these issues are listed. The issues are interconnected and have cross-reinforcing tendencies, e.g., overfishing intensifies conflicts between small and large-scale fisheries leading to the use of destructive gears and increased habitat degradation. The management interventions are also interconnected, although only the link to the main issues being addressed is illustrated. Apart from providing a summary, Fig. 11 in essence presents a systems matrix of generic elements which should be considered in advancing coastal fisheries management efforts in South and Southeast Asia.

Beyond the reflection and debate, Fig. 11 illustrates the need for effective action on a wide front at various levels of the institutional hierarchy. The management interventions outlined in this paper show scope for action at the local, national and international levels. Much of the overall success will depend on national institutional capabilities. The strengthening and upgrading of these capabilities and effective implementation of the interventions outlined are in turn dependent on the resources that can be mobilized for such purposes. In the context of the development needs of these countries, there is competition for resources given other equally pressing developmental and social needs. The reviews given by Holdgate et al. (1982), Tolba and El-Kholy (1992), and FAO (1995b) identify positive and negative international trends affecting the environment, food, agriculture and fisheries particularly relevant to this study. High population growth, external debt burden, declining commodity prices, market access difficulties and the shrinking international aid 'pot' are minuses for the ability of most countries to devote sufficient resources to the problems at hand. The positive developments are increased economic growth (although this can lead to more pollution problems), environmental awareness, democratization and regional collaboration. Thus, the ultimate solutions to the multiplicity of issues impacting coastal fisheries are also premised on addressing poverty and promoting overall development in South and Southeast Asia.

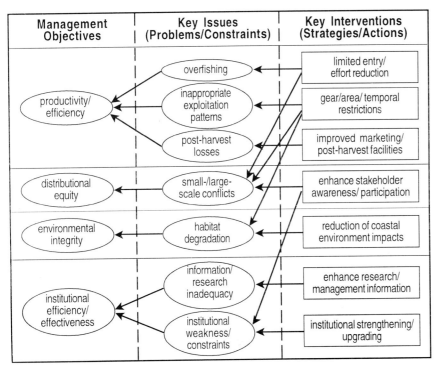

Fig. 11. Summary of management objectives, key management issues and constraints, and interventions (strategies and actions) for the coastal fisheries of the developing countries of tropical Asia. Management interventions have crosscutting benefits/implications, but only the connections to the main issue being addressed are illustrated.

Acknowledgments

Thanks are due Mr. Len Garces and Ms. Bing Santos for assistance in assembling some of the information given in the tables, and for locating important references used in this study.

References

ADB. 1995. Key indicators of developing Asian and Pacific countries. Vol. XXVI. Asian Development Bank, Manila. 417 p.

ADB. 1996. Asian development outlook - 1996 and 1997. Asian Development Bank, Manila. 245 p.

Alverson, D., M. Freeberg, S. Murawski and J.G. Pope. 1994. A global assessment of fisheries bycatch and discards. FAO Fish. Tech. Pap. (339): 233 p.

Anderson. L.G. 1987. Expansion of the fishery management paradigm to include institutional structure and function. Trans. Am. Fish. Soc. 116(3): 396-404.

Aoyama, T. 1973. The demersal fish stocks and fisheries of the South China Sea. FAO and UNDP, Rome. SCS/DEV/73/3: 80 p.

APFIC. 1996. Summary report of the APFIC Symposium on Environmental Aspects of Responsible Fisheries, Seoul, Republic of Korea, 15-18 October 1996. RAP Publ. 1996/42. Asia Pacific Fisheries Commission, Bangkok, Thailand. 27 p.

APO. 1988. Fishing industry in Asia and the Pacific. Asian Productivity Organization, Tokyo. 481 p.

Armstrong, D., R. Ferro, D. MacLennan and S. Reeves. 1990. Gear selectivity and the conservation of fish. J. Fish. Biol. 37(A): 261-262.

Beverton, R. and S. Holt. 1957. On the dynamics of exploited fish populations. Fishery Invest., London, Ser. 2, 19: 533 p.

Bohnsack, J.A. 1994. Marine reserves: they enhance fisheries, reduce conflicts, and protect resources. Naga, ICLARM Q. 17(3): 4-7.

Butcher, J. 1994. Harvesting the sea. An ecological history of the marine fisheries of Southeast Asia. Research School of Pacific and Asian Studies, Australian National University. 116 p.

Butcher, J. 1996. The marine fisheries of the Western Archipelago: towards an economic history, 1850 to the 1960s. p. 24-39. In D. Pauly and P. Martosubroto (eds.) Baseline studies of biodiversity: the fish resources of Western Indonesia. ICLARM Stud. Rev. 23, 312 p.

Caddy, J.F. and R. Mahon. 1995. Reference points for fishery management. FAO Fish. Tech. Pap. (347): 82 p.

Caddy, J.F. Fisheries management after 2000: will new paradigms apply? Rev. Fish Biol. Fish. (In press).

Christensen, V. and D. Pauly. 1995. Fish production, catches and the carrying capacity of the world oceans. Naga, the ICLARM Q. 18(3): 34-40.

Christensen, V. and D. Pauly. 1996. Ecological modeling for all. Naga, the ICLARM Q. 19(2): 25-26.

Chou, L.M. 1994. Marine environmental issues of Southeast Asia: state and development. Hydrobiologia 285: 139-150.

Chua, T. E. and D. Pauly (eds.) 1989. Coastal area management in Southeast Asia: policies, management strategies and case studies. ICLARM Conf. Proc. 19, 254 p.

Chullasorn, S. and P. Martosubroto. 1986. Distribution and important biological features of coastal fish resources in Southeast Asia. FAO Fish Tech. Pap. (278): 84 p.

Clark, J.R. 1992. Integrated management of coastal zones. FAO Fish. Tech. Pap. (327): 167 p.

Dalzell, P., P. Corpuz, R. Ganaden and D. Pauly. 1987. Estimation of maximum sustainable yield and maximum economic rent from the Philippine small pelagic fisheries. Bureau of Fisheries and Aquatic Resources Tech. Pap. Ser. 10(3), 23 p.

Dwiponggo, A., T. Hariati, S. Banon, M.L. Palomares and D. Pauly. 1986. Growth, mortality and recruitment of commercially important fishes and penaeid shrimps in Indonesian waters. ICLARM Tech. Rep. 17: 91 p.

Dwiponggo, A. 1988. Recovery of overexploited demersal resources and growth of its fishery on the north coast of Java. Indones. Agric. Res. Dev. J., 10(3): 65-72.

FAO. 1978. Some scientific problems of multispecies fisheries. Report of the Expert Consultation on Management of Multispecies Fisheries. FAO Fish. Tech. Pap. (181): 42 p.

FAO. 1992. Review of the state of world fishery resources. Part 1. The marine resources. FAO Fish. Circ. (710), Rev. 8, Part 1. 114 p.

FAO. 1994. FAO yearbooks, Fishery Statistics. Vol. 78, 702 p.

FAO. 1995a. Review of the state of world fishery resources: marine fisheries. FAO Fish Circ. (884): 105 p.

FAO. 1995b. The state of world fisheries and aquaculture. FAO, Rome. 57 p.

FAO/SEAFDEC/SIFR. 1994. Status of fishery information and statistics in Asia. Proceedings of the Regional Workshop on Fishery Information and Statistics in Asia, Bangkok, Thailand, 18-22 January 1994. Vol. I and II.

Flewwelling, P. 1995. An introduction to monitoring, control and surveillance for capture fisheries. FAO Fish. Tech. Pap. (338): 217 p.

Fortes, M.D. 1988. Mangrove and seagrass beds of East Asia: habitats under stress. Ambio 17(3): 207-213.

Fortes, M.D. 1995. Seagrasses of East Asia: environmental management perspectives. RCU/EAS Technical Reports Series No. 6.

Froese, R. and D. Pauly. 1996. FishBase 96: concepts, design and data sources. ICLARM, Manila. 179 p.

Gomez, E.D. 1988. Overview of environmental problems in the East Asian Seas region. Ambio 17(3): 166-169.

Gomez, E.D., E. Deocadiz, M. Hunspreugs, A.A. Jothy, Kuan Kwee Jee, A. Soegiarto and R.S.S. Wu. 1990. State of the marine environment in the East Asian Seas Region. UNEP Regional Seas Reports and Studies No. 126. 63 p.

Gulland, J. 1983. Fish stock assessment: a manual of basic methods. FAO/Wiley, New York. 223 p.

Hamley, J.M. 1975. Review of gillnet selectivity. J. Fish. Res. Board Can. 32(11): 1943-1969.

Hilborn, R. and C.J. Walters. 1992. Quantitative fisheries stock assessment: choice, dynamics and uncertainty. Chapman and Hall, New York. 570 p.

Holdgate, M., M. Kassas and G. White (eds.) 1982. The world environment, 1972-1982. A Report by UNEP. Tycooly International, Dublin.

Holmgren, S. 1994. An environmental assessment of the Bay of Bengal region. Bay of Bengal Programme, Madras. BOBP/REP/67: 256 p.

Hotta, M. 1996. Regional review of the fisheries and aquaculture situation and outlook in South and Southeast Asia. FAO Fish. Circ. (904): 45 p. FAO, Rome.

Ingles, J. and D. Pauly. 1984. An atlas of the growth, mortality and recruitment of Philippine fishes. ICLARM Tech. Rep. 13:127 p.

IPFC. 1987a. Papers presented at the Symposium on the exploitation and management of marine fishery resources in Southeast Asia, Darwin, Australia, 16-19 February 1987. RAPA Rep. 1987/10: 552 p.

IPFC. 1987b. Report of the symposium on the exploitation and management of marine fishery resources in Southeast Asia. Darwin, Australia, 16-19 Feb. 1987. RAPA Rep. 1987/9: 39 p.

Khoo, H.W., S. Selvanathan and H.A.M.S. Halidi. 1987. Capture fisheries, p. 89-109. In T.E. Chua, L.M. Chou and M.S.M. Sadorra (eds.) The coastal environmental profile of Brunei Darussalam: resource assessment and management issues. ICLARM Tech. Rep. 18, 193 p.

Koe, L.C.C. and M.A. Aziz. 1995. Regional programme of action on land-based activities affecting coastal and marine areas in the East Asian Seas. UNEP, Bangkok. RCU/EAS Tech. Rep. Ser. (5): 117 p.

Lawson, R. 1978. Incompatibilities and conflicts in fisheries planning in Southeast Asia. Southeast Asian J. Soc. Sci. 6(1-2): 115-136.

Lilburn, B.V. 1987. Formulation of fisheries management plans. p. 507-527. In IPFC. Papers presented at the symposium on the exploitation and management of marine fishery resources in Southeast Asia. Darwin, Australia. 16-19 February 1987. RAPA Rep. 1987/10: 552 p.

Longhurst, A. and D. Pauly. 1987. Ecology of tropical oceans. Academic Press, New York. 407 p.

Marr, J.C. 1976. Fishery and resource management in Southeast Asia. RFF/PISFA Pap. 7, 62p.

Marr, J.C. 1981. Southeast Asian marine fishery resources and fisheries. pp. 75-109. In L.S. Chia and C. MacAndrews (eds.). Southeast Asian Seas: Frontiers for Development. McGraw-Hill, Singapore.

McManus, J.W. 1986. Depth zonation in a demersal fishery in the Samar Sea, Philippines, p. 483-486. In Maclean, J.L., Dizon, L.B., Hosillos, L.V. (eds.) The First Asian Fisheries Forum. Asian Fisheries Society, Manila, Philippines.

McManus, J.W. 1988. Coral reefs of the ASEAN region: status and management. Ambio 17(3): 189-193.

McManus, J.W. 1989. Zonation among demersal fishes in Southeast Asia: the southwest shelf of Indonesia, p. 1011-1022. In Proceedings of the Sixth Symp. On Coastal and Ocean Management/ASCE , 11-14 July 1989. Charleston, South Carolina.

McManus, J.W. 1996. Marine bottom fish communities from the Indian Ocean coast of Bali to mid-Sumatra, p. 91-101. In D. Pauly and P. Martosubroto (eds.) Baseline studies of biodiversity: the fish resources of Western Indonesia. ICLARM Stud. Rev. 23: 312 p.

Olver, C.H., B.J. Shuter and C.R. Minns. 1995. Towards a definition of conservation principles for fisheries management. Can. J. Fish. Aquat. Sci. 52: 1584-1594.

Pauly, D. 1979. Theory and management of tropical multispecies stocks: a review, with emphasis on the Southeast Asian demersal fisheries. ICLARM Stud. Rev. 1, 35 p.

Pauly, D. and G.I. Murphy (eds.). 1982. Theory and management of tropical fisheries. Proceedings of the ICLARM/CSIRO Workshop on the Theory and Management of Tropical Multispecies Stocks, 12-21 January 1981, Cronulla, Australia. ICLARM Conf. Proc. 9, 360 p.

Pauly, D. and T.E. Chua. 1988. The overfishing of marine resources: socioeconomic background in Southeast Asia. Ambio 17(3): 200-206.

Pauly, D. 1989. Fisheries resources management in Southeast Asia: why bother? p. 1-10. In T.E. Chua and D. Pauly (eds.) 1989. Coastal area management in Southeast Asia: policies, management strategies and case studies. ICLARM Conf. Proc. 19, 254 p.

Pauly, D. 1994. From managing fisheries to managing ecosystems. ICES, Copenhagen. ICES Inf. 24:7.

Pauly, D. 1995. Anecdotes and the shifting baseline syndrome of fisheries. Trends in Ecol. Evol. 10 (10): 430.

Pauly, D. and P. Martosubroto. 1996. Baseline studies of biodiversity: the fish resources of Western Indonesia. ICLARM Stud. Rev. 23, 321 p.

Pauly, D. 1996a. Fleet-operational, economic, and cultural determinants of by-catch uses in Southeast Asia, p. 285-288. In Solving by-catch: considerations for today and tomorrow. Alaska Sea Grant College Prog. Rep. No. 96-03. University of Alaska, Fairbanks.

Pauly, D. 1996b. One hundred tons of fish and fisheries research. Fish. Res. 25(1): 25-38.

Pauly, D. 1997. Small-scale fisheries in the tropics: marginality, marginalization, and some implications for fisheries management. In E.K. Pikitch, D.D. Huppert and M.P. Sissenwine (eds.) Global trends: fisheries management. American Fisheries Society Symposium 20, Bethesda, Maryland.

Pope, J., A. Margetts, J. Hamley and E. Akjüz. 1975. Manual of methods for fish stock assessment, part III. Selectivity of fishing gear. FAO Tech. Pap. 41 (Rev. 1), 65 p.

Raja, B.T.A. 1980. Current knowledge of fisheries resources in the staff area of the Bay of Bengal. Bay of Bengal Programme, Madras. BOBP/WP8, 23 p.

Ricker, W.E. 1975. Computation and interpretation of biological statistics of fish populations. Bull. Fish. Res. Board. Can. (191), 382 p.

Rothlisberg, P., D. Staples and B. Hill. 1988. Factors affecting recruitment in penaeid prawns in tropical Australia. p. 241-248. In A. Yañez-Arancibia and D. Pauly (eds.) IOC/FAO Workshop on Recruitment in Tropical Coastal Demersal Communities. IOC Works. Rep. No. 44.

Sainsbury, K. 1984. Optimum mesh size for tropical multispecies trawl fisheries. J. Cons. CIEM 41: 129-139.

Sardjono, I. 1980. Trawlers banned in Indonesia. ICLARM Newsl. 3(4): 3.

Sen Gupta, R., M. Ali, A.L. Bhuiyan, M.M. Hossain, P.M. Sivalingam, S. Subasinghe and N.M. Tirmizi. 1990. State of the marine environment in the South Asian Seas Region. UNEP Reg. Seas Rep. Stud. No. 123.

Shindo, S. 1973. General review of the trawl fishery and the demersal fish stocks of the South China Sea. FAO Fish. Tech. Pap. 120, 49 p.

Silvestre, G.T., R.B. Regalado and D. Pauly. 1986. Status of Philippine demersal stocks-inferences from underutilized catch rate data, p. 47-96. In D. Pauly, J. Saeger and G. Silvestre (eds.) Resources, management and socio-economics of Philippine marine fisheries. Tech. Rep. Dep. Mar. Fish. Tech. Rep. 10, 217 p.

Silvestre, G.T. and D. Pauly. 1986. Estimate of yield and economic rent from Philippine demersal stocks, 1946-1984. Paper presented at the WESTPAC Symposium on Marine Science in the Western Pacific, Townsville, Australia, 1-6 December 1986.

Silvestre, G.T. 1990. Overexploitation of demersal stocks in Lingayen Gulf, Philippines, p. 973-876. In R. Hirano and I. Hanyu (eds.) The Second Asian Fisheries Forum, Asian Fisheries Society, Manila, Philippines.

Silvestre, G.T., M. Soriano and D. Pauly. 1991. Sigmoid selection and the Beverton and Holt yield equation. Asian Fish. Sci. (4): 85-98.

Silvestre, G.T. and H.J.H. Matdanan. 1992. Brunei Darussalam capture fisheries: A review of resources, exploitation and management, p. 1-38. In G. Silvestre, H.J.H. Matdanan, P.H.Y. Sharifuddin, M.W.R.N. De Silva and T.E. Chua (eds.). The coastal resources of Brunei Darussalam: status, utilization and management. ICLARM Conf. Proc. 34, 214 p.

Silvestre, G.T. 1995. Fisheries management and the selectivity of fishing operations. FAO Consultation of Experts and Industry on Selective Fishing for Responsible Exploitation of the Resources in Asia, Beijing, 12-17 October 1995. Inf. Pap. (12): 25 p.

Silvestre, G.T. 1996. Integrated management of coastal fisheries: lessons from initiatives in San Miguel Bay, Philippines. ICLARM, Manila, 13 p.

Sivasubramaniam K. 1985, Marine fishery resources of the Bay of Bengal. Bay of Bengal Programme, Madras. BOBP/WP/36: 66 p.

Smith, I.R., D. Pauly and A.N. Mines. 1983. Small-scale fisheries of San Miguel Bay, Philippines: options for management and research. ICLARM Tech. Rep. 11, 80 p.

Soysa, C.H., L.S. Chia and W.L. Collier (eds.) 1982. Man, land and sea: coastal resource use and management in Asia and the Pacific. Agricultural Development Council, Bangkok, 320 p.

Staples, D. 1991. Penaeid prawn recruitment: geographic comparison of recruitment patterns within the Indo-West Pacific Region. Mem. Queensland Mus. 31: 337-348.

Stephenson, R.L. and D.E. Lane. 1995. Fisheries management science: a plea for conceptual change. Can. J. Fish. Aquat. Sci. 52: 2051-2056.

Thomson, D. 1988. The world's two marine fishing industries — how they compare. Naga, ICLARM Q. 11(3): 17.

Tiews, K. (ed.). 1973. Fisheries resources and their management in Southeast Asia. German Foundation for International Development, Federal Research Board for Fisheries and FAO. Berlin (West), 511 p.

Tolba, M. and O. El-Kholy (eds.). 1992. The world environment, 1972-1992. Two decades of challenge. Chapman and Hall for UNEP, London.

WCED. 1987. Our common future. World Commission for Environment and Development. Oxford University Press, Oxford.

Williams, M. 1996. The transition in the contribution of living aquatic resources to food security. Food, Agriculture, and the Environment Discussion Paper 13, 41 p. International Food Policy Research Institute, Washington D.C.

Wilkinson, C.R., S. Sudara and L.M. Chou (eds.) 1994. Proceedings, Third ASEAN-Australia Symposium on Living Coastal Resources. Vol. 1: Status reviews. Australian Institute of Marine Science, Townsville, Australia. 454 p.

WRI. 1995. People and the environment. WRI in collaboration with UNEP and the UNDP.

Yanagawa, H. and P. Wongsanga. 1993. Review of fishery production, provisional estimation of potential yield and the situation of fisheries in the Southeast Asian region - 1976 to 1989. SEAFDEC Spec. Publ. (18), 114 p.

The Coastal Fisheries of Bangladesh

M.G. KHAN
Marine Fisheries Survey
Department of Fisheries
Chittagong 4100, Bangladesh

M. ALAMGIR
Fisheries Research Institute
Marine Fisheries and Technology Station
Cox's Bazar, Bangladesh

M.N. SADA
Marine Fisheries Survey
Department of Fisheries
Chittagong 4100, Bangladesh

Khan, M.G. and M. Alamgir and M.N. Sada. 1997. The coastal fisheries of Bangladesh, p. 26-37. *In* G. Silvestre and D. Pauly (eds.) Status and management of tropical coastal fisheries in Asia. ICLARM Conf. Proc. 53, 208 p.

Abstract

This contribution provides an overview of coastal fisheries in Bangladesh and covers the coastal environment, capture fisheries characteristics and trends, and management issues relevant to the sector. The country's coastal fisheries accounted for 28% of the annual fisheries production of 1 087 000 t in 1993-1994. The industrial and artisanal fisheries operate in shallow, coastal waters; the former mainly in 40-100 m depth and the latter mainly between the shoreline and 40 m depth.

A multiplicity of issues adversely affect the coastal fisheries of Bangladesh. Excessive fishing effort and growth overfishing impact currently fished resources (particularly shrimp and demersal finfish). Competition and conflict among artisanal and industrial gears are intense. About 80% of the catch of trawlers is not landed but discarded at sea. Moreover, shrimp seed collection in nearshore waters is causing severe damage to coastal resources. In addition to fisheries-specific issues, a number of cross-sectoral issues (e.g., mangrove denudation, siltation) impact the coastal zone and the long-term sustainability of coastal fisheries. Management measures in response to these issues are briefly discussed.

Introduction

The fisheries sector in Bangladesh is important as a source of food, livelihood and foreign exchange. The country's annual fish production has increased from 640 000 t in 1975-1976 to about 1 087 000 t in 1993-1994. Per capita fish consumption has declined during the same period from 33.4 kg to 21.0 kg because production has not kept pace with the rapid population increase. Nevertheless, the fisheries sector still supplied 80% of animal protein consumed in the country in 1993-1994.

The marine capture fisheries sector generated only 28% of fisheries production in 1993-1994. The sector is subdivided into industrial and artisanal fisheries, the former involving solely the use of trawlers and the latter involving the use of relatively simple gear such as gillnets and bagnets. The marine capture fisheries in Bangladesh operate in shallow, coastal waters principally between the shoreline and 100 m depth. This contribution attempts to provide an overview of the marine capture fisheries of Bangladesh and the main issues impacting their management.

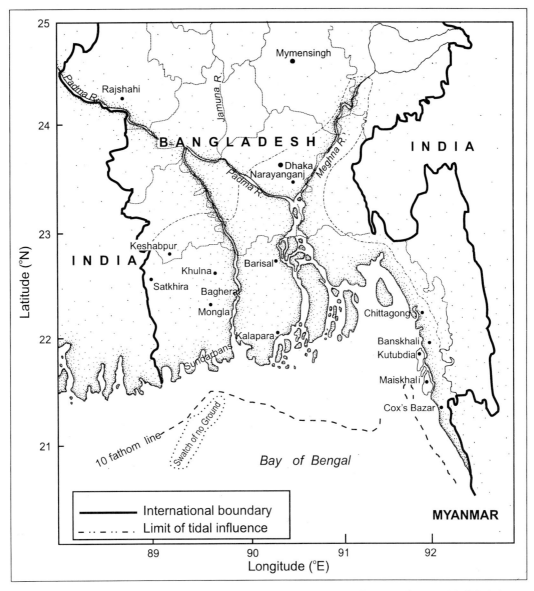

Fig 1. Map of Bangladesh showing the limit of salt penetration in the coastal zone. (Modified from Holmgren 1994).

Table 1. Extent of EEZ and depth distribution of the continental shelf off Bangladesh.

Depth zone (m)	Area (km²)
≤10	24 000
10 - 24	8 400
25 - 49	4 800
50 - 74	5 580
75 - 99	13 410
100 - 199	10 250
All shelf	66 440
Total EEZ	164 000

Marine Environment

Bangladesh has a land area of 144 000 km² and is bounded by India on the west, north and northeast, by Myanmar (Burma) on the east and southeast, and by the Bay of Bengal on the south (Fig. 1). The country's exclusive economic zone (EEZ) spans 164 000 km² and the shelf area covers roughly 66 440 km². Table 1 gives the depth distribution of the shelf area. The coastal waters are very shallow, with depths less than 10 m covering 24 000 km². The shelf area down to about 150 m appears to be very smooth with very few obstacles to bottom trawling. The continental edge occurs at depths between 160 m and 180 m. Its slope is very precipitous and thus, it appears presently not possible to trawl in waters deeper than 180 m.

28

The southwest monsoon, characterized by hot and humid winds blowing from the Bay of Bengal, occurs from May to around August/September. About 80% of the total annual rainfall is recorded during this period. The northeast monsoon blows from November to around March/April, bringing cool, dry air from the continental areas. Between the two monsoons, winds are variable and unstable and cyclones may occur. Coastal waters are characterized by a prolonged low saline regime due to river discharges (which peak during the southwest monsoon). A strong (3 m) semidiurnal tide mixes the highly turbid coastal waters which receives sediments from as far as the Himalayas (FAO 1968; Eysink 1983). Monsoon seasonality governs surface current patterns (Lamboeuf 1987).

Fig. 2 gives typical surface and subsurface distributions of salinity, temperature and oxygen off Bangladesh during the summer. The surface salinity distribution shows a strong gradient in a north-south direction. This is reflected by the subsurface salinity distribution, which shows a 32‰ frontal zone or 'tongue' due to river discharges. Surface water temperature varies from 29.5°C to 38.8°C with higher values prevailing offshore. Thermocline depth is around 30-40 m in summer (compared to about 70 m during winter months). In the more offshore areas the upper layer is more homogenous, at least down to the thermocline. Oxygen content declines rapidly with depth, the 1 ml·l^{-1} isoline being situated around 80 m depth in summer (compared to about 60 m in winter). Oxygen deficits are prominent in the subsurface layer of the outer shelf area.

Primary production in the Bay of Bengal is known to be high during the northeast monsoon, i.e., 0.15-1.45 gC·m^{-2}·day^{-1}, but similar information during the southwest monsoon is lacking. Zooplankton biomass in the area has been reported to vary between 0.98 and 3.90 gC·m^{-2}·day^{-1} (Sivasubramaniam 1985).

Coral reefs are quite limited off Bangladesh due to high river discharge and turbidity. Four species of *Acropora* (*A. pulchra, A. horrida, A. humilis* and *A.*

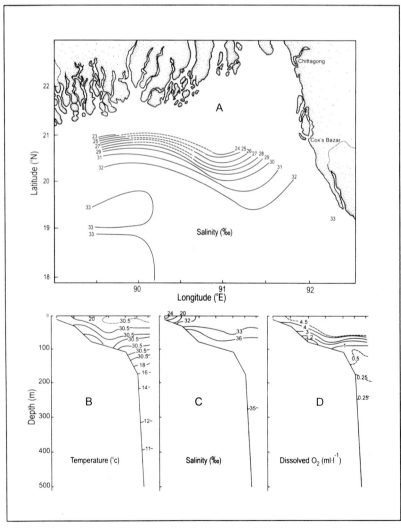

Fig 2. Surface (A) and subsurface (B-D) distribution of temperature, salinity and oxygen off Bangladesh. (Source: Saetre 1981).

variabilis) and ten other coral reef genera have been reported from offshore islands (Mahmood and Haider 1992). Information about seagrasses in Bangladesh is lacking, although seafronts of newly-formed islands and some low-lying coastal areas are often carpeted with seagrass. *Halodule uninervis* has been reported from sandy littoral zones (Islam 1980).

Mangroves are useful buffers against cyclones and tidal surges. The country has 587 400 ha of natural mangroves, with a further 100 000 ha replanted with various mangrove species (Mahmood 1995). Mangroves also provide a habitat for numerous crustaceans and finfishes (Mahmood et al. 1994). However, the mangroves of Bangladesh are under severe stress due to a reduced freshwater flow (resulting from construction of dams, barrages and embankments) and conversion to shrimp ponds (Mahmood 1995). Numerous studies have shown the relationship between fish/shrimp yields and the extent of mangrove cover (Martosubroto and Naamin 1977; Turner 1977; Pauly and Ingles 1986). Mangrove destruction in Bangladesh may be partly responsible for the observed decline of *Penaeus*

Table 2. Commercially important fishes and shrimps exploited in Bangladesh.

Scientific name	Common name
Demersal/small pelagics	
Pampus argenteus	Silver pomfret
P. chinensis	Chinese pomfret
Pomadasys argenteus	White grunter
Lutjanus johnii	Red snapper
Mene maculata	Moonfish
Tenualosa ilisha	Hilsa shad
Polydactylus indicus	Indian salmon
Lepturacanthus savala	Ribbonfish/hairtail
Arius spp.	Catfish
Johnius belangerii	Croaker
Otolithes ruber	Croaker
Nemipterus japonicus	Japanese threadfin bream
Upeneus sulphureus	Goatfish
Saurida tumbil	Lizardfish
Ilisha filigera	Bigeye ilisha
Sphyraena barracuda	Great barracuda
Congresox talabonoides	Indian pike conger
Larger pelagics	
Euthynnus affinis	Eastern little tuna
Katsuwonus pelamis	Skipjack tuna
Thunnus maccoyii	Southern bluefin tuna
Thunnus obesus	Bigeye tuna
Thunnus tonggol	Longtail tuna
Auxis rochei	Bullet tuna
Auxis thazard	Frigate tuna
Shrimps	
Penaeus monodon	Giant tiger
P. semisulcatus	Tiger shrimp
P. japonicus	Tiger shrimp
P. indicus	White shrimp
P. merguiensis	Banana/white shrimp
Metapenaeus monoceros	Brown shrimp
M. brevicornis	Brown shrimp
M. spinulatus	Brown shrimp
Parapenaeopsis sculptilis	Pink shrimp
P. stylifera	Pink shrimp

monodon post larvae in various areas and the declining shrimp catch rates of fishers.

Capture Fisheries

The capture fisheries of Bangladesh exploit a complex, multispecies resource. A single trawl haul, for instance, usually catches over a hundred species (Khan 1983; White and Khan 1985). A number of species, however, are targetted by the fishery. Table 2 gives a list of the commercially important fish and shrimp exploited in Bangladesh waters. A number of studies (particularly demersal trawl surveys – see Appendix III) have been undertaken to examine the development potential and status of these resources. In the case of trawlable (dominantly demersal) stocks, estimates of standing stock vary widely, from the 40 000-55 000 t given by Penn (1983) to the 260 000-370 000 t figure given by West (1973). The current consensus based on reassessment of these and related studies (Chowdhury and Khan 1981; Saetre 1981; Khan 1983; White and Khan 1985; Lamboeuf 1987) is a trawlable standing stock of 150 000-160 000 t in the coastal waters off Bangladesh. The composition and depth distribution of this stock is shown in Table 3 using the results of Lamboeuf (1987). The trawlable stock is dominated by sciaenids, catfishes, threadfin breams, carangids and goatfishes. Roughly 53% (85 400 t) of the standing stock consists of commercially important demersals, while 16% (25 700 t) consists of commercially important pelagics.

Studies of the shrimp resources provide standing stock estimates of 1 500-9 000 t of shrimps (West 1973; Penn 1983; White and Khan 1985; Van Zalinge 1986; Mustafa et al. 1987). Recent work suggests a maximum sustainable yield (MSY) figure of 7 000-8 000 t of penaeid shrimps (Khan et al. 1989). Available information on the magnitude of the pelagic resources of the country are quite limited. The only available estimate is that of Saetre (1981) who gives a pelagic standing stock figure of 90 000-160 000 t based on acoustic survey results. This figure is believed to be a very preliminary and conservative estimate which requires further research. Information pertaining to the country's tuna and large pelagic resources are quite patchy, emanating mostly from exploratory fishing trials. More reliable estimates await the results of assessment efforts further offshore.

A multiplicity of gear is used by fishers in Bangladesh to exploit the multispecies resources. Table 4 gives a summary of the fishing gear used in the country, together with their common target species/groups and depth of operation or deployment. The gear is subdivided into industrial and artisanal categories. The industrial gear consists of fish and shrimp trawlers which were introduced in 1974 and 1978,

Table 3. Estimate of biomass by family/group and depth strata off Bangladesh. (Adapted from Lamboeuf 1987).

Family/group	Common name	Biomass (t)	Relative abundance (%)	Distribution by depth strata (%)			
				10-20 m	20-50 m	50-80 m	80-100 m
1. Sciaenidae	Croakers	20 670	12.8	66.5	28.5	3.1	1.9
2. Ariidae	Catfishes	18 729	11.6	50.8	31.1	10.6	7.5
3. Nemipteridae	Threadfin breams	7 117	4.4	0.1	3.3	15.2	81.4
4. Carangidae	Jacks, scads	5 039	3.2	21.4	28.3	24.0	26.3
5. Mullidae	Goatfishes	4 811	3.0	2.4	47.8	35.7	14.1
6. Synodontidae	Lizardfishes	4 663	2.9	10.3	25.1	23.4	41.2
7. Trichiuridae	Hairtailfishes	4 043	2.5	20.8	48.5	16.2	14.5
8. Leiognathidae	Ponyfishes	3 998	2.5	24.5	69.5	4.1	1.9
9. Pomadasyidae	Grunters	3 415	2.1	81.0	15.3	2.5	1.2
10. Clupeidae	Sardines	3 109	1.9	43.7	45.1	9.7	1.5
11. Scombridae	Mackerels	1 836	1.1	10.5	10.8	21.6	57.1
12. Priacanthidae	Bullseyes	1 433	0.9	0.1	1.6·	7.6	90.7
13. Stromateidae	White Chinese pomfrets	1 348	0.8	44.7	34.8	18.2	2.3
14. Cephalopods	Squids, cuttlefishes	1 296	0.8	12.8	10.7	27.7	48.8
15. Engraulidae	Anchovies	1 082	0.7	36.5	45.4	16.4	1.7
16. Gerreidae	Silver-biddies	959	0.6	2.5	50.7	37.2	9.6
17. Harpadontidae	Bombay duck	783	0.5	65.5	34.5	0	0
18. Lutjanidae	Snappers	356	0.2	24.0	41.4	14.9	19.7
19. Rajidae	Rays, skates	6 714	4.2	88.1	10.8	0.9	0.2
20. Others	-----	69 679	43.3	-	-	-	-
Commercially important demersals		85 366	53.0	38.6	26.4	12.7	22.3
Commercially important pelagics		25 676	15.9	19.6	29.2	22.0	29.2
Total		161 080	100	35.0	28.5	14.9	21.6

respectively. From a few units in 1974, fish trawlers expanded to a high of 137 units in 1980 and declined to 100 units in 1985. Currently, there are 13 fish trawlers ranging in size from 17.5 m to 28 m (350-1 200 hp). Fish trawlers have headrope lengths of 18-32 m with 60-65 mm codend mesh size. From 4 shrimp trawlers operating in 1978, the number increased to 100 units in 1984 (White and Khan 1985). At present, there are 39 shrimp trawlers with overall length from 20.5 m to 44.5 m. Shrimp trawlers use riggers and operate twin nets with 45-50 mm codend mesh size. Trawlers commonly operate in waters 40-100 m deep. In the early 1990s, trawlers landed an average of 3 500 t of shrimps and 100 000 t of fish annually (Table 5). It is estimated that 80% of the fish catch is not landed by trawlers but discarded at sea (White and Khan 1985). Table 5 shows the annual shrimp catch and fishing effort of trawlers from 1981 to 1991.

The artisanal fisheries utilize a variety of gear to exploit the multispecies mix. These include five types of gillnets, three types of set bagnets, trammel nets, bottom longlines, beach seines and other gear deployed in shallow estuarine and coastal areas (Table 4). With the exception of bottom set gillnets, artisanal gear is commonly deployed in depths of only up to 30 m. With the exception of drift gillnets, marine set bagnets and bottom longline operations, artisanal fishing involves the use of non-motorized boats. It is estimated that the artisanal fishery in the early 1990s used about 4 900 boats (using roughly 5 500 artisanal gear units) and landed about 247 000 t of fish and shrimps annually. About 55% of the annual production was contributed by drift gillnets and 30% by estuarine set bagnets.

Fig. 3 illustrates the common areas of operation or deployment of the more important fishing gear used in the coastal waters off Bangladesh. The pushnets, estuarine set bagnets and beach seine are used in the shallow, estuarine parts throughout the country's coast. The bottom longlines, trammel nets and marine set bagnets are deployed with increasing depth (up to about 30 m). Drift gillnets are also deployed in these areas, targeting principally small pelagics (i.e., *Tenualosa ilisha, T. kelee, T. toli*). Trawlers predominate in the deeper areas from 30 m to 100 m depth. Given that the size of fish and shrimps increases with depth, they sequentially exploit many species. Management of fisheries, therefore, requires consideration of this overlapping of target species and their sequential exploitation.

Moreover, fishing in shallow areas down to 30 m depth is quite intense and competition between artisanal and industrial gear has increased (Islam et al. 1993; Paul et al. 1993; Rahman 1993; Khan et al. 1994).

Table 4. Fishing gear used in Bangladesh with common target species/group and depth of operation.

Fishery/gear	Target species/group	Depth of operation (m)
A. Industrial		
1. Trawl		
a. Shrimp trawl	*Metapenaeus monoceros*	40 - 100
b. Fish trawl	*Penaeus monodon*	
	P. semisulcatus,	40 - 100
	P. merguiensis, sciaenids, catfish, Indian salmon (*Polynemus indicus*), sharks and rays, pomfret	
B. Artisanal		
1. Gillnet		
a. Drift gillnet	*Tenualosa ilisha*	down to 30
b. Fixed gillnet	*Tenualosa ilisha*	8 - 10
c. Large mesh drift gillnet	Sharks	down to 30
d. Bottom set gillnet	Indian salmon	down to 80
e. Mullet gillnet	Grey mullet	5 - 10
2. Set bagnet		
a. Estuarine set bagnet	Brown and pink shrimp, Bombay duck (*Harpadon nehereus*), sciaenids, anchovies, clupeids, hairtail	5 - 20
b. Marine set bagnet	Brown and pink shrimp, hairtail, Bombay duck, anchovies, clupeids	10 - 30
c. Large mesh set bagnet	*Lates calcarifer* (sea perch)	10 - 30
3. Trammel net	White, tiger, and brown shrimp, sciaenids, catfish	8 - 20
4. Bottom longline	Sciaenids	10 - 30
5. Beach seine	Small brown and pink shrimp, clupeids, anchovies, sciaenids hairtail	8 - 10
6. *Char pata jal*	Brown, white and tiger shrimp	down to 10
7. Cast net	Brown, white and tiger shrimp	down to 10
8. Push net	Larvae of *P. monodon*	down to 10
9. Fixed bagnet	Larvae of *P. monodon*	down to 5
10. Dragnet	Larvae of *P. monodon*	down to 2

Table 5. Annual shrimp catch and fishing effort of trawlers in Bangladesh during the period 1981-1982 to 1990-1991. (Sources: Mustafa and Khan 1993; Khan and Latiff 1995).

Fishing season	Standard effort (days)			Shrimp catch (t)			Catch per effort (kg·day⁻¹)
	Shrimp trawlers	Fish trawlers	Total	Shrimp trawlers	Fish trawlers	Total	
1981-82	2 987	795	3 782	1 340	357	1 697	449
1982-83	4 510	2 514	7 024	2 004	1 116	3 120	444
1983-84	6 087	3 575	9 662	3 441	2 020	5 461	565
1984-85	6 267	1 892	8 159	4 239	1 279	5 518	676
1985-86	5 941	502	6 444	3 716	318	4 034	626
1986-87	6 449	479	6 928	4 178	310	4 488	648
1987-88	6 239	344	6 583	3 339	184	3 523	535
1988-89	6 615	330	6 945	4 661	232	4 893	705
1989-90	5 460	86	5 546	3 086	48	3 134	565
1990-91	4 437	62	4 499	3 384	47	3 431	763

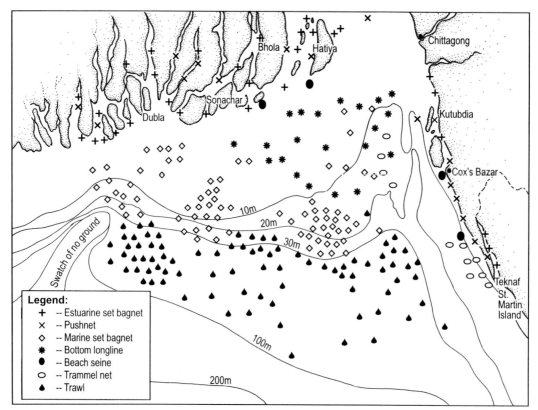

Fig 3. Depth distribution and area of operations of dominant fishing gears used in the coastal waters off Bangladesh. (Source: Khan et al. 1994).

Sequential exploitation of target resource groups off Bangladesh may be best illustrated by the case of penaeid shrimps (particularly *Penaeus monodon*). Fig. 4 illustrates graphically the various stages of the life cycle history of penaeid shrimps and the variety of gear used to exploit them at each stage. Push nets or larval nets exploit the postlarval stage to supply the wild seed demand of the expanding shrimp aquaculture industry. Estuarine set bagnets exploit the juvenile shrimp, while beach seines and marine set bagnets exploit the subadults. Trammel nets and trawlers are used to harvest the adult shrimps in deeper waters.

Paul et al. (1993) provided estimates of the size distribution of *P. monodon* catches resulting from the sequential fisheries operating in Bangladesh coastal waters (Fig. 5). This study noted that over 2 billion *P. monodon* postlarvae were caught by seed collectors annually (i.e., 99.6% of the total number harvested, the rest coming from other types of gear). The shrimp seed collection wreaks tremendous damage on the larvae and juveniles of other species, apart from the danger of growth and recruitment overfishing for the target *P. monodon* stock. Fig. 5 also illustrates the considerable growth overfishing resulting from estuarine set bagnet operations. Suggestions for the gradual phaseout of estuarine set bagnets have been made (Islam et al. 1993; Paul et al. 1993; Khan et al.

1994). However, the socioeconomic dislocation of 200 000 people directly dependent on the estuarine set bagnet fishery requires serious consideration and more innovative approaches (e.g. in the area of alternative livelihood development). The problem of growth overfishing illustrated here for *P. monodon* is one which impacts other species exploited in Bangladesh, given the sequential fisheries operating therein and the intense fishing effort in the shallow, coastal areas.

Management Issues and Opportunities

A number of comprehensive assessments of coastal capture fisheries and the environment of Bangladesh have been conducted in recent years (BOBP 1993; Rahman 1993; Mahmood 1995). These studies raise a host of management issues and also make various recommendations. In this contribution, we have chosen to simply highlight some of the issues which we believe require the most immediate attention. The studies referred to above give a more detailed treatment.

Excessive fishing effort in nearshore waters (particularly between the shoreline and 40 m depth) is the primary issue requiring attention. The increasing fisher population and intense competition in nearshore waters between artisanal gears has led to a decline in their incomes. In the case of industrial

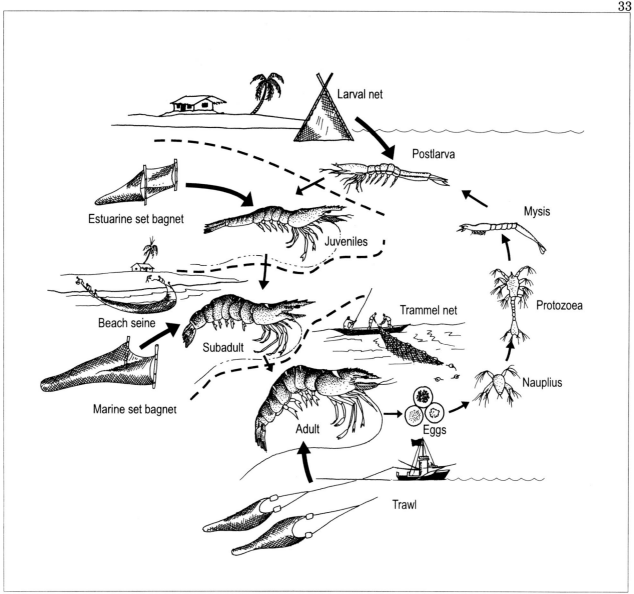

Fig 4. Graphical illustration of penaeid shrimp life cycle and fishing gears used to exploit them at various stages in Bangladesh. (Source: Khan and Latiff 1995).

fisheries, the existing number of trawlers is deemed sufficient to exploit the available shrimp and demersal fish stocks and no expansion in trawler effort is foreseen. The incursion of trawlers in nearshore, artisanal fishing grounds increases competition and conflict in these areas and requires increased enforcement efforts.

The intense fishing pressure on currently exploited resources may be gleaned from the results of recent assessments. Table 6 gives the population parameters of species caught by estuarine set bagnets obtained by Khan et al. (1992). Estimates of asymptotic length (L_∞), growth constant (K) and natural mortality rate (M) are consistent with those available in other literature. Note the very high fishing mortality (F) and exploitation rates (E) of most species, as well as the relatively small length at first capture (L_c). Apart from

Acetes indicus and *Setipinna taty*, these species are evidently fished heavily. Table 7 gives the results of various assessments conducted for trawl-caught species in Bangladesh. There is also a preponderance of very high F and E values for the species investigated.

Growth overfishing is a serious problem and results principally from the heavy concentration of push nets and estuarine set bagnets (and marine set bagnets to some extent) in nearshore waters (see Figs. 3 to 5 and L_c values in Table 6). Push nets supply shrimp seeds to the expanding shrimp aquaculture industry. The development of hatcheries to supply shrimp seeds and the improvement of handling and transport of these seeds could reduce push net effort and the damage it entails. In the case of estuarine set bagnets, area and seasonal closures for operation of the gear and the development of alternative livelihood projects

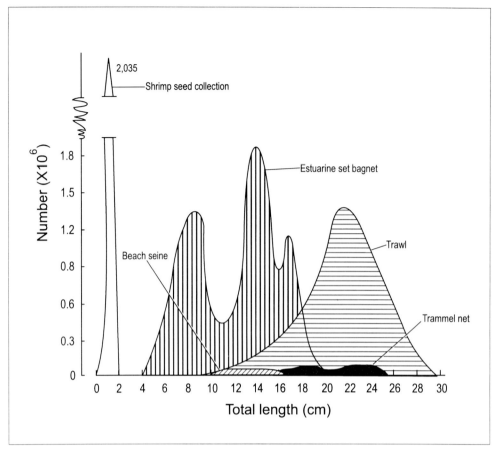

Fig 5. Size distribution of giant tiger shrimp (*Penaeus monodon*) catches by various gears used in Bangladesh. (Source: Khan et al. 1994).

Table 6. Population parameters of some species common in the catch of estuarine set bagnets in Bangladesh. (Source: Khan et al. 1992).

Species	L_∞ (cm)[a]	K (year⁻¹)	M (year⁻¹)	F (year⁻¹)	L_c (cm)[b]	E (=F/Z)
Crustaceans						
Penaeus monodon	31.4	0.72	1.42	8.38	13.8	0.86
P. indicus	22.8	0.55	1.30	3.70	5.9	0.74
Metapenaeus monoceros	19.8	0.44	1.17	3.65	5.9	0.76
M. brevicornis	15.6	0.31	1.00	4.24	4.8	0.81
M. spinulatus	20.1	0.39	1.08	5.90	5.3	0.84
Parapenaeopsis sculptilis	16.9	0.76	1.75	4.15	15.3	0.70
P. stylifera	14.4	1.66	3.06	3.00	2.8	0.50
Acetes indicus	5.0	0.73	2.40	1.10	2.0	0.31
Macrobrachium rosenbergii	35.5	0.34	0.84	1.96	7.3	0.70
Palaeomon styliferus	15.4	0.63	1.59	3.20	3.74	0.67
Fish						
Raconda russelliana	23.6	0.43	1.10	2.10	2.9	0.66
Setipinna taty	21.3	0.53	1.28	0.80	15.8	0.28
Stolephorus tri	16.8	0.65	1.59	9.00	3.4	0.85
Harpadon nehereus	34.9	0.38	0.91	3.75	6.3	0.80
Lepturacanthus savala	93.0	0.29	0.58	2.62	22.6	0.82
Eleutheronema tetradactylum	38.1	0.10	0.85	3.50	5.3	0.87
Polynemus paradiseus	21.6	0.52	1.28	4.72	2.7	0.79
Sillaginopsis panijus	43.3	0.38	0.86	2.70	13.1	0.76
Sillago sihama	27.4	0.39	0.99	3.00	5.1	0.75

[a] Asymptotic total length in the von Bertalanffy growth equation.

[b] Mean total length at first capture.

Table 7. Population parameters of some trawl-caught species in Bangladesh. (Source: DOF internal reports).

Species	L_∞ (cm)[a]	K (year^{-1})	M (year^{-1})	F (year^{-1})	L_c (cm)[b]	E(=F/Z)
Crustaceans						
Penaeus monodon (F)	30.5	1.14	1.94	4.89	17.5	0.71
P. monodon (M)	31.5	1.35	2.14	3.58	15.7	0.62
Metapenaeus monoceros (M)	15.7	1.60	2.91	2.98	8.9	0.50
M. monoceros (F)	18.5	1.65	2.84	1.68	9.5	0.37
Fish						
Pampus argenteus	30.5	1.66	2.35	2.90	-	0.55
Upeneus sulphureus	22.0	1.10	2.96	7.63	-	0.72
Nemipterus japonicus	25.0	1.06	1.94	1.81	-	0.48
Saurida tumbil	39.0	0.97	1.66	0.88	-	0.35
Pomadasys argenteus	56.9	0.38	0.81	0.79	-	0.51
Lepturacanthus savala	105.0	0.85	1.33	0.73	-	0.65
Harpadon nehereus	38.3	0.42	0.94	0.60	-	0.38
Lutjanus johnii	64.7	0.28	0.59	2.11	-	0.78
Ariomma indica	22.0	1.12	2.10	3.43	-	0.62

[a] Asymptotic total length in the von Bertalanffy growth equation.
[b] Mean total length of first capture.

Table 8. Typical coastal transect showing main activities and issues relevant to effective integrated coastal zone and coastal fisheries management in Bangladesh.

Major zones	Terrestrial			Coastal			Marine	
	Upland (>18% slope)	Midland (8-18% slope)	Lowland (0-<8% slope)	Interface (1 km inland from HHWL-30 m depth)	Nearshore (30 m-200 m depth)		Offshore (>200 m depth-EEZ)	Deepsea (beyond EEZ)
Main resource uses/activities	Logging Mining Urban development Agriculture	Logging Mining Urban development Industries Agriculture Brick/boulder extraction	Logging Agriculture Urban development Industries Tourism	Mangrove forestry Aquaculture Ports/marine transport Artisanal fishing	Artisanal fishing Commercial/industrial fishing Marine transport Pearl collection		Commercial/industrial fishing	Marine transport
Main environmental issues/impacts on the coastal zone	Siltation Erosion Flooding Agrochemical loading Water pollution Toxic mine tailings	Siltation Erosion Flooding Toxic mine tailings	Siltation Erosion Flooding Agrochemical loading Sewage pollution Industrial pollution	Reduction of biodiversity Habitat degradation Overfishing Beach erosion Organic loading Oil spills/slicks	Reduced biodiversity Overfishing Oil spills/slicks		Overfishing Oil spills/slicks	Oil spills/slicks

in the context of integrated community development have been suggested (Khan et al. 1994).

The evident need for the capture fisheries sector in Bangladesh is to reduce fishing effort (particularly in nearshore waters) and rationalize current exploitation and utilization of fishery resources. Given the country's development context it is a clear challenge to find means to achieve this. Briefly, the following points deserve attention in this regard:

- introduction of a zoning scheme governing artisanal and industrial gear deployment to minimize conflicts and optimize utilization of exploited resources;
- development of small and large pelagics fisheries to relocate effort away from the heavily fished shrimps and demersals;
- reduction of by-catch and discards, particularly in the industrial fisheries;
- enhancement of participatory management and an integrated, community-based development approach;
- improvement of the information and extension capabilities of the Department of Fisheries in support of fisheries management efforts;
- computerization and improvement of the capture fisheries statistical system;
- improvement of research, enforcement and monitoring capabilities consistent with requirements of the Marine Fisheries Ordinance; and
- improvement of the information and extension capabilities of the Department of Fisheries in support of fisheries management efforts; and
- reactivation of the Bangladesh National Fishermens' Cooperative Society to enhance support, management and rehabilitation programs.

Apart from these evident fisheries-specific issues, a number of cross-sectoral issues require attention for long-term sustainability of coastal fisheries. Table 8 summarizes the main activities and issues which impact the coastal zone (and ultimately coastal fisheries) in Bangladesh. The multiple issues impacting the country's coastal zone require multiple, but integrated, interventions. Integrated coastal zone management (ICZM), therefore, requires increased emphasis and support to resolve the multiple impacts (both actual and potential) on the coastal ecosystem. Various ICZM initiatives in the country emphasize the need for improved collaboration and coordination among relevant government agencies (e.g., Ministry of Fisheries and Livestock, Marine Mercantile Depart-

ment, Ministry of Land, Ministry of Environment and Forests, Ministry of Defense) as well as an upgrading of their capabilities to effectively resolve coastal zone impacts.

References

BOBP. 1993. Studies of interactive marine fisheries of Bangladesh. Bay of Bengal Programme, Madras, India. 117 p.

Chowdhury, W. and M.G. Khan. 1981. A report on results of the fisheries survey cruise no.3 (Oct. 1981) of the R.V. Anusandhani to the shrimp grounds in the Bay of Bengal of Bangladesh. Marine Fisheries Exploration and Biological Research Station, Chittagong.

Eysink, W.D. 1983. Basic consideration on the morphology and land accretion potentials in the estuary of the Lower Meghna River. Tech. Rep. 15. 39 p. Land Reclamation Project. Bangladesh Water Development Board.

FAO. 1968. Report to the government of East Pakistan on oceanography of the North Eastern Bay of Bengal useful to the fisheries development and research in East Pakistan based on the work of Dr. Z. Popovioi. PAK/F1 102 p. United Nations Development Programme.

Holmgren, S. 1994. An environmental assessment of the Bay of Bengal Programme, Madras. BOBP/REP/67: 256 p.

Islam, A.K.M.N. 1980. A marine angiosperm from St. Martin's Island. Bangladesh J. Bot. 9(2): 177-178.

Islam, M.S., M.G Khan, S.A. Quayum, M.N.U. Sada and Z.A. Chowdhury. 1993. The estuarine set bagnet fishery. In Studies of interactive marine fisheries of Bangladesh. BOBP/WP/89. 19-50 p. Bay of Bengal Programme. Madras, India.

Khan, M.G. 1983. Results of the 13th cruise (July 1983) with the R.V. Anusandhani of the demersal fish and shrimp ground of the Bay of Bengal, Bangladesh. Marine Fisheries Research, Management and Development Project, Chittagong.

Khan, M.G. and M.A. Latif. 1995. Potentials, constraints and strategies for conservation and management of open brackishwater and marine fishery resources. Paper presented at the National Seminar on Fisheries Resources Development and Management organized by MOFL, Bangladesh, with FAO and ODA, 29 October to 31 November 1995. Dhaka, Bangladesh. 19 p.

Khan, M.G., M.G. Mustafa, M.N.U. Sada and Z.A. Chowdhury. 1989. Bangladesh offshore marine fishery resources studies with the special reference in the penaeid shrimp stock 1988-89. Annual report: Marine Fisheries Survey, Management and Development Project. GOB, Chittagong. 213 p.

Khan, M.G., M.S. Islam, M.G. Mustafa, M.N.U. Sada and Z.A. Chowdhury. 1994. Biosocioeconomic assessment of the effect of the estuarine set bagnet on the marine fisheries of Bangladesh. BOBP/WP/94. 28 p. Bay of Bengal Program, Madras, India.

Khan, M.G., M.S. Islam, S.A. Quayum, M.N.U. Sada and Z.A. Chowdhury. 1992. Biology of the fish and shrimp population exploited by the estuarine set bagnet. Paper presented at the BOBP Seminar, 12-15 January 1992, Cox's Bazar, Bangladesh. 20 p.

Lamboeuf, M. 1987. Bangladesh demersal fish resources of the continental shelf. FAO/BGD. FE:DP/BGD/80/075. 26 p. Marine Fisheries Research Management and Development Project.

Mahmood, N. 1995. On the fishery significance of the mangroves of Bangladesh. Paper presented at the Workshop on Coastal Aquaculture and Environmental Management, 25-28 April 1995. Cox's Bazar. Organized and sponsored by Institute of Marine Science, Chittagong University (IMS, CU) UNESCO.

Mahmood, N., N.J.U. Chowdhury, M.M. Hossain, S.M.B. Haider and S.R. Chowdhury. 1994. Bangladesh, p. 75-129. *In* An environmental assessment of the Bay of Bengal region. Swedish Centre for Coastal Development and Management of Aquatic Resources, Bay of Bengal Programme. BOBP/Rep/67.

Mahmood, N. and S.M.B. Haider. A note on corals of the St. Martin's Island. Bangladesh. Chittagong University Studies. (In press).

Martosubroto, P. and N. Naamin. 1977. Relationship between tidal forests (mangroves) and commercial shrimp production in Indonesia. Mar. Res. Indones. No. 8: 81-86.

Mustafa, M.G. and M.G. Khan. 1993. The bottom trawl fishery, p. 89-106. *In* studies of interactive marine fisheries of Bangladesh. BOBP/WP/89. 117 p. Bay of Bengal Programme, Madras, India.

Mustafa, M.G., M.G. Khan and M. Humayon. 1987. Bangladesh Bay of Bengal penaeid shrimp trawl survey results, *R.V. Anusandhani*, November 1985, January 1987. UNDP/FAO/GOB. Marine Fisheries Research, Management and Development Project, Chittagong. 15 p.

Paul, S.C., M.G. Mustafa, Z.A. Chowdhury and M.G. Khan. 1993. Shrimp seed collection. *In* Studies of interactive marine fisheries of Bangladesh. BOBP/WP/89, 3-17 p. Bay of Bengal Programme, Madras, India.

Pauly, D. and J. Ingles. 1986. The relationship between shrimp yields and intertidal vegetation (mangroves). p. 277-283. *In* IOC/FAO Workshop on Recruitment in Tropical Coastal Demersal Communities, Ciudad del Carmen, Campeche, Mexico, 21-25 April 1986.

Penn, J.W. 1983. An assessment of potential yield from the offshore demersal shrimp and fish stock in Bangladesh water (including comments on the trawl fishery 1981-1982). FE:DP/BGD/81. 634. Field Doc. 22 p. Fishery Advisory Service (Phase-II) Project, Rome, FAO.

Rahman, A.K.A. 1993. Marine small scale fisheries in Bangladesh. Regional Office for Asia and the Pacific, Bangkok. 55 p.

Saetre, R. 1981. Survey on the marine fish resources of Bangladesh, November-Dececember 1979 and May 1980. Report on surveys in the *R.V. Dr. Fridtjof Nansen*. Institute of Marine Research, Bergen. 67 p.

Sivasubramaniam, K. 1985. Marine fishery resources of the Bay of Bengal. Bay of Bengal Programme, Madras, India. 66 p.

Turner, R.E. 1977. Intertidal vegetation and commercial yields of penaeid shrimps. Trans. Am. Fish. Soc. 106: 411-416.

Van Zalinge, N.P. 1986. Bangladesh shrimp resources. Field Doc. BAD/80/025.

West, W.Q.B. 1973. Fishery resources of the upper Bay of Bengal. Indian Ocean Programme. FAO/UNDP/IOFC/DEV/73/28. 42 p.

White, T.F. and M.G. Khan. 1985. Marine fishery resources survey. Demersal Trawling Survey Cruise Rep. 1. FAO/BGD/80/025/CRI. 67 p. Chittagong.

The Marine Fisheries of Indonesia, with Emphasis on the Coastal Demersal Stocks of the Sunda Shelf

B. E. PRIYONO

Directorate General of Fisheries
Department of Agriculture
Jakarta, Indonesia

B. SUMIONO

Agency for Agricultural Research and Development
Institute for Marine Fisheries
Jakarta, Indonesia

PRIYONO, B.E. and B. SUMIONO. 1997. The marine fisheries of Indonesia, with emphasis on the coastal demersal stocks of the Sunda shelf, p. 38-46. *In* G. Silvestre and D. Pauly (eds.) Status and management of tropical coastal fisheries in Asia. ICLARM Conf. Proc. 53, 208 p.

Abstract

Indonesia, with a combined fisheries catch and aquaculture production of 2.8 million t·year⁻¹, has the tenth largest fishery in the world. A potential for future development through mariculture still exists in the eastern part of the archipelago. In coastal waters, especially in the western part, the increasing number of small fishing vessels, with either inboard or outboard motors, has caused overfishing of coastal resources. This problem has led the Indonesian government to put emphasis on developing offshore fisheries, notably for large pelagics.

Following trawl surveys conducted in the early 1960s, demersal trawling -- mainly for shrimp and high-value fish -- started in the mid-1960s in the Malacca Strait. From there it expanded throughout the waters of the western part of the archipelago. In the east, trawl surveys by Indonesian vessels in the Arafura Sea began in 1964. A ban on trawling in all parts of Indonesia (except in the Arafura Sea) was implemented in 1980. This paper briefly assesses some effects of the ban and of other coastal fisheries management policies in Indonesia.

Introduction

Indonesia is an archipelagic country located in Southeast Asia. Consisting of over 17 000 large and small islands, it straddles the equator between 6°N and 11°S latitude (Fig. 1). When it claimed a 200 nautical mile exclusive economic zone (EEZ) in 1980, Indonesia became the world's largest archipelago, stretching some 3 000 miles along the equator.

With combined fishery catches and aquaculture production of 2.8 million t·year⁻¹, Indonesia is the tenth largest fishing nation in the world. The following account, updating some issues raised in Martosubroto and Badrudin (1984) and Bailey et al. (1987), reviews key aspects of the fisheries sector in Indonesia, with some emphasis on the coastal fisheries of the western archipelago (i.e., the Sunda Shelf area).

Indonesian Marine Environment

The total water area claimed by Indonesia covers 5.8 million km², of which 3.1 million km² are territorial and archipelagic waters with the remainder being claimed as part of the Indonesian EEZ. These waters consist of productive continental shelves (the Sunda Shelf in the west and the Sahul Shelf in the east) and deeper, less productive waters in between (Dalzell and Pauly 1989).

The demersal fishing grounds (including shrimping grounds) and the fishing grounds for small pelagic fishes are found in the shallow parts of the shelves, usually down to 50 m depth (Martosubroto 1996; Martosubroto et al. 1996). These include a variety of habitat types such as soft (muddy or sandy) bottoms and hard, rocky areas generally interspersed with coral

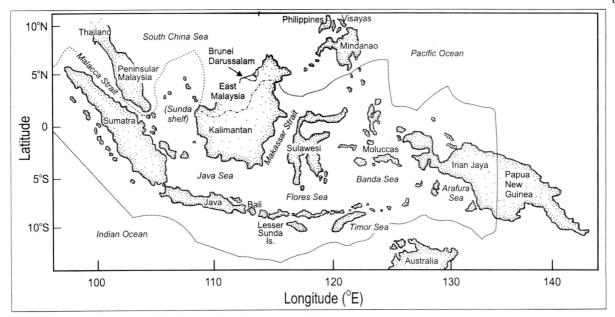

Fig. 1. Indonesian territorial waters and exclusive economic zone.

reefs. The oceanic pelagic species are found in deep offshore and oceanic waters where water depth exceeds 100 m.

Sea conditions in Indonesia are governed by the alternating southwest (December to February) and northeast (June to August) monsoons. During the southwest monsoon, northeasterly currents intensify in the western part of the South China Sea dominating surface water flows, while the Pacific Ocean waters from the Flores and Celebes Seas flow into the Java Sea. During the northeast monsoon, these surface flows are reversed (Wyrtki 1961; Roy 1996; Sharp 1996).

In general, average annual variations of sea surface temperatures are less than 2°C in western Indonesian waters, with slightly higher differences, of 3°C to 4°C, in the Banda, Arafura and Timor Seas, and South of Java (Soegiarto and Birowo 1975; Roy 1996).

On the Sunda Shelf and Sahul Shelf, primary production is generally high due to the influence of river water discharges. Primary production is also high during seasonal localized upwellings, notably along the south coast of Java and especially the Bali Strait, where they support a relatively large stock of *Sardinella lemuru*, which in turn supports an important fishery (Dwiponggo 1974; Ghofar and Mathews 1996; Venema 1996).

Coral reefs are of great economic importance for Indonesia, particularly in the eastern part, where they contribute to the food resources of the local population with a supply of fish, invertebrates and plants. Coral reefs are also used as a source of building materials. This, combined with other stresses, has led to the degradation of many Indonesian coral reefs. Overexploitation has also reduced other valuable resources, e.g., spiny lobster, ornamental fish and sea turtle populations. A particularly insidious and recent stress is cyanide fishing for the Hong Kong live fish market. This has resulted in the decline of live coral cover and of large coral reef fishes, such as Napoleon or humphead wrasses, *Cheilinus undulatus* (Johannes 1995; Johannes and Riepen 1995; Erdmann and Pet-Soede 1996; and see p. 44).

Live coral cover, which determines reef carrying capacity for fishes, is declining. Only 7% of coral reefs still have an excellent coral cover (i.e., 75%-100%), 34% good (50%-75%), 49% fair (25%-50%) and 11% poor (0-25%).

Indonesia has approximately 3.8 million ha of mangrove. Of these, over 2.9 million are located in Irian Jaya; 417 000 ha in Sumatra; 275 000 ha in Kalimantan; and 34 300 ha in Java. Mangroves are distributed throughout coastal areas, often near estuaries, and between tidal swamp forests and seagrass beds.

In many areas, mangroves have been considerably reduced due to logging for timber and firewood, reclamation for port and industrial sites, and transformation into brackishwater fishponds. Most of the original mangrove forests have disappeared from the north coast of Java, and the areas have been converted to shrimp and fish ponds. This may have effects on coastal shrimp fisheries, as there is a relationship between the surface area of mangroves and shrimp yields in Indonesia (Martosubroto and Naamin 1977), similar to that reported from other parts of the world (Turner 1977; Pauly and Ingles 1986).

Capture Fisheries Sector

Small-scale fisheries

Small-scale fisheries are defined in Indonesia as including all fishing units which use boats powered by sail or outboard engines. Fishers who operate fixed or mobile gear without a boat are classified as small-scale, whatever the size of their gear. While the small-scale fishing operations tend to be labor-intensive and limited to coastal waters, the sector still contributes about 90% of the total marine fishery catches of Indonesia. However, intense fishing pressure and competition between small-scale fishers and trawl operators have caused severe overexploitation of the coastal resources throughout much of the western archipelago. This situation led to serious social unrest in the 1970s, which resulted in the trawling ban of 1980 (Sardjono 1980).

The Directorate General of Fisheries (DGF) classifies Indonesian small-scale craft into three categories:

1. Dug-outs (*jukung*) - made of hollowed-out logs. These comprised 32% of fishing boats in 1993. More than 77% of these crafts are found in the eastern part of Indonesia, i.e., the Moluccas and Irian Jaya, though they also occur in Sulawesi and the lesser Sunda Islands.

2. Nonpowered plank-built boats, divided into:
   a) small - up to 7 m long;
   b) medium - from 7 to 10 m; and
   c) large - more than 10 m.
The total number of boats in these categories is about 16 000, or 15% of all fishing boats in Indonesia.

3. Boats powered by outboard engines, including designs with a long, trailing propeller shaft, and gasoline or diesel engine from 2 to 15 hp. About 21% of fishing boats in Indonesia use outboard motor; 43% of these operate along the north coast of Java.

In general, small-scale craft for daily trips have a crew of two to three, though small-scale fishers often operate alone. During the 1970s, 96-98% of fishing boats were small-scale, but the number has dropped since, due to the gradual increase of boats powered with outboard engines (Table 1). In 1993, there were 329 962 small-scale boats, or 85% of the total number of fishing boats (Table 2).

The gear most commonly used for small-scale fishing are gillnets (29% of all gear), drift gillnets, seine nets, cast nets and traps, though specialized traditional gear is also used for shellfish and seaweed collection. The most popular gear for catching shrimps are trammel nets and shrimp (monofilament) gillnets, especially where the trawl ban has made larger

Table 1. Number of fishing boats, 1968-1993. (Source: DGF 1985, 1995).

| Year | Small-scale | | | | Medium- and large-scale | | Total |
	Without engine	Outboard engine	Subtotal	%	With inboard engine	%	
1968	278 206	-	278 206	98.0	5 707	2.0	283 913
1969	275 314	-	275 314	98.1	5 319	1.9	280 633
1970	289 402	2 798	292 200	98.9	3 236	1.1	295 436
1971	277 662	2 652	280 314	98.4	4 524	1.6	284 838
1972	286 463	2 877	289 340	98.0	5 941	2.0	295 281
1973	230 615	5 019	235 634	97.0	7 248	3.0	242 882
1974	257 164	5 931	263 095	97.3	7 274	2.7	270 369
1975	242 221	6 771	248 992	96.8	8 160	3.2	257 152
1976	228 244	7 746	235 990	96.0	9 735	4.0	245 752
1977	228 228	9 601	237 829	95.7	10 715	4.3	248 544
1978	222 121	13 226	235 347	94.9	12 766	5.1	248 113
1979	225 804	17 343	243 147	94.3	14 758	5.7	257 905
1980	226 866	26 523	253 389	93.2	18 467	6.8	271 856
1981	225 949	43 831	269 780	91.3	25 847	8.7	295 627
1982	215 466	55 265	270 731	90.1	29 818	9.9	300 549
1983	220 706	57 490	278 196	90.6	28 861	9.4	307 057
1984	219 929	61 789	281 718	89.9	31 922	10.2	313 640
1985	220 823	61 867	282 690	89.3	33 756	10.7	316 446
1986	219 130	62 809	281 939	88.6	36 156	11.4	318 095
1987	222 233	70 380	292 613	87.6	41 459	12.4	334 072
1988	220 138	71 154	291 292	87.2	42 910	12.8	334 202
1989	218 553	71 122	289 675	86.4	45 413	13.6	335 088
1990	225 359	73 144	298 503	86.5	46 542	13.5	345 045
1991	231 659	75 416	307 075	86.6	47 709	13.4	354 784
1992	229 377	77 779	307 156	85.6	51 750	14.4	358 906
1993	247 745	82 217	329 962	84.7	59 536	15.3	389 498

Table 2. Number of fishing boats, by sector and size, 1993. (Source: DGF 1995).

Boat type	Number of boats
Small-scale fishery:	
Dug out boats	123 760
Plank-built boats	
- small	73 452
- medium	41 822
- large	8 711
Boats with outboard engines	82 217
Subtotal	329 962
Medium-scale fishery:	
Boats with inboard engines	
- less than 5 GT	43 396
- 5-10 GT	9 791
- 10-20 GT	2 812
- 20-30 GT	1 558
Subtotal	57 557
Large-scale fishery:	
Boats with inboard engines	
- 30-49 GT	1 170
- 50-100 GT	351
- 100-200 GT	213
- more than 200 GT	245
Subtotal	1 979
Total	389 498

shrimps more accessible to small-scale fishers. The use of such gear has expanded rapidly in recent years along the southeast and north coast of Java, the east coast of Sumatra and in the Malacca Strait.

Medium-scale fisheries

Medium-scale fisheries in Indonesia include boats of generally less than 30 GT that use inboard engines (about 15% of the Indonesian fishing fleet in 1993). The owners of medium-scale fishing boats tend to have access to few or none of the shore-based amenities such as ice plants, cold storage facilities or workshops, the use of which characterize the large-scale fisheries. Investment levels for medium-scale fishing units (boat and gear) are commonly in the range of Rp.5-20 million. Individual operators own one or, at most, a few fishing units.

Medium sized vessels are based mainly in the Malacca Strait, the Moluccas and Irian Jaya. After the trawl ban of 1980, the number of medium boats and artisanal trawlers increased, i.e., modern vessels with sizes ranging from 10 to 35 GT (Tables 1 and 2).

Medium-scale fishers use gear such as skipjack poles and lines, purse seines, gillnets and otter trawls (until the 1980 trawl ban). Many of them also catch shrimp using trammel nets along the south coasts of Java. Thus, medium-scale fisheries in Indonesia cannot be differentiated from other fisheries on the basis of the gear type they use.

Large-scale fisheries

Indonesia's large-scale fisheries operate in isolation from medium and small scale fisheries, due to their strong export orientation (which limits competition in the local market) and the offshore location of their fishing operations. Most of these fisheries target high-value commodities, especially shrimp and tuna. They use shrimp nets (equipped with by-catch excluder devices which have gradually replaced standard demersal trawls since 1992) and tuna longlines, i.e., large-scale gear deployed mainly in the eastern part of Indonesia.

Trawling for shrimp in the Arafura Sea began in 1969 with nine trawlers, ranging in size from 90 to 600 GT, and from 260 to 1 200 hp. By the end of 1982, the number of shrimp trawlers in the Moluccas and Irian Jaya had peaked at 188 units. In the 1990s, there were 87 trawlers equipped with the by-catch excluder device (Sujastani 1984), operated by eight joint-venture companies.

Aside from shrimp, tuna and skipjack are among the most important export commodities from Indonesian fisheries. About 80% of the landings come from the eastern part of Indonesia, mainly Irian Jaya, the Moluccas, and North Sulawesi.

To estimate fisheries resource potential, a study was conducted by the DGF in cooperation with other institutions such as the Research Institute for Marine Fisheries (RIMF). The results of the study indicate a potential yield estimate of about 580 000 t·year[-1] of fishery resources, including tunas, skipjacks, shrimps, small pelagic and demersal fishes (Table 3). The landings in 1990 were about 48% of the estimated potential yield.

Development of the Sunda Shelf Trawl Fisheries

Trawl fisheries started commercially in 1966 in the Malacca Strait, particularly in the area surrounding the estuary of the Rokan River, with Bagansiapi-api as its base. This fishery was characterized by wooden sampan-like motorized vessels of 5-20 GT, employing a single gulf-type shrimp trawl of 12-15 m headrope length. This fishery developed rapidly and by the end of 1971, over 800 vessels were engaged in it.

The development of trawl fishery in Indonesia may have been influenced by western Peninsular Malaysia. The ancestors of many Chinese fishers in Riau Province, Indonesia, have migrated from there and still maintain contact with their relatives in Malaysia. Another strong input came from Thailand (Eiamsa-ard and Amornchairojkul, this vol.) which had long influenced Indonesian fisheries developments (Butcher 1996).

Table 3. Estimate of potential yield and landings. (Source: DGF 1991).

Fish resources	Potential yield (t·year⁻¹)	1990 landings (t)
1. Tuna and skipjack		
- Tuna	178 400	75 700
- Skipjack	295 000	150 600
2. Shrimp		
- Penaeid	111 755	75 600
- Nonpenaeid	4 488	2 398
3. Small pelagics	4 042 000[a]	1 115 000
4. Other resources		
- Carangidae	66 037	-
- Demersal fishes	916 000[b]	1 200 000
- Seaweeds	148 750	118 395
Total	5 762 430	2 737 693

[a] Acknowledged to be an overestimate.
[b] Acknowledged to be an underestimate.

The number of Chungking trawlers (of type 15 GT) operating from Bagansiapi-api had increased to 227 in 1976. In the following years, the trawl fishery spread throughout western Indonesia, via southeastern Sumatra to the north and south coasts of Java and to southern Sulawesi. The sizes of the trawlers gradually increased from 15 to 35 GT and those of their engines from 66 to 120 hp. Polyethylene nets were used, with headrope length ranging from 13.5 to 22.5 m, and codend mesh size of 2 cm.

Data from the Provincial Fisheries Offices of the Malacca Strait provinces of Aceh, North Sumatra, and Riau showed that in the early to mid-1970s about 20% of trawler catch was shrimp (Martosubroto et al. 1996). Along the south coast of Java, the contribution of shrimps to trawler landings was about 28%.

The most important shrimp caught by trawl are banana, or *jerbung* (*Penaeus merguiensis, P. indicus, P. chinensis*); tiger, or *windu* (*P. monodon, P. semisulcatus, P. latisulcatus*); endeavour, or *dogol* (*Metapenaeus monoceros, M. ensis, M. elegans*); rainbow, or *krosok* (*Parapenaeopsis sculptilis, P. stylifera*); and pink (*Solenocera subnuda, Solenocera* spp.). The first three groups are well defined in DGF fisheries statistics, while rainbow and pink shrimps belong to the 'other shrimps' category.

The ban on trawling in Indonesia has reduced shrimp and demersal fish yields. In the first year after the trawl ban, shrimp catches in artisanal fisheries decreased by about 60%, while the catches of demersal fish declined by about 40%. In the third year (1983), shrimp catches and exports were still lower than in 1979, the year immediately preceding the ban (Naamin and Martosubroto 1984; Chong et al. 1987).

Some demersal fish (e.g., of the genera *Harpadon, Setipinna, Leiognathus* and *Cynoglossus*), which had formed a major component of the by-catch of trawlers before the ban, are not common in the catches of trammel nets (which is the gear that many fishers use as a substitute for trawls).

Similarly, there was a significant reduction of the by-catch of trawlers operating in the Arafura Sea because of their use of by-catch excluder devices, although these did not significantly affect the catch of shrimps, their target group.

Monitoring of demersal stocks has been done through trawl surveys, performed mainly by RIMF with or without foreign partners and using either research trawlers or chartered commercial trawlers. The contributions in Pauly and Martosubroto (1996) review the knowledge that has been gained from trawl surveys conducted between 1976 to 1980 in western Indonesia, so we shall not duplicate these contributions (but see Dwiponggo et al. 1986; Silvestre and Pauly, this vol., Appendix III, this vol.)

Management Issues

The coastal zone faces a wide array of problems due to multiple uses that are not always compatible. Table 4 summarizes management issues relevant to fisheries in Indonesia. These issues or problems are mostly site-specific, particularly in highly urbanized areas. Most are consistent with those identified by Silvestre and Pauly (this vol.), but some need to be elaborated upon, such as unsustainable aquaculture development, overfishing and destructive fishing.

Large areas for brackishwater culture production have been unproductive because of disease and

Table 4. Typical transect across an Indonesian coastline, illustrating range of activities and issues that must be addressed by integrated coastal management schemes.

Major zones	Terrestrial			Coastal		Marine	
	Upland (>18% slope)	Midland (8-18% slope) depth)	Lowland (0-<8% slope)	Interface (1 km inland from HHWL-30 m)	Near shore (30 m-200 m depth)	Offshore (>200 m depth-EEZ)	Deep sea (beyond EEZ)
Main resource uses/activities	Logging Mining Agriculture Freshwater aquaculture	Mining Agriculture Freshwater aquaculture	Urban development Industries Agriculture Tourism Sand/gravel mining	Mangrove forestry Brackishwater fish culture Ports Fish processing Marine recreation and ecotourism Artisanal fisheries	Artisanal fisheries Industrial fishing Marine transport Mariculture Recreational fishing, sailing, swimming and diving Artificial reef	Marine transport Industrial fishing Offshore development	Marine transport Industrial fishing Offshore development
Main environmental issues/impacts on the coastal zone	Siltation Erosion Flooding Toxic mine tailings Sewage	Siltation Erosion Flooding Water pollution Sewage	Siltation Erosion Industrial pollution Agriculture run-off	Habitat degradation Denudation of coastal wetlands Endangered wildlife Coastal erosion Sedimentation	Reduced biodiversity Gear conflicts Oil spills Destructive fishing practices	Overfishing Oil spills Poaching Destructive fishing practices	Oil spills Poaching Capture of whales and dolphins

44

declining water quality (pollution). In 1989, there were 270 000 ha of brackishwater ponds in Indonesia, with 155 136 farmers involved in brackishwater culture. Since 1987, total shrimp and prawn production contributed substantially to the fishing industry.

Fisheries data from 1969 to 1990, consisting of total catch and total effort of Arafura trawler fisheries, showed that catch per effort is declining significantly while total effort is increasing. Such data indicate that the prawn stock population in the area is overexploited. Previous studies in the Arafura Sea (Naamin and Farid 1980; Priyono 1991, 1994) also showed that the shrimp resources are fully exploited.

In the Bali Strait, *Sardinella lemuru* is the main target fish. The fishers there use several traditional boats, medium and large sizes of *perahu* with outboard engines. 'Payao' and purse seine gear are commonly used by the fishers. Analysis of the potential of *S. lemuru* resources showed that total existing catch is more than the estimated MSY, which means that the fishing operation may be in peril (DGF 1992).

Marine fishing development in Malacca Strait (Indonesian side) is dominated by smaller boats, mostly concentrated in the coastal areas. From 1986 to 1990, the major fishing gear used by the Indonesian fishers in the Malacca Strait were gillnets and seine nets. In 1992 and 1993, handlines became the second most commonly used gear.

Using data on catch and purse seine standard effort from 1976 to 1990, a trend of consistently increasing total catch and decreasing total effort during the period indicates that the total existing effort exceeds the effort required to generate MSY, which indicates overcrowded conditions.

The use of sodium cyanide for catching aquarium fish is rapidly growing. Humphead or Napoleon wrasse, *Cheilinus undulatus*, and the light fin grouper (or polka dot grouper) *Cromileptus altifalis* are the main target species. Prime, plate-size specimens sell to Hong Kong consumers for as much as $180·kg⁻¹. The industry currently exports about 25 000 t of live reef fishes annually, with about 60% caught from the wild. The estimated wholesale value is about US$1 billion. The premium prices paid for these kinds of fish encourage the use of fishing practices that cause widespread devastation of coral reefs.

Conclusion

Integrated planning and coastal zone management could still reverse the evident degradation and decline of coastal resources in Indonesia.

Indonesia's marine fisheries sector has experienced rapid technological changes over the past decades. The most prominent has been the tremendous increase in the number of powered boats and the decline in the number of nonmotorized craft. The ban on trawl fishing in Indonesia has not caused as much social friction and dislocation as expected. Although catches of some commodities have been affected, the income of traditional fishers has increased. The ban seems to have achieved its main objectives. However, what has been gained in the short term will be lost in the long term if the pattern of degradation is not reversed (Chong et al. 1987; Martosubroto and Chong 1987).

The importance of effective fisheries resource management is clearly understood in Indonesia. Nonetheless, most government efforts have been directed towards resource development through expanded use of more productive fishing gear and boats, rather than through effort controls. The most serious management problem facing policymakers is related to the coastal resources exploited by the vast majority of fishers. Existing management regulations attempt to protect both vulnerable resources and small-scale fishers' rights of access to fishing grounds. Over the long term these objectives will turn out to be incompatible, and hard choices will have to be made.

Action is needed to address the issues in aquaculture, overfishing and irresponsible fishing so that the advantages of fundamental productivity can be effectively exploited. In formulating the action plans, two broad options should be considered: (1) a 'status quo' option in which there will be no increase in marine and coastal development in overcrowded conditions; and (2) an 'agri-business concept' for fisheries development, with emphasis on the shrimp brackishwater culture and marine fishing in the offshore areas and the EEZ.

In the short term, option 1 has the least implications for government funding and organizational arrangements. However, it fails to exploit the potential benefits of improved cross-sectoral management and ignores long-term productivity and tax revenue losses resulting from illicit activities and negative cross-sectoral impacts. Adopting this option without any improvement on monitoring, control and surveillance (MCS) and enforcement (to limit poachers) runs the risk of destruction of marine resources and may foreclose future sustainable development options in the coastal waters. Thus, for long-term planning, option 1 is not preferred. Option 2 considers the very basic problem in fisheries development, i.e., poverty alleviation of the coastal community. Therefore, the Indonesian government should improve the income-generating programs for these communities. If profit is a function of reducing the cost of production, then the government can run a simultaneous program to increase this profit margin (Fig. 2).

If option 2 is selected, the government should improve MCS and enhancement activities to improve fisheries management as a whole, so that the goal of sustainable development can be reached in the long term.

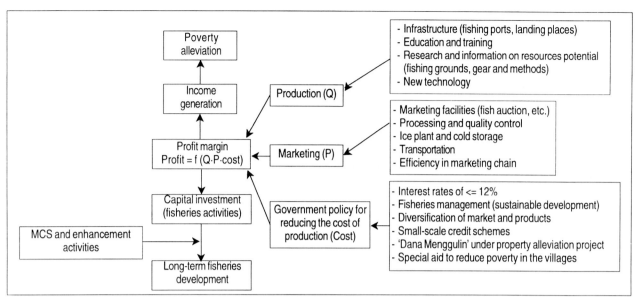

Fig. 2. A concept of sustainable development of fisheries for Indonesia.

References

Bailey, C., A. Dwiponggo and F. Marahudin. 1987. Indonesian marine capture fisheries. ICLARM Stud. Rev. 10, 196 p.

Bell, J.D. and R. Galzin. 1984. Influence of live coral cover on coral-reef fish communities. Mar. Ecol. Prog. Ser. 15: 265-274.

Butcher, J. 1996. The marine fisheries of the western archipelago: towards an economic history, 1850 to the 1960s, p. 24-39. *In* D. Pauly and P. Martosubroto (eds.) Baseline studies of biodiversity: the fish resources of western Indonesia. ICLARM Stud. Rev. 23, 321 p.

Chong, K.C., A. Dwiponggo, S. Ilayas and P. Martosubroto. 1987. Some experiences and highlights of the Indonesia trawl ban: bioeconomics and socioeconomics, p. 458-477. *In* Papers presented at the Symposium on the Exploitation and Management of Marine Fishery Resources in Southeast Asia, 16-26 February 1987, Darwin, Australia. RAPA Rep. 1987/10.

Dalzell, P. and D. Pauly. 1989. Assessment of the fish resources of southeast Asia, with emphasis on the Banda and Arafura seas. Neth. J. Sea Res. 24(4): 641-650.

DGF (Directorate General Fisheries). 1985. Fisheries Statistics.

DGF (Directorate General Fisheries). 1995. Fisheries Statistics.

Dwiponggo, A. 1974. The fishery for and preliminary study on the growth rate of 'lemuru' (oil sardine) at Muntjar, Bali Strait. Symposium on Coastal and High Sea Pelagic Resources. Pro. Indo-Pac. Fish Counc. 15 (Section III): 221-240.

Dwiponggo, A., T. Hariati, S. Banon, M.L. Palomares and D. Pauly. 1986. Growth, mortality and recruitment of commercially important fishes and penaeid shrimps in Indonesian waters. ICLARM Tech. Rep. 17, 91 p.

Erdmann, M.V. and L. Pet-Soede. 1996. How fresh is too fresh? The live reef food fish trade in eastern Indonesia. Naga, ICLARM Q. 19(1): 4-8.

Ghofar, A. and C.P. Mathews. 1996. The Bali Straits lemuru fishery, p. 146. *In* D. Pauly and P. Martosubroto (eds.) Baseline studies of biodiversity: the fish resources of western Indonesia. ICLARM Stud. Rev. 23, 321 p.

Johannes, R.E. 1995. Fishery for live reef food fish trade is spreading human death and environmental degradation. Coast. Manage. Trop. Asia (September): 8-9.

Johannes, R.E. and M. Riepen. 1995. Environmental, economic and social implications of the live reef fish trade in Asia and the western Pacific. The Nature Conservancy, Jakarta Selatan, Indonesia.

Martosubroto, P. and K.C. Chong. 1987. An economic analysis of the 1980 Indonesian trawl ban. IARD J. 9(1-2): 1-12.

Martosubroto, P. and M. Badrudin. 1984. Notes on the status of the demersal resources off the north coast of Java, p. 33-36. *In* Report of the Fourth Session of the Standing Committee on Resources Research and Development of the Indo-Pacific Fisheries Commission, Jakarta, Indonesia, 23-29 August 1984. FAO Fish. Rep. (318).

Martosubroto P. and N. Naamin. 1977. Relationship between tidal forests (mangroves) and commercial shrimp production in Indonesia. Mar. Res. Indones. 8: 81-86.

Martosubroto, P., T. Sujastani and D. Pauly. 1996. The mid-1970s demersal resources in the Indonesian side of the Malacca Strait, p. 40-46. *In* D. Pauly and P. Martosubroto (eds.) Baseline studies of biodiversity: the fish resources of western Indonesia. ICLARM Stud. Rev. 23, 321 p.

Martosubroto, P. 1996. Structure and dynamics of the demersal resources of the Java Sea, 1975-1979, p. 62-76. *In* D. Pauly and P. Martosubroto (eds.) Baseline studies of biodiversity: the fish resources of western Indonesia. ICLARM Stud. Rev. 23, 321 p.

Naamin, N. and A. Farid. 1980. A review of the shrimp fisheries in Indonesia, p. 33-48. *In* Report of the Workshop on the Biology and Resources of Penaeid Shrimps in the South China Sea area - Part I. SCS/GEN/80/26. 162 p. South China Sea Fisheries Development and Coordination Program, Manila.

Naamin, N. and P. Martosubroto. 1984. Effect of gear changes in the Cilacap shrimp fishery, p. 25-32. *In* Report of the Fourth Session of the Standing Committee on Resources Research and Development of the Indo-Pacific Fisheries

46

Commission, Jakarta, Indonesia, 23-29 August 1984. FAO Fish. Rep. (318).

Pauly, D. and J. Ingles. 1986. The relationship between shrimp yields and intertidal vegetation (mangroves). p. 277-283. *In* A. Yañez-Arancibia and D. Pauly (eds.) Proceedings of the IREP/OSLR Workshop on the Recruitment in Tropical Coastal Demersal Communities, 21-25 April 1986, Campeche, Mexico, IOC (UNESCO) Workshop Rep. No. 44 (Suppl.).

Pauly, D. and P. Martosubroto (eds). 1996. Baseline studies of biodiversity: the fish resources of western Indonesia. ICLARM Stud. Rev. 23, 321 p.

Roy, C. 1996. Variability of sea surface features in the western Indonesian archipelago: inferences from the COADS dataset, p. 15-23. *In* D. Pauly and P. Martosubroto (eds.) Baseline studies of biodiversity: the fish resources of western Indonesia. ICLARM Stud. Rev. 23, 321 p.

Sarjono, I. 1980. Trawlers banned in Indonesia. ICLARM Newsl. 3(4): 3.

Sharp, G.D. 1996. Oceanography of the Indonesian archipelago and adjacent areas, p. 7-14. *In* D. Pauly and P. Martosubroto (eds.) Baseline studies of biodiversity: the fish resources of western Indonesia. ICLARM Stud. Rev. 23, 321 p.

Soegiarto, A. and S. Birowo (eds). 1975. Atlas oseanologi perairan Indonesia dan sekitarnya. Book 1. Lembago Oseanologi Nasional, Jakarta.

Sujastani, T. 1984. The by-catch excluder device, p. 91-95. *In* Report of the Fourth Session of the Standing Committee on Resources Research and Development of the Indo-Pacific Fisheries Commission, Jakarta, Indonesia, 23-29 August 1984. FAO Fish. Rep. (318).

Turner, R.E. 1977. Intertidal vegetation and commercial yields of penaeid shrimps. Trans. Am. Fish. Soc. 106: 411-416.

Venema, S.C. 1996. Results of surveys for pelagic resources in Indonesian waters with the *R.V. Lemuru*, December 1972 to May 1976, p. 102-122. *In* D. Pauly and P. Martosubroto (eds.) Baseline studies of biodiversity: the fish resources of western Indonesia. ICLARM Stud. Rev. 23, 321 p.

Wyrtki, K. 1961. Naga report: scientific results of marine results investigations of the South China Sea and the Gulf of Thailand, 1959-1961. Vol. 2. Scripps Institution of Oceanography, La Jolla, California.

Status of Fisheries in Malaysia — An Overview

A. ABU TALIB

Department of Fisheries
Wisma Tani, Jln Sultan Salahuddin
50628 Kuala Lumpur, Malaysia

M. ALIAS

Fisheries Research Institute
11960 Batu Maung, Pulau Penang, Malaysia

ABU TALIB, A. and M. ALIAS. 1997. Status of fisheries in Malaysia — an overview, p. 47-61. *In* G. Silvestre and D. Pauly (eds.) Status and management of tropical coastal fisheries in Asia. ICLARM Conf. Proc. 53, 208 p.

Abstract

The marine fisheries sector in Malaysia provides a significant contribution to the national economy in terms of income, foreign exchange and employment. In 1994, marine fisheries production contributed to 90% of total fish production and was valued at US$1.04 billion. This paper reviews the status of the marine environment and coastal resources in Malaysia with emphasis on coastal fisheries resources and the development of trawl fisheries. Trawl fisheries were initiated in 1963 and the first trawl survey was conducted in 1970.

Fishery resources are being intensively exploited and the sustainability of these resources is affected by various activities in the coastal zone. Currently the government is implementing several management strategies and measures to ensure the sustainability of coastal fishery resources.

Introduction

The fisheries sector in Malaysia plays an important role in the national economy in terms of income, foreign exchange and employment. Marine fisheries alone contributed 1 066 000 t in 1994, or 90% of total fish production. This production, valued at RM 2.6 billion[1] (US$1.04 billion), constitutes about 1.5% of the national Gross Domestic Product (GDP). The demand for fish, as the main source of protein, is expected to increase from an annual consumption of 630 000 t to over 1 580 000 t by year 2010 (DOF 1995b). The exports of fishery products increased from RM577 million (US$231 million) in 1989 to RM740 million (US$296 million) in 1991, consisting of 175 000 t of fishery products.

While the demand for fish continues to increase, the task of managing fisheries resources on a sustainable basis has become increasingly complex. The threat of overexploitation as well as the decrease in recruitment due to the destruction of aquatic habitats and the degradation of the aquatic environment have become serious problems that need to be solved. Some mangrove forest, which provide important breeding, nursery and feeding grounds for fish, prawns and a variety of invertebrates, have been converted to agriculture and aquaculture or have been decimated by pollution or diversion of drainage water. Coastal areas are threatened by severe erosion along large parts of the coastline. Continued economic growth and industrialization are exerting considerable pressure on the sensitive coastal ecosystems. Effluents from industrial and domestic discharges, land reclamation as well as illegal dumping and accidental spills contribute to the degradation of water quality in the aquatic environment.

---

[1] 1RM = 0.4004 US$ in 1994.

Marine Environment

Malaysia is a confederation comprising the Malay peninsula (or Peninsular Malaysia) situated at the southeastern tip of the Asian mainland and the States of Sabah and Sarawak in the northwestern part of Borneo Island (Fig. 1). It has a land area of 332 600 km², making it the third largest nation in the ASEAN region, after Indonesia and Thailand. The continental shelf area up to 200 m depth is 373 500 km². With the extension of maritime jurisdiction, Malaysian claims expanded to 138 700 km² or about 42% of its total land area.

The Strait of Malacca is one of the busiest navigational passage in Southeast Asia. Given such a strategic location, issues such as sovereignty, security, sea lanes and environmental protection are of considerable importance to Malaysia's maritime development (Kent and Valencia 1985). Malaysian waters are generally divided into four sub-areas: the west coast and the east coast of Peninsular Malaysia, the coast of Sarawak and the coast of Sabah.

The depth of water off the west coast of Peninsular Malaysia seldom exceeds 120 m within the exclusive economic zone (EEZ). The deepest part is at the northern tip of the Strait of Malacca or the eastern part of the Bay of Bengal where the two water bodies meet. The east coast of Peninsular Malaysia has a relatively flat sea bottom and water depth less than 100 m. Deep waters occur off the north coast of Sarawak, with soundings in excess of 1 000 m. In the west and north of Sabah, a deep water trench comes in from the northeast and runs parallel to the coast. The trench has steep sides and the water depth exceeds 2 000 m.

Throughout the year, the flow of the current in the Strait of Malacca is in the northwest direction through the central part of the Strait. During the northeast monsoon, from October to April, the current flows from the south going into the South China Sea and passes the extremity of the Malay peninsula. During the southwest monsoon period, from May to September, water flows westward in the Java Sea and northwestward through the Karimata Strait towards the South China Sea then passes directly into the Malacca Strait (Hydrographic Department 1964). Current speed exceeding 1 knot may be experienced throughout the year in the Malacca Strait.

Surface currents in the South China Sea flow in accordance with the monsoon-driven wind system. The southwesterly current of the northeast monsoon is

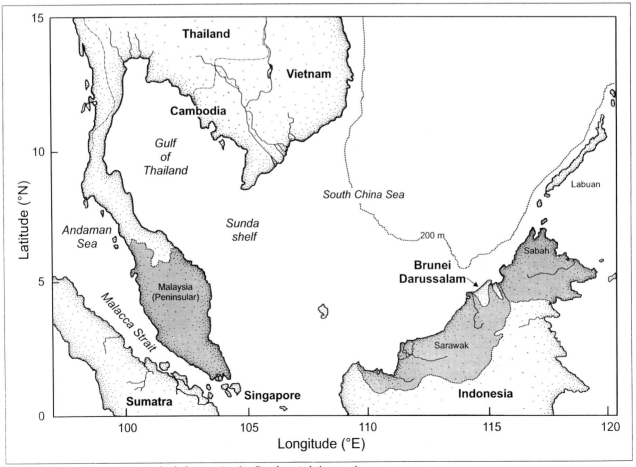

Fig. 1. Location of Malaysia (shaded area) in the Southeast Asian region.

stronger and more constant than the northeasterly current of the southwest monsoon. This behavior is due to stronger and more constant winds (Hydrographer 1975). The southwesterly flow of the northeast monsoon prevails from October to February. Current flow reaches its greatest rate and constancy from December to January. The northeasterly current of the southwest monsoon occurs in May. The current attains its greatest rate and constancy from June to August-September. During the monsoons, the winds fluctuate considerably in strength owing to variations in the distribution of atmospheric pressure. The northeast monsoon winds are stronger and less liable to interference than the southwest monsoon winds. The direction of water movement over the eastern side of the South China Sea is controlled, to a large extent, by the position of eddies, which occur in most months, and by the flow of water to or from the Sulu Sea and through the broad straits between Peninsular Malaysia and Borneo. These features are controlled by varying oceanographic and meteorological conditions outside the area. Moreover, the currents are generally weak with mean rates less than 0.5 knot in most regions. When monsoons are fully developed (July to August and December to February) the mean rates increase to between 0.5 to 1 knot. On rare occasions the current may run up to 3 knots between Peninsular Malaysia and Borneo and 4 knots through some of the passages linking the South China Sea to the Sulu Sea.

The sea surface temperatures are relatively constant throughout the year, ranging from 26 to 32°C (FRI 1987). Water temperature varies from 26 to 29°C during the northeast monsoon. The off monsoon period has water temperatures of 28 to 30°C. The average surface water temperature during the southwest monsoon period ranges from 29 to 32°C (Hamid 1989).

Surface salinity in the west coast of Peninsular Malaysia ranges from 27‰ to 33‰ during the northeast monsoon. At depths of 13 m to 21 m salinity ranges from 25‰ to 33‰ (Samsuddin 1995). From 100 to 1 000 m the salinity off Sarawak waters averages 34‰ during the southwest monsoon period (Saadon et al. 1988). At 2 000 m depth the vertical profile of salinity increases from 35‰ at the surface to 36‰ while temperature drops from 29°C at the surface to 5°C (Hamid 1993).

In most cases, the thermocline is found at 20 m to 50 m (FRI 1987; Saadon et al. 1988). The typical water temperature above the thermocline ranges from 28°C to 30°C, except in a few areas in the west coast of Peninsular Malaysia. In some shallow waters (less than 50 m), a thermocline may be lacking.

On the east coast of Peninsular Malaysia, particulate organic carbon ranges from 3.9 gC·m^{-2} on the surface to 6.0 gC·m^{-2} at 50 m depth (Ichikawa 1986), and from 3.7 gC·m^{-2} to 6.0 gC·m^{-2} respectively in Sarawak waters (Ichikawa and Law 1988).

The average surface concentration of chlorophyll a in the waters off the east coast of Peninsular Malaysia is 0.08 mg·m^{-3}; with an average value of 0.208 mg·m^{-3} considering all depth strata (Raihan and Ichikawa 1986). Similarly low content of chlorophyll a is observed in Sarawak waters ranging from 0.049 mg·m^{-3} to 0.150 mg·m^{-3} (Shamsudin et al. 1988).

Table 1 gives a synoptic description of activities and issues in the coastal area of Malaysia. Mangrove forests are an important source of commercial timber, i.e., the Matang Mangrove Forest Reserve in Perak which has been sustainably managed for timber since the 1920s. Other activities in the lowland areas include urban development, aquaculture, agriculture and industries. The marine sector also contributes significantly to the country's economy from the traditional marine-related uses such as commercial fishing and shipping to more dynamic high-growth industries such as petroleum and gas. The problems associated with these activities are generally site-specific and most obvious in urbanized coastal areas.

Mangrove forests cover a total area of 641 000 ha. More than half of the mangrove forests are found in eastern Sabah with the remaining primarily distributed along the west coast of Peninsular Malaysia and in northern and southwestern Sarawak. About 446 000 ha of these forests are gazetted as forest reserves while the remaining are state land forests (Chew 1996). The total area of mangrove forests in Peninsular Malaysia is estimated to be around 103 000 ha and is mainly located on the west coast (Tang et al. 1990). The major mangrove areas in Peninsular Malaysia include the Larut Matang mangroves in the state of Perak (40 000 ha), the mangroves in Johor (25 700 ha), Kelang in Selangor (22 500 ha) and Merbok in Kedah (9 000 ha) (Choo et al. 1994). The Larut Matang mangroves is reported to be the largest mangrove forest in Peninsular Malaysia and cited as the best managed mangrove forest in the world (Gong et al. 1980). However, mangrove forests are being extensively exploited. The threat to Malaysia's mangroves lies in land conversion or deforestation for agriculture, industry and aquaculture.

The coral reef areas of Peninsular Malaysia are found around the islands located off the coast of Terengganu (Pulau Redang, Pulau Perhentian and Pulau Tenggol) and Pahang/Johor (Pulau Tioman groups) in the east, the Payar group of islands in the north of Kedah, and the Sembilan group of islands off Perak in the west coast. There are also fringing reefs along the mainland coast, particularly on the east coast, with small isolated reefs occurring along the west coast. The more extensive and exotic coral reef areas are located in the north coast of Sabah and the Spratleys.

The biggest threat to coral reefs has been their exploitation for commercial and tourism purposes, coupled with siltation and sedimentation caused by development projects (Liew and Hoare 1982). These activities have subjected corals to stress and leaching which have resulted in the deterioration of coral reefs. The Department of Fisheries has taken steps to conserve and rehabilitate the country's coral reefs by classifying many of the islands where these corals occur as marine parks.

In a recent review of seagrass distribution off the west coast of Peninsular Malaysia (Kushairi 1992), five species were noted in shallow waters (0.2 m to 1.8 m), i.e., *Halophila ovalis*, *H. minor*, *Halodule uninervis*, *H. pinifolia* and *Enhalus acoroides*. The seagrass beds are prone to erosion, fecal contamination and heavy metal pollution (Phang 1990; Kushairi 1992).

In terms of water quality of Malaysian coastal waters, loading of suspended solids has been a major environmental problem for the past few years. Activities associated with continuous and intensive land clearing, uncontrolled development, and mining and logging in the catchment areas may be responsible for the increased siltation (Table 1). The high fecal coliform levels come from both domestic and animal waste, especially pig waste. Indiscriminate use of pesticides has resulted in organochlorines contaminating some of Malaysia's river systems and aquatic life.

A high quantity of oil and grease have also been found along the coast of Perak. Heavy metal concentrations of lead, copper and cadmium exceed the proposed standard of 0.05 mg·l^{-1} for Pb, 0.01 mg·l^{-1} for Cu and 0.005 mg·l^{-1} for Cd. Despite the elevated concentration of these metals in the water, studies indicate that the levels of these elements found in fish and shellfish samples do not pose a threat to public health (Shahunthala 1989).

Effluents of palm oil and rubber processing contain very high concentrations of organic material, suspended solids and nutrients like nitrogen and phosphorus. Land development, agriculture and high population density are among the main causes of water pollution in Malaysia through increased sediment load, pesticide run-offs, discharges from agro-based industries and sewage.

Development activities that resulted in a severe destruction of marine resources caused several national agencies, like the Department of Fisheries, the Department of Environment and the National Working Group for Mangroves, to initiate conservation and rehabilitation measures for proper utilization and management of marine resources.

Fishery Resources and Exploitation

Total landings have stabilized in most of the coastal fisheries and it is believed that the yield has already reached its maximum. The local supply of fish to the local market has been greatly reduced and fish has to be imported from neighboring countries to overcome the shortage. This can only serve as a temporary remedy. In the long run fisheries resources have to be managed for the increasing nutritional needs and economic improvement. In addition to venturing into aquaculture and deep-sea fishing, sustainable exploitation of the resources in coastal waters is very important.

Resources

A demersal survey conducted in 1986-1987 in the Malaysian EEZ beyond 25 nm from the coastline revealed a potential yield of 167 000 t of demersal resources (Table 2). The most promising area was off the east coast of Peninsular Malaysia followed by areas off Sarawak and Sabah. Pelagic resources have a potential yield of 171 000 t, mostly off Sarawak and the east coast of Peninsular Malaysia

An acoustic survey conducted in 1986-1987 in the EEZ beyond 12 nm from the coastline indicated a potential yield of 255 000 t·year^{-1}. However, the greatest concentrations of fish were found close to shore, particularly outside the monsoon period. An offshore-onshore migration pattern between the two monsoon seasons was also observed.

A new fishery product in Malaysia is oceanic tuna. Tuna such as skipjack and yellowfin are found in deeper waters off the Sarawak and Sabah coasts. With the expansion of maritime jurisdiction, tuna is seen as an important fishery resource to be exploited. In 1984, Malaysian and Thai fishers operating in the Gulf of Thailand and off the east coast of Peninsular Malaysia landed an estimated 84 000 t of tuna. The dominant species in this catch was longtail tuna. The potential yield of longtail tuna from the east coast of Peninsular Malaysia was estimated to be about 50 000 t and off Sabah and Labuan about 50 000 t. A recent study conducted in the waters off Sarawak found skipjack and yellowfin tuna. It is believed that the area serves as an important migratory route to and from the adjacent waters (Hamid 1993).

As traditional coastal resources are being fully exploited, 'new' non-conventional products such as shellfish, jellyfish and bivalves are emerging. Landings of shellfish and bivalves have increased but landings of jellyfish have declined. This sometimes creates

Table 1. Synoptic transect of activities and issues relevant to coastal fisheries management in Malaysia.

Major zones	Terrestrial			Coastal		Marine	
	Upland (> 18% slope)	Midland (8-18% slope)	Lowland (0-<8% slope)	Interface (1 km inland from HHWL- 30 m depth)	Nearshore (30 m - 200 m depth)	Offshore (>200 m depth- EEZ)	Deepsea (beyond EEZ)
Main resource uses/activities	Upstream development Mining Logging Resort development	Logging	Land use conversion to rubber and palm oil plantation Palm oil mills and pineapple processing Residential, industrial and tourism development Charcoal processing Petrochemical/industrial estate Pig farming Aquaculture	Sand mining Mangrove conversion to aquaculture, agriculture Industrial and tourism development Dredging and reclamation Port and shipyard activities Artisanal fishing Tourism	Commercial fishing Marine transport	Offshore development Marine transport	Marine transport
Main environmental issues/Impacts on the coastal zone	Limited water supply Flashfloods High sediment load Pollution	Erosion	Agrochemical pollution Domestic and sewage pollution Industrial pollution Organic pollution Landuse conflict of industries sited in aquaculture areas	Sedimentation and water turbidity Coastline erosion and siltation Destruction of mangrove forest Deterioration of water quality due to chemical effluents Landuse conflict of industries sited in aquaculture areas Aquaculture development above carrying capacity Oil pollution Use of destructive fishing methods Overfishing Illegal collection of corals Coral reefs damaged by anchorage of boats	Conflicts between traditional and commercial fisheries Competition for the use of the coastal zone Oil pollution	Sewage and solid waste disposal Limited water supply Sedimentation Water pollution	Oil pollution

Table 2. Estimates of potential yield and commercial landings for offshore grounds in 1994.

Area	Potential yield estimate (t·year⁻¹)				Offshore landings in 1994 (t)
	Pelagic	Demersal	Tuna	Total	Total
East Coast of Pen. Malaysia	54 600	82 200	50 000	186 800	69 506
West Coast of Pen. Malaysia	16 950	11 300	-	28 250	28 370
Sarawak	81 550	62 300	-	143 850	28 429
Sabah and Labuan	17 750	10 900	50 000	78 650	9 231
Total	170 850	166 700	100 000	437 550	135 536

conflicts among fishers, especially if the resources command good prices, as in the high-profit bivalve fishery in the west coast of Peninsular Malaysia.

Capture Fishery Sector

Marine capture fisheries in Malaysia can be divided into two categories, coastal or inshore fisheries and deep-sea fisheries. The coastal fishery is the most important subsector of the fisheries industry from the socioeconomic point of view. The fishing vessels operate within 30 nm from the coastline. They range from traditional craft to commercial vessels of less than 70 Gross Registered Tonnage (GT). A large proportion of the licensed traditional vessels are below 9.9 GT. In 1994 inshore fisheries contributed 930 000 t or about 87% of the total marine fish landings (Table 3). The west coast of Peninsular Malaysia contributed 47%, east coast of Peninsular Malaysia 27%, Sabah coast 19% and Sarawak coast 7% of the total landings of the coastal fisheries. Using surplus production models, the maximum sustainable yield (MSY) for the west coast of Peninsular Malaysia was estimated to be about 430 000 t (Alias 1994). Hence, the present exploitation level has exceeded the estimated MSY level in the area.

Deep-sea fishing vessels operate beyond 30 nm from shore. In 1994 deep-sea fisheries contributed 13% (136 000 t) of the total marine fish landings in the country. The east coast of Peninsular Malaysia contributed 51%, west coast of Peninsular Malaysia 21%, Sarawak 21%, Labuan 5% and Sabah 2% of the deep-sea fish landings. The fishing vessels in operation are fairly large (more than 70 GT) and commercial gear such as trawl and purse seines are used. In 1994 a total of 31 403 fishing vessels were licensed, of which 520 units were deep-sea fishing vessels.

Marine capture fisheries also use various types of fishing gear. The fishing gear are classified into commercial and traditional. The commercial fishing gear consist of trawl and purse seine while the traditional fishing gear are drift/gillnets, bagnets, hook and lines, trammelnets, liftnets and traps. The number of vessels declined from 43 492 units in 1980 to 31 403 in 1994 (Table 4). The number of offshore/commercial fishing vessels (70 GT and above) increased in 1988-1989, following government encouragement of offshore fishing.

Total landings from marine capture fisheries were reported to be 1.07 million t in 1994. Commercial fishing gear landed 789 000 t or 74% and traditional fishing gear landed 278 000 t or 26% (Table 5). From 1987 to 1994 marine capture fisheries increased their landings at an average annual rate of around 10%.

The landings of demersal fish show a fluctuating trend (Table 6). Trawl landings in 1994 were about 588 000 t. The west coast of Peninsular Malaysia is the main contributor. The majority of trawls are also registered in the west coast. The east coast of Peninsular Malaysia and the coast of Sabah contributed about half, while the Sarawak coast contributed a quarter of the amount landed on the west coast of Peninsular Malaysia. In general, the landings of demersal fish from all coasts were stable from 1990 to 1994. In terms of fishing gear, trawl is the main gear used for the exploitation of the resource, except for the Sarawak and the Sabah coasts where traditional gear is still very much in use.

Table 3. Marine landings in 1994 from the inshore/coastal and offshore fishing areas in Malaysia. (See text).

Area	Inshore/coastal[a] landings (t)	Offshore[b] landings (t)	Total (t)
West Coast of Peninsular Malaysia	431 932	28 370	460 302
East Coast of Peninsular Malaysia	255 271	69 506	324 777
Sarawak	67 195	28 429	95 624
Sabah	175 651	9 231	184 882
Total	930 049	135 536	1 065 585

[a] Inshore/coastal grounds are <30 nm from shore.

[b] Offshore grounds are >30 nm from shore.

Table 4. Number of licensed fishing vessels in the commercial and traditional fisheries sector in Malaysia from 1980 to 1994. (Source: Annual fisheries statistics 1980-1994).

Year	Commercial[a]	Traditional[a]	Total
1980	(6 969)	(23 551)	43 492
1981	(6 870)	(23 520)	44 595
1982	(6 589)	(21 741)	41 497
1983	(5 813)	(19 882)	40 168
1984	(6 263)	(19 410)	41 079
1985	(5 590)	(17 781)	38 933
1986	(5 404)	(17 072)	40 383
1987	(6 042)	(16 096)	35 429
1988	8 951	28 851	37 802
1989	9 162	30 808	39 970
1990	8 906	30 635	39 541
1991	8 727	24 174	32 901
1992	8 426	24 124	32 550
1993	8 234	23 341	31 575
1994	7 923	23 480	31 403

[a] Figures in bracket are for Peninsular Malaysia only, as the breakdown into traditional or commercial category for Sabah and Sarawak is unavailable for 1980-1987.

Table 5. Marine landings by the commercial and traditional fisheries sector in Malaysia from 1980 to 1994. (Source: Annual fisheries statistics 1980-1994).

Year	Marine landings (t)		
	Commercial	Traditional	Total
1980	515 590	254 590	770 180
1981	481 680	315 669	797 349
1982	433 747	286 669	720 416
1983	479 672	295 821	775 493
1984	429 118	227 154	656 272
1985	408 707	200 046	608 753
1986	391 828	222 046	613 874
1987	667 277	191 869	859 146
1988	667 277	191 851	859 128
1989	676 123	206 369	882 492
1990	765 242	186 065	951 307
1991	707 343	204 590	911 933
1992	771 371	252 145	1 023 516
1993	766 574	280 776	1 047 350
1994	787 877	277 708	1 065 585

Table 6. Landing and effort data for Malaysian trawl fishery, 1968-1994. (Source: Annual fisheries statistics 1968-1994).

Year	Trawl landings (t)					Licensed trawlers				
	Pen. Malaysia		Sarawak	Sabah	Total	Pen. Malaysia		Sarawak	Sabah	Total
	West	East				West	East			
1968	-	6 973	-	-	-	-	-	-	230	-
1969	-	8 518	-	-	-	-	-	-	300	-
1970	-	17 419	-	26 009	-	-	-	176	315	-
1971	-	19 274	-	26 721	-	-	-	243	275	-
1972	-	16 779	7 715	27 127	-	-	-	433	320	-
1973	-	27 965	24 792	31 192	-	-	-	472	325	-
1974	-	42 856	35 672	32 309	-	-	-	479	350	-
1975	-	35 599	33 540	33 020	-	-	-	530	330	-
1976	175 595	45 084	43 821	30 700	295 200	3 039	1 276	569	340	5 224
1977	226 521	34 603	45 582	34 900	341 606	3 029	1 256	713	580	5 578
1978	235 347	49 672	42 439	40 100	367 558	3 321	1 183	744	560	5 808
1979	201 761	47 439	58 839	40 200	348 239	3 316	1 175	485	590	5 566
1980	193 974	33 306	52 283	33 300	312 863	3 347	1 240	780	516	5 883
1981	199 934	34 337	58 839	38 000	331 110	3 414	1 160	861	620	6 055
1982	194 477	31 689	52 283	39 800	318 249	3 365	1 535	765	444	6 109
1983	195 932	29 786	42 008	45 500	313 226	3 236	1 355	708	789	6 088
1984	158 269	31 936	37 592	50 200	277 997	3 487	950	631	908	5 976
1985	161 628	27 567	30 116	48 600	267 911	3 281	1 371	723	1 054	6 429
1986	192 902	35 682	30 258	46 900	305 742	3 281	837	732	856	5 706
1987	311 132	78 556	38 034	13 839	441 561	3 336	967	637	1 103	6 043
1988	266 975	127 224	36 710	13 228	444 137	3 257	943	727	852	5 779
1989	298 852	137 435	51 918	15 310	503 515	3 331	1 002	1 010	1 040	6 383
1990	324 676	192 049	42 236	17 262	576 223	3 187	1 031	992	1 048	6 258
1991	259 276	184 552	53 045	32 671	529 544	3 224	1 059	603	1 058	5 944
1992	291 771	177 329	54 034	49 261	572 395	3 294	1 015	599	1 043	5 951
1993	270 343	194 910	48 309	48 380	561 942	3 155	997	587	1 010	5 749
1994	275 330	187 995	59 222	65 381	587 928	3 137	939	557	1 046	5 679

The landings of the pelagic and semi-pelagic fish in 1994, of 856 100 t, were more than four times greater than the demersal fish landed (DOF 1995a). About 43% comes from the west coast of Peninsular Malaysia, 32% from the east coast of Peninsular Malaysia, while 15% and 10% come from the coasts of Sabah and Sarawak, respectively. The great majority of the pelagic and semi-pelagic fish are landed by commercial fishing gear, which contributed about 70%. The traditional fishing gear is as important as the commercial gear for this fishery along the coast of Sabah.

54

Trawl fisheries

Trawling was introduced into Malaysia indirectly, through a German project in Thailand. The success of the Thais provided the impetus for Malaysian fishers to venture into trawling operations in 1963. In the early stages, only vessels of 50 GT were allowed to operate but later 25 to 30 GT boats were also given permission to operate beyond 7 miles from shore. Later still, small boats below 25 GT were allowed to operate and this made investment more attractive because small trawlers had better access to the richer inshore areas. However, in early 1964 the Malaysian Government banned trawling on the ground that it would deplete the fish stocks in inshore waters and ruin the livelihood of the traditional fishers. The ban resulted in illegal trawling. The highly profitable nature of fish trawling, coupled with the need to increase fish landings to meet the growing demand for food, made it impossible to enforce the ban. The trawl ban was lifted in October 1964 and a pilot trawling scheme was launched with the goal of devising methods for better control and regulation of trawling. A survey of the coastal waters off the east coast of Peninsular Malaysia was carried out by Malaysia in cooperation with Thailand and the Federal Republic of Germany in 1967 to assess the potential demersal fish resources in the area. Following this survey, the government introduced a pilot scheme in mid-1968 to promote the development of trawl fishing in the area. By 1975 there were about 910 trawlers operating around the coastal waters off the east coast of Peninsular Malaysia. In Sabah, trawling was first introduced in the early 1960s and within five years there were 350 trawlers recorded in the area. The mainstay of the coastal fisheries in Sabah was still prawn capture, mainly within the 3 nautical mile-belt. By the mid-1980s, coastal fisheries had an upturn as the exploitation of demersal fish became as profitable as the prawn fisheries. Presently there are 1 548 licensed trawlers operating in the area.

Trawl fishing is firmly established in Malaysia but remains under the strict control stipulated by the Fisheries Act of 1963 and 1985. Currently trawls yield about 55% of the total marine landings in the country compared to only 34% in 1974 (Annual fisheries statistics 1974, 1994). Table 7 details the major events with impact on trawl fisheries development in Malaysia.

Currently, the otter bottom trawl net is the main fishing gear used to harvest the demersal finfish and penaeid prawn resources in the waters off Malaysia. Of the 580 000 t total marine landings from trawl fishery in 1994, 540 000 t (92%) were finfish and cephalopods while the remainder (8%) consists of penaeid prawns. A large number of fish species are landed by the trawls. While demersal finfish still remain the dominant catch of trawl nets, the development of 'high opening' trawl nets have resulted in the ability to catch pelagic finfish, mainly the Indo-Pacific mackerel *Rastrelliger brachysoma,* on the west coast of Peninsular Malaysia. Here, penaeid prawns constitute an important component of the catches of trawls operating in inshore waters. Although penaeid prawns landed by trawl nets contributed only 4.4% to the total marine fisheries landings in 1994, they are the mainstay of the trawl fishery due to their high commercial value and market demand. The rapid development and concentration of the trawl fishery within the coastal waters has resulted in intensive exploitation of coastal demersal finfish and penaeid prawn resources. While the total landings of the coastal demersal finfish and penaeid prawn fisheries may appear to be stable or even increasing, the disappearance of certain species, notably *Lactarius lactarius*, from the coastal waters of Peninsular Malaysia indicates overexploitation of the resource. This is further evidenced by the relatively high catch of 'trash fish', usually consisting of the juveniles of commercial fish. The trawl landing data from 1968 to 1994 is presented in Table 6.

The highest number of trawls licensed in coastal fisheries (6 429 units) was recorded in 1985. Since then it has been fluctuating at a somewhat lower level. The distribution of vessels by area in 1994 shows that 55% of them were located on the west coast of Peninsular Malaysia, 18% in Sarawak, 16% on the east coast of Peninsular Malaysia and 10% in Sabah.

Trawl Surveys in Malaysia

To further develop commercial trawl fisheries, the status of the industry and the state of the resources have to be determined. This is possible by collecting commercial catch statistics and by conducting regular and systematic surveys. With this information the catch trends resulting from the operation of trawlers can be studied and further expansion and rational management of the trawl fisheries can be planned.

Trawl surveys were first initiated on the east coast of Peninsular Malaysia with assistance from the Federal Republic of Germany under a bilateral agreement. The first trawl survey was conducted in 1970, followed by surveys in 1971, 1972 and 1974. These surveys covered coastal waters 10 to 60 m deep and 5 to 60 nm offshore, with the survey area stretching from the Thai/Malaysian border in the north to Johore in the south.

The survey area was divided into sub-areas on the east and west coast (Figs. 2 and 3). All trawl surveys were carried out with a standard German otter trawl net with a codend mesh size of 40 mm. The same

Table 7. Key events with impact on the trawl fisheries of Malaysia.

Date	Events, facts and related information	Source
Before 1962	Trawling trials using large seagoing vessels (*S.T. Tongkol* - 1926, *Kembong* - 1954 and *Mannihine* - 1955) in offshore trawling waters off the East Coast of Peninsular Malaysia were conducted and trawling deemed unviable due to economic reasons.	Mohamed (1984)
1963	A few Malaysian fishers from the northwest coast of Peninsular Malaysia go to Thailand to observe trawling operation and bring back 2 types of trawl nets (otter trawl and beam trawl).	Selvadurai and Andrew (1977)
Early 1964	Ban on trawling imposed by the Government because it would deplete inshore fish stocks and ruin the livelihood of traditional fishers. Physical conflicts between traditional fishers and trawlers are reported.	
October 1964	Pilot project on the use of otter trawl launched in Pulau Langkawi (in waters deeper than 15 fathoms using 42 boats from Penang and Kuala Kedah). The ban is lifted and trawling licenses issued on a limited scale to fishers from fishing centers such as Kuala Kedah, Penang and Pulau Pangkor.	Lam and Pathansali (1977)
1965	The Government legalizes trawling in Kedah.	
1966	The Government legalizes trawling in Penang.	
July 1966	Ten boats in Pulau Pangkor converted to trawlers. Within three months, the number increases to 40.	Mohamed (1984)
January 1966 - August 1967	Analysis of trawl fishery in Peninsular Malaysia at its nascent stage indicates that the catch rate is very high.	Selvadurai and Andrew (1977)

1967	Trawling subjected to stipulations listed under Fisheries (Maritime) Regulation 1967, made under the Fisheries Act of 1963. The terms and conditions are not only for codend mesh size but also the distance from the shore where trawling may be carried out, and the places where the catch should be landed.
	
	The minimum mesh size shall not be less than 1 inch internal measure at the cod end.
	  The operational zones are:
	- 100 gross tons (GT) and above shall be used only water beyond 12 miles;
	- 25 GT and above shall be used only in waters beyond seven miles;
	- 25 GT and below shall be used only in waters beyond three miles.
	  Time restriction for operation are:
	- vessel of 25 GT and below can only operate within 06:00-18:00 hours;
	- vessel above 25 GT are permitted to operate day and night.
	
	Fish caught by trawlers can be landed only at 49 specified places spread out between east and west coast of Peninsular Malaysia.
		Fisheries (Maritime) Regulation

1967	A survey of the coastal waters off the east coast of Peninsular Malaysia is carried out jointly by Malaysia, Thailand and the Federal Republic of Germany to assess potential demersal fish resource.	Mohamed (1984)
1968	Trawling becomes popular among fishers along the west Coast of Peninsular Malaysia. Authorities decided to issue trawling licenses to fishers' cooperatives but not to individual fishers.	Selvadurai and Andrew (1977)
1968	Trawling starts in Penang waters. Initially, five licenses are granted to a trawler society in Telok Bahang, Penang which bought five trawlers from Pangkor Island.	Lam and Pathansali (1977)
Mid 1968	The Government introduces a pilot scheme to promote development of trawl fishing off the east coast of Peninsular Malaysia. About 7 000 t landed by the estimated 300 trawlers in the area.	
1969 -1970	Costs and earnings survey of Malaysian trawl fisheries conducted by Peace Corps. Volunteers illustrates changes in the trawling industry between the nascent stage and the present.	
1970	The Government initiates a program, with assistance from the Federal Republic of Germany, to survey the coastal waters off the west coast of Peninsular Malaysia.	Lam et al. (1975)

Date	Events, facts and related information	Source
	A total of 4 surveys were carried out (in 1970, 1971, 1972 and 1974) covering coastal waters 10-60 m deep and 5-60 nautical miles from the shore.	
1971	Trawling in inshore waters highly successful that within a year, a total of 1 709 trawl licenses are issued in Perak State.	Mohamed (1984)
1971-1974	Study on economics of fish trawling in Penang concludes that the demersal resources in the West Coast of Peninsular Malaysia are overexploited.	Selvadurai and Andrew (1977)
1973	Landings by trawlers increase four-fold from 67 323 t in 1970 to 276 234 t in 1973. The 1973 figure is about 34% of total fish landings in the country.	DOF (1995a)
1974	Trawlers account for 39% of the total fishing gear in operation and described as having "over-expanded".	Saharuddin (1995)
1975	Five-year German Technical Aid Program commences to develop, modernize and spearhead fishery development through improved technology. Efforts targeted to modernize shrimp trawling in Sarawak by deploying double-rig/outrigger vessels.	Mohamed (1984)
1977	Introduction of large-scale (corporate) fishing in offshore areas in the South China Sea.	
April 1980	Malaysia proclaims its Exclusive Economic Zone (EEZ).	
May 1985	Malaysia enforces the EEZ Act.	
May 1985	Fisheries Act of 1985 is gazetted introducing more comprehensive provisions on enforcement, marine fishery management and conservation within the extended jurisdiction. The main focus is allocation of fishing grounds through the zoning concept. The license limitation program allocates fishing grounds by type of fishing gear, size of vessel and ownership status as follows: Zone A: Within 5 miles from the shoreline - reserved only for artisanal, owner-operated vessels. Zone B: Between 5 and 12 miles - reserved for owner-operated trawlers and purse seines of less than 40 GT. Zone C: Between 12 and 30 miles - reserved for trawlers and purse seines greater than 40 GT, wholly owned and operated by Malaysian fishers. Zone D: Beyond 30 miles - reserved for deep sea fishing vessels of 70 GT and above. Foreign fishing through joint-ventures or charter are restricted to this zone.	
1985	Government launches 'National Fisheries Development Plan' under the 1984 'National Agriculture Policy'. Main components attempt to restrict fishing effort in inshore areas and develop fishing in offshore grounds through commercialization and modernization of fishing activities.	DOF (1994)
1987	Deep sea fishing campaign is launched to bring about awareness of potential resources in the EEZ.	DOF (1994, 1995b)
1994	Licensed trawl fishing vessels recorded as many as 5 679 units which landed 587 928 t or 55% of total marine landings in the country.	DOF (1995a)

survey methodology was followed in the surveys conducted in the 1980s and 1990s. On the west coast of Peninsular Malaysia at least 11 demersal fish resource surveys have been carried out since 1970. The fishing ground covered during the surveys was relatively shallow water of 10 to 60 m. Standard German otter trawls with codend mesh size of 40 mm were used in all the surveys. The west coast of Peninsular Malaysia is also known as an important prawn fishing ground. A total of six prawn surveys have been carried out since 1972. During the first survey, German trawl nets with a codend mesh size of 30 mm and 35 mm were used. These nets were replaced by locally designed prawn trawl nets with 40 mm codend mesh size in the next five surveys. The first survey carried out on the coast of Sarawak and Sabah was in March/May 1972 using a standard German otter trawl with codend mesh size of 40 mm. The survey area ranged between 5 nm and 60 nm from the shoreline with depths of 10 m to 60 m. The whole coast was divided into six sub-areas as shown in Fig 3. At least six prawn trawl surveys have been conducted in the area since 1980. A double rigger trawler using 31.5 mm codend mesh size trawl gear was employed during the

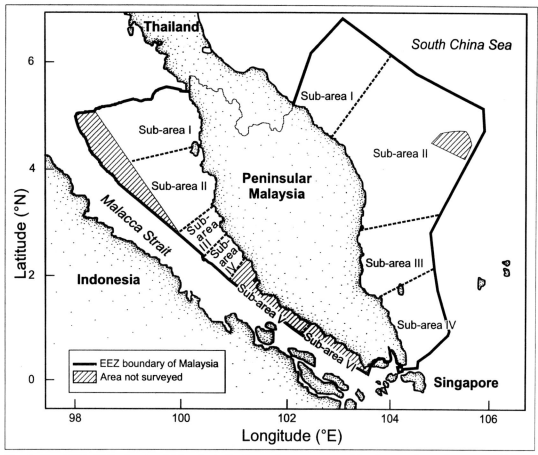

Fig. 2. The EEZ off Peninsular Malaysia indicating sub-areas covered by demersal trawl surveys. (See Table 7 and text).

surveys. A survey of demersal resources in the EEZ was also conducted in 1986/87 by the FAO vessel the *R.V. Rastrelliger*. The survey covered areas beyond 30 nm from shore. A summary of the information from these surveys is given in Appendix III to this volume.

The average catch rates obtained from surveys conducted in the east coast of Peninsular Malaysia indicate that these have declined from 516 kg·hour^{-1} in 1970 (Pathansali et al. 1974) to 255 kg·hour^{-1} in 1972 (Lam et al. 1975). The rate further declined to 72 kg·hour^{-1} in 1988 (Ahmad Adnan 1990). The recent survey carried out in 1995 showed a catch rate of 29 kg·hour^{-1} (Table 8). The same situation has also been observed for resources on the west coast of Peninsular Malaysia. Seasonal variability and natural fluctuations occur in fish populations, but the impact of fishing itself is unmistakable.

Ahmad Adnan (1990) reviewed the changes in the composition of major species or fish groups in terms of changes of dominance, based on the surveys conducted in Malaysia. He noted that the most dominant fish during the survey in 1970 were the thread-fin breams (*Nemipterus* spp.) (Table 9). However, after 1990 the dominance of this group appeared to have declined. The decreasing dominance of sharks, sciaenids, carangids and leiognathids has also been observed, while squids, priacanthids and synodontids show increasing dominance. A very clear decline in dominance is also seen for sciaenids, which were third in 1970 and became almost negligible in 1995. On the other hand, squids and lizardfish have shown marked increases. Squids, which ranked 14th in 1970, have become the dominant group in the 1990s. Another group that shows a major increase in their degree of dominance, particularly since the 1980s, are the priacanthids.

The survey data are kept in a DBase IV database and in a simple database developed by an individual researcher using the BASIC programming language. Also, the SEAFDEC research center in Terengganu has developed a database for biological and survey data using INFORMIX, while the Department of Fisheries has developed an integrated database comprising most of the information about fisheries (licensing, enforcement, catch/effort, aquaculture, etc.) using ORACLE.

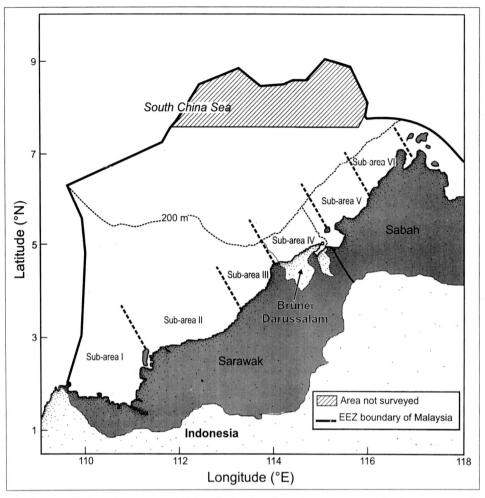

Fig. 3. The EEZ off Sarawak and Sabah indicating the sub-areas covered by demersal trawl surveys.

Table 8. Mean catch rates (kg·hour) of demersal trawl surveys in Malaysia. (See also Figs. 2 and 3).

Year	Peninsular Malaysia		Sarawak	Sabah[d]
	West coast	East coast		West coast
1967	-	438.4	-	-
1970	131.1[a]	515.6	-	-
1971	-	166.6	-	-
1972	141.7[a]	254.8	150-250	450-550
1973	125.1[b]	-	210-220	-
1974	92.1[a]	238.1	-	-
1975	-	-	200	-
1977	-	-	178.4	-
1978	69.4[a]	-	-	-
1979	-	-	141.9	-
1980	59.0[a]	-	154.0	-
1981	55.0[a]	160.0	162.0	-
1982	-	-	171.0	-
1983	-	129.2	220.0	-
1984	41.2[a]	90.5	79.0	-
1986	67.5[a]	-	-	-
1988	51.4[a]	71.8	-	-
1989	?	-	-	-
1990	-	-	-	-
1991	36.7[a]	69.9	235.5[c]	-
1992	25.9[a]	-	-	-
1993	-	-	-	208.0
1995	-	29.3	-	-

[a] Sub-area I and II only.
[b] Sub-area III, IV, V and VI only.
[c] Sub-area I only.

[d] The east coast of Sabah was not surveyed.
? Indicate inaccessible/unstandardized data.

Table 9. Ranking of dominant species groups caught during demersal fish surveys in the east coast of Peninsular Malaysia from 1970 to 1995. (Source: Ahmad Adnan 1996).

Survey year	1970	1971	1972	1974	1981	1983	1984	1988	1990	1991	1995
No. of hauls	151	150	144	97	78	115	120	120	60	62	100
15 dominant groups in the first survey:											
Nemipterus spp.	1	2	2	2	1	1	1	2	4	5	4
Carangidae	2	1	3	6	5	4	3	3	6	8	8
Sciaenidae	3	10	1	15	12	11	-	-	-	-	-
Mullidae	4	4	6	4	2	3	4	1	3	2	2
Tachysuridae	5	5	9	1	11	5	14	8	13	14	9
Leiognathidae	6	15	-	-	-	-	-	-	-	-	12
Lutjanidae	7	6	7	10	-	7	5	7	8	6	5
Scolopsis spp.	8	8	10	8	7	10	9	11	15	-	11
Rays	9	3	5	5		2	2	5	1	7	6
Gerridae	10	14	14	-	14	-	-	-	-	-	-
Sharks	11	7	12	11	-	12	13	9 .	7	-	-
Pentapodidae	12	-	8	12	9	-	-	-	-	11	-
Synodontidae	13	13	11	9	6	9	6	10	11	4	3
Loligoidea	14	11	4	3	4	8	7	6	2	1	1
Diagramma punctatum	15	9	13	-	-	15	15	-	-	-	13
New entries into list of dominant groups:											
Abalistes stellaris	-	12	15	-	15	14	11	13	-	-	15
Pomadasys spp.	-	-	-	7	-	-	10	-	-	-	-
Sepioidea	-	-	-	13	8	-	12	-	10	-	15
Thenus orientalis	-	-	-	13	13	-	-	15	14	-	10
Brachyura	-	-	-	14	-	-	-	-	-	-	-
Caesionidae	-	-	-	-	10	-	-	-	-	-	-
Serranidae	-	-	-	-	15	-	-	-	-	10	-
Priacanthus spp.	-	-	-	-	-	6	8	4	5	3	7
Sphyraena spp.	-	-	-	-	-	-	-	14	-	-	-
Siganus spp.	-	-	-	-	-	13	-	-	-	-	-
Lethrinus spp.	-	-	-	-	-	-	-	-	-	9	-
Rachycentron canadum	-	-	-	-	-	-	-	-	12	13	-
Aluteridae	-	-	-	-	-	-	-	-	-	15	-

A module for biological and survey data has yet to be developed and integrated into this database.

Management Issues and Opportunities

Fishery resources in the coastal areas are not only being intensively exploited but also threatened by pollution and habitat degradation. Currently, various management strategies and measures have been applied to ensure the sustainability of inshore fisheries production. The Malaysian Government has implemented the following measures through its legal and institutional framework:

1. *Direct limitation of fishing effort* - to ensure that the current high fishing pressure on the limited coastal fisheries resources will not be increased thus mitigating overexploitation;

2. *Closed fishing area* - commercial fishing vessels are prohibited from fishing in waters less than 5 nm from shore;

3. *Management zones* - Four fishing zones have been established through a licensing scheme based on specific fishing gear, class of vessel and ownership. These zones are:

- Zone A - less than 5 nm from shore, reserved solely for small-scale fishers using traditional fishing gear and owner-operated vessels;
- Zone B - beyond 5 nm where owner-operated commercial fishing vessels of less than 40 GT using trawl nets and purse seine nets are allowed to operate;
- Zone C_1 - beyond 12 nm where commercial fishing vessels of more than 40 GT using trawl nets and purse seine nets are allowed to operate;
- Zone C_2 - beyond 30 nm where deep-sea fishing vessels of 70 GT and above are allowed to operate.

4. *Setting-up of Marine Parks* - to conserve marine resources as well as the nursery/breeding/feeding habitat of marine life in the area. Thirty five islands have been designated as Marine Parks all over the country with a non-fishing zone of 3 miles around these islands;

5. *Artificial Reef Project* - to enhance fish resource in the coastal area. Apart from provision of artificial habitat for juvenile fish, the

60

reefs also prevent the encroachment of trawlers into shallow waters. A total of 54 tire reefs, 10 boat reefs and 10 concrete reefs have been constructed. An experimental reef using PVC pipes has also been set up to study the effect of the artificial reef;

6. *Set-up a good MCS system* - the Monitoring, Control and Surveillance (MCS) system provides effective and efficient scientific data acquisition for resource evaluation. It also provides for the effective monitoring and control of fisheries enforcement activities to ensure that only authorized fishing vessels conduct their fishing activities within designated areas in Malaysian waters;

7. *Deep-sea Fishing* - the fisheries resources within inshore water appear to have reached their maximum level of exploitation, but the declaration of the EEZ has enabled the Malaysian Government to actively encourage the development of offshore fisheries.

Several ordinances and acts were introduced to regulate the fishing industry. The first one was the Fisheries Ordinance of 1909. This ordinance was subsequently amended in 1912, 1924, 1926 and was finally repealed in 1951. The Fisheries Rules of 1951 was enforced on 10 August 1951. During this time there were also seven Fisheries Ordinances/ Enactments enforced by the various states. The fishing industry at that time consisted mainly of traditional fishing and regulation was minimal. The introduction of trawling in the coastal waters in the 1960's created much conflict between traditional fishers and trawl operators. This led to the formulation of the Fisheries Act of 1963 that provided a more comprehensive legal framework to manage fisheries in Malaysian waters. This Act was subsequently repealed and replaced by the Fisheries Act of 1985 with the following objectives:

a. to integrate and strengthen the legal framework relating to marine and inland fisheries;
b. to protect the natural living resources;
c. to protect the interest of the fishers;
d. to ensure equitable allocation of fishery resources; and
e. to strengthen administrative activities to reduce conflict among the fishing communities.

Presently fish resources in the nearshore area are heavily exploited causing a reduction in catch rates. There has been a drastic reduction in the licensing of small vessels (of less than 9.9 GT) because of government policies. In contrast, there has been a significant increase in the number of large vessels (more than 70 GT), which contribute about 11% of the total value of fish landings. The fishers have also been given the opportunity to leave the fishing industry and participate in other economic activities such as fish processing and aquaculture. This will help sustain fish production in inshore waters as well as increase income in the industry.

To ensure successful long term fisheries management, there is a need to look at the marine environment as a multiple-use area (see Table 1). Pollution due to sewage, solid waste and chronic oil spills requires careful monitoring and regulation in order to conserve marine habitats. Malaysia needs to explore new species with market potential for mariculture. The present monoculture practice is open to diseases and market instability. The cutting of mangroves for aquaculture, agriculture activities and factories require stringent control and close monitoring as this ecosystem is the breeding and feeding ground of many commercial species.

Acknowledgments

Thanks are due the Director General, Department of Fisheries, Malaysia, Y. Bhg. Dato' Shahrom bin Hj. Abdul Majid for his kind permission to present this paper. Thanks are also due to the Director of Fisheries Management and Protection Branch and the Director of Research, Fisheries Research Institute. The authors are also grateful for the assistance rendered by staff of ICLARM, Manila, in the preparation of this report.

References

Ahmad Adnan, N. 1990. Demersal fish resources in Malaysian waters (16). Fifth trawl survey of the coastal waters off the east coast of Peninsular Malaysia (June-July 1981) Fish. Bull. Dep. Fish Malays. Bul. Perikanan Jabatan Perikanan Malaysia. 60, 36 p.

Ahmad Adnan, N. 1996. Changes in the species composition of demersal fish based on trawl surveys in the east coast of Peninsular Malaysia. Unpublished report. Department of Fisheries, Malaysia.

Alias, M. 1994. Mechanization of fishing boats and its consequences on the management of the fishery in the West Coast of Peninsular Malaysia, p. 1-27. In Proceedings of Fisheries Research Conference, DOF, Malaysia, Kuala Terrenganu. 27 p.

Chew, Y. F. 1996. Wetland resources in Malaysia. In State of the Malaysian environment. CAP-SAM National Conference, 5-9 January 1996. Penang, Malaysia. Unpublished.

Choo, P.S., I. Ismail and H. Rosly. 1994. The west coast of Peninsular Malaysia, p. 33-54. In S. Holmgren (ed). An environmental assessment of the Bay of Bengal Region. BOBP/REP/67. Bay of Bengal Programme and Swedish

Centre for Coastal Development and Management of Aquatic Resources, Madras.

DOF. 1994. Perikanan seratus tahun: menuju era dinamik. Ministry of Agriculture, Malaysia.

DOF. 1995a. Annual fisheries statistics, 1968-94. Department of Fisheries, Ministry of Agriculture, Malaysia, Kuala Lumpur, Malaysia.

DOF. 1995b. Review of national fisheries management policy - implications for the Malaysian fisheries sector. Paper presented at the Workshop on Fisheries Management in Malaysia: Alternatives and Directions, 6-7 November 1995. MIMA (Malaysian Institute of Maritime Affairs), Kuala Lumpur, Malaysia.

FRI. 1987. Deep-sea fisheries resources survey within the Malaysian Exclusive Economic Zone. Fisheries Research Institute, Penang, Malaysia.

Gong, W. K., J.E. Ong, C. H. Wong and G. Dhanarajah. 1980. Productivity of mangrove trees and its significance in a managed mangrove ecosystem in Malaysia. Paper presented at the Asian Symposium on Mangrove Environmental Research and Management, UNESCO and University of Malaya, 25-29 August 1980.

Hamid, Y. 1989. Physical oceanography and plankton survey in the east coast of Peninsular Malaysia. *In* A.A.Jothy (ed.), Proc. Res. Sem. DOF Malaysia. Kuala Lumpur. 18 p.

Hamid, Y. 1993. On biological aspects of skipjack (*Katsuwonus pelamis*) and yellowfin (*Thunnus albacares*) off the Sarawak waters. FAO/UNDP Indo-Pacific Tuna Development and Management Programme. Unpublished report.

Hashim, A. 1993. Socio-economic issues in the management of coastal fisheries in Malaysia. Paper presented at the IPFC (Indo-Pacific Fishery Commission) Symposium on Socio-economic Issues in Coastal Fisheries Management, 23-26 November 1993, Bangkok, Thailand.

Hydrographic Department, Admiralty. 1964. Malacca Strait Pilot. 3rd ed. London.

Hydrographer of the Navy. 1975. China Sea Pilot. Vol. 2. London.

Ichikawa, T. 1986. Particulate organic carbon in the waters off the Terengganu coast, p. 87-92. *In* A.K.M. Mohsin, M.I. Mohamed and M.A. Ambak (eds.) *Ekspedisi Matahari '85*: A study on the offshore waters of the Malaysian EEZ. Occas. Pap. 3. Faculty of Fisheries and Marine Science, Universiti Pertanian Malaysia.

Ichikawa, T. and A.T. Law. 1988. Particulate organic carbon of the south-eastern portion of the South China Sea, p. 37-42. *In* A.K.M. Mohsin and M.I. Mohamed (eds.) *Ekspedisi Matahari '87*: A study on the offshore waters of the Malaysian EEZ. Occas. Publ. 8. Faculty of Fisheries and Marine Science, Universiti, Pertanian Malaysia.

Kent , G. and M.J. Valencia. 1985. Marine policy in Southeast Asia. University of California Press, Berkeley.

Kushairi, R. 1992. The areas and species distribution of seagrasses in Peninsular Malaysia. Paper presented at the 1st National Natural Res. Symp., FSSA, Universiti Kebangsaan Malaysia, Kota Kinabalu, Sabah, 23-26 July 1992.

Lam, W.C., S.A.L. Mohammed Shaari, A.K. Lee and W. Weber. 1975. Demersal fish resources in Malaysian waters. Second trawl survey off the west coast of Peninsular Malaysia (30 November 1971-11 January 1972). Fish Bull. 7. Ministry of Agriculture, Malaysia. 29 p.

Lam, W.C. and D. Pathansali. 1977. An analysis of Penang trawl fisheries to determine the maximum sustainable yield. Fish. Bull. No. 16. Ministry of Agriculture, Malaysia. 10 p.

Liew, H.C. and R. Hoare. 1982. The effects of sediment accumulation and water turbidity upon the distribution of Scleractinian corals at Cape Rachado, Malacca Strait. *In* Proc. of the Conf. on Trends in App. Biol. in Southeast Asia. 11-14 October 1979, Penang, Malaysia. SBS, USM.

Mohammed Shaari, S.A.L. 1976. Demersal fish resources surveys and problems of fisheries resources development, p. 1-26. *In* Fish. Bull. 15. Ministry of Agriculture and Rural Development, Kuala Lumpur, Malaysia.

Mohamed, M.I. 1984. The development of fishing technology in Malaysia. Bull. Jentera 3(1): 20 p.

Pathansali, D. G. Rauck, A.A. Jothy, S.A.L. Mohammed Shaari and T.B. Curtin. 1974. Demersal fish resources in Malaysian waters. Min. of Agric. and Fish. Malays. Fish. Bull. 1, 46 p.

Phang, S.M. 1990. Seagrass - A neglected natural resource in Malaysia. Proc. 12th Ann. Sem. Mal. Mar. Sci. Kuala Lumpur.

Raihan, A. and T. Ichikawa. 1986. Chlorophyll *a* content off the Terengganu coast, p. 117-120. In *Ekspedisi Matahari '85*: A study on the offshore waters of the Malaysian EEZ. Occas. Publ. 8. Faculty of Fisheries and Marine Science, Universiti, Pertanian Malaysia.

Saadon, M.N., L.H. Chark and M. Uchiyama. 1988. Salinity and temperature vertical profiles of the Sarawak coasts, p. 29-36. *In* A.K.M. Mohsin and M.I. Mohamed (eds.) *Ekspedisi Matahari '87*: A study on the offshore waters of the Malaysian EEZ. Occas. Publ. 8. Faculty of Fisheries and Marine Science, Universiti Pertanian Malaysia.

Saharuddin, A.H. 1995. Development and management of Malaysian marine fisheries: Technical conservation measures. Mar. Policy 19(2): 115-126 p.

Samsuddin, B. 1995. Kajian oseanograpi perikanan pantai perairan Kedah/Pulau Langkawi (internal report).

Selvadurai, S. and L.K.K. Andrew. 1977. Marine fish trawling in Peninsular Malaysia: A case study of costs and earnings. Fish. Bull. No. 19. Ministry of Agriculture, Malaysia. 30 p.

Shahunthala, D.V. 1989. Heavy metal levels in Malaysian fish. Fish. Bull. 58, Ministry of Agriculture, Malaysia.

Shamsudin, L.B., K.B.B. Kaironi and M.N.B. Saadon. 1988. Chlorophyll *a* content off the Sarawak waters of the South China Sea, p. 87-90. *In* A.K.M. Mohsin and M.I. Mohamed (eds.) *Ekspedisi Matahari '87*: A study on the offshore waters of the Malaysian EEZ. Occas. Publ. 8. Faculty of Fisheries and Marine Science, Universiti Pertanian Malaysia. 233 p.

Tang, H. T., Haron Hj. Abu Hassan and E.K. Cheah. 1990. Mangrove forest of Peninsular Malaysia - A review of management and research objectives and priorities. Unpublished report.

Overview of Philippine Marine Fisheries[1]

N.C. BARUT
M.D. SANTOS
Bureau of Fisheries and Aquatic Resources
Department of Agriculture
Quezon City, Philippines

L.R. GARCES
International Center for Living Aquatic Resources Management
MCPO Box 2631, Makati City 0718, Philippines

BARUT, N.C., M.D. SANTOS and L.R. GARCES. 1997. Overview of Philippine marine fisheries, p. 62-71. *In* G. Silvestre and D. Pauly (eds.) Status and management of tropical coastal fisheries in Asia. ICLARM Conf. Proc. 53, 208 p.

Abstract

Marine fisheries are an important source of protein, livelihood and export earnings for the Philippines. In 1994, total marine fisheries catch was 1.67 million t (62% of total fisheries production) valued at about US$1.65 billion. Of this total, 277 thousand t were demersal fishes, 885 thousand t small pelagics, 305 thousand t tunas and 203 thousand t other species or groups. Current catches have leveled off since 1991 (at a level near estimated maximum sustainable yield) and existing fishing effort is clearly too high.

This paper reviews the status of marine fisheries and the development of trawl fisheries in the Philippines. The combined effects of excessive fishing effort and environmental degradation have contributed to the depletion of fishery resources, particularly coastal demersal and small pelagic fishes.

Introduction

Fisheries are an important component of the agricultural sector in the Philippines. In 1994, they accounted for 4.3% of gross domestic product and 18% (US$2.5 billion) of gross value added by the agriculture, fishery and forestry sectors. Fisheries provide livelihood to about 1 million individuals or about 5% of the country's labor force. Fishery exports were about 172 000 t in 1994 valued at US$578 million[2]. Fish consumption in the Philippines is high at about 28.5 kg/capita/year.

Marine fisheries landings in 1994 were about 1.67 million t valued at about US$1.65 billion. This represented roughly 62% of the total fisheries production in the country; the rest was contributed by the aquaculture and inland fisheries sectors. The municipal (i.e., small-scale) fisheries sector contributed 47% (787 000 t) of marine fisheries catches, while the balance (885 000 t) came from the commercial fisheries sector.

The increased demand for fish from a rapidly growing population and increasing exports has substantially increased fishing pressure on the marine fishery resources in the past two decades. In February 1996 the National Fisheries Workshop on Policy Planning and Industry Development pointed to resources depletion and environmental degradation as the key issues facing the fisheries sector. Declining catch rates in many traditional fishing grounds and the leveling off of marine landings since 1991 support these conclusions. The sections below provide an overview of the status of Philippine marine fisheries.

---

[1] ICLARM Contribution No. 1390.
[2] In 1994: US$1 = P23.75.

Marine Environment and Activities

The Philippines is an archipelago consisting of more than 7 100 islands. It extends about 2 000 km in a south-north direction, between 4°05' and 4°30' N latitude, from the northeast coast of Borneo to 150 km off Taiwan (Fig. 1). The total territorial water area, including the Exclusive Economic Zone (EEZ), is about 2.2 million km². The shelf area, down to 200 m covers 184 600 km².

Pioneering expeditions that contributed significantly to information on Philippine hydrography were those of the *Nuestra Señora de Buena Esperanza* (1587), *Desire* (1588), *Cygnet* (1688), *Elizabeth* (1762), *Atrevida* and *Descibierta* (1792), *Santa Lucia* and *Magallanes* (1800), *Rhone* (1819), *Samarang* (1843),

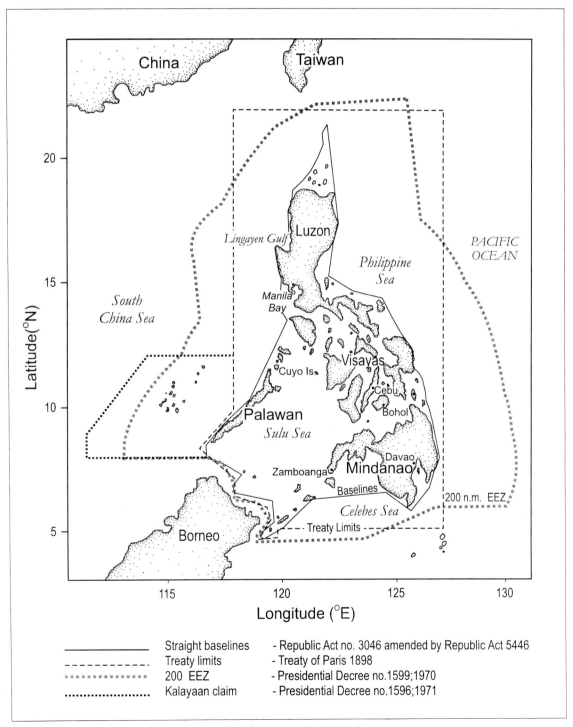

	Straight baselines	- Republic Act no. 3046 amended by Republic Act 5446
	Treaty limits	- Treaty of Paris 1898
	200 EEZ	- Presidential Decree no.1599;1970
	Kalayaan claim	- Presidential Decree no.1596;1971

Fig. 1. Marine jurisdictional boundaries of the Republic of the Philippines.

64

Royalist (1850-1854), *Rifflemann* and *Nassau* (1868-1869) (Sebastian 1951). Oceanographic data were also made available through similar explorations during the late 19th to the mid-20th centuries. The Danish *Galathea* collected oceanographic information in Manila Bay, Dinagat and Surigao between 1845 and 1847. In 1875, the British *H.M.S. Challenger* Deep Sea Expedition conducted investigations in the Sulu Sea and the Visayas. From 1907 to 1909 the marine life of the islands was surveyed extensively by the *Albatross* Philippine expedition which was sponsored by the US Bureau of Fisheries. Other explorations that followed were the German survey vessel *Planet* (1907-1912), the German cruiser *Emden* (1927), the Danish research vessel *R.S.S. Dana II* (1929), the Dutch *H.M.S. Willebrord Snellius* (1903), the Japanese *Manshu*, *Yamato*, and *Musashi* (1934-1942) and the US naval vessel *Cape Johnson* (1945). In 1947-1950, the *S.F. Baird*, under the Philippine Fishery Program of the US Fish and Wildlife Service, conducted an extensive investigation of the Celebes Sea, Sulu Sea, South China Sea and the waters of the Philippine Sea covering an area of more than 2 074 500 km². In 1951-1952, the *Galathea* made an intensive biological survey of the Philippine Deep.

The waters east of the Philippines are affected by the major large-scale ocean currents of the Pacific. The major current system affecting the Philippines is the North Equatorial Current which flows westward across the Pacific, hits the eastern coast of the country and splits into northward and southward branches. The northward branch flows along the east coast of the Visayas and Luzon, moving to Taiwan and Japan as the Kuroshio Current. The southward branch becomes the Mindanao Current, moving southward along the east coast of Mindanao. The influence of the strong seasonally reversing monsoon winds on water circulation is more pronounced on the western side of the Philippines. During the northeast monsoon a cyclonic pattern of surface water movement develops in the South China Sea with a northwesterly flow along the western coasts of Palawan and Luzon. During the southwest monsoon water movement in the South China Sea is generally northeasterly, flowing out through the straits between Luzon and Taiwan (Wyrtki 1961; Soegiarto 1985). The work of Wyrtki (1961) described the water masses in the area. Historical hydrographic data indicated that Philippine waters are very similar to those of the western Pacific.

The country's marine environment is distinctly tropical in character, with relatively warm and less saline waters. Sea surface temperatures are generally above 28°C in summer and only a few degrees lower during the cold months. Salinity variations are very small, especially in the eastern parts of the country. These variations increase during the moisture-laden southwest monsoon in the western parts of the

country. The waters are typically poor in nutrients, with small upwellings, gyres and mixing processes occasionally enhancing local productivity. Nutrient depletion can extend to depths of 50-100 m in open waters. The thermocline depth is usually about 150 m and varies seasonally. Nutrient concentrations and biological productivity are highest over the shelves, declining rapidly with depth and distance from the coast.

Water quality studies in the country are limited to highly localized and pollution-prone areas. Evident in all these studies is the deterioration of water quality brought about by mine tailings, agricultural runoffs, siltation, domestic sewage and oil spills. Almost all results indicate abiotic and biotic parameters (e.g., pH, salinity, turbidity, dissolved oxygen, heavy metal content, coliform count) exceeding standards set by the Environmental Management Bureau of the Department of Environment and Natural Resources.

Coral reefs abound in shallow water areas not subjected to low salinity from freshwater inflows, sedimentation and physical perturbations, with about 27 000 km² of coral reef area within the 30 m depth contour. There are more than 70 genera and 400 species of hard corals documented, as well as about a thousand associated fish species (Gomez et al. 1994). Reef areas contribute substantially to fisheries productivity, with fish yields ranging from 5 to 37 t·km⁻² (Alcala and Gomez 1985). Extensive coralline or hard bottoms are found around Palawan, Sulu, the Visayas and the central part of the country's Pacific coast. A large portion of Philippine coral reefs have been subjected to serious degradation which reduced their productivity (Yap and Gomez 1985). Major destructive factors are sedimentation and siltation from coastal development and activities, illegal and destructive methods of fishing and overfishing (Gomez et al. 1994). Coral cover data from various surveys of Philippine reefs indicate that 5% are in excellent condition (more than 75% living coral cover), 25% in good condition (50%-75% living coral cover) and the rest in fair and poor condition (below 50% living coral cover) (Gomez 1991).

Mangrove communities are integral and important components of the coastal ecosystem. These are categorized into mangrove swamps composed mainly of large trees and associates, and *nipa* swamps which are characterized by stemless palm growths. Forty-one species of mangroves have been identified in the Philippines. These yield by-products such as timber and other building materials, high grade charcoal, tannins, resins, dyes and medicines. Mangrove areas are under pressure for conversion to other uses, notably aquaculture and human settlement. In 1965, mangrove areas covered about 4 500 km². Ten years later only about 2 500 km² were left. Sixty percent of this decline was due to conversion into aquaculture ponds for milkfish and prawns (Primavera 1991). By 1981 an aggregate cover of only 1 460 km² was intact. This prompted government and nongovernment agencies to suspend permits for mangrove conversion to fishponds, accelerate reforestation activities and spur community-based management.

Seagrass communities regulate water flow and wave energy together with coral reefs and mangroves. There are 16 seagrass species recorded in the Philippines, second only to Western Australia among the 27 countries of the Indo-Pacific region. Extensive seagrass beds have been identified in Bolinao, Palawan, Cuyo Island, Cebu, Bohol, Siquijor, Zamboanga and Davao. Seagrass communities in the country manifest signs of degradation due to the combined effects of natural calamities, predation, aquaculture, deforestation, siltation and destructive fishing methods (Fortes 1990).

Seaweed beds, like coral reef, mangrove and seagrass communities, play a vital role in the coastal environment. They provide feeding and nursery grounds for different types of marine macro and microorganisms and interact with seagrasses to control ocean wave action. Aside from its ecological function this group of marine macrobentic algae is also an important human food source. There are 190 species of seaweed recorded in the Philippines. About 150 species are considered economically important but only a few are cultivated (particularly *Eucheuma* spp.). Other species under the genus *Sargassum* and *Gracilaria* are harvested from natural beds. To date,

the Philippines is the world's leading supplier of *Eucheuma*, producing about 80% of total world supply. There are about 80 000 seaweed farmers with 350 000 dependents that rely on the seaweed industry in the country (Dacay 1992).

Marine Capture Fisheries

Philippine marine fisheries are conventionally subdivided into municipal (small-scale) and commercial fisheries on the basis of vessel gross tonnage. Municipal fisheries include capture operations using boats less than 3 GT and those that do not involve the use of watercraft. A license is issued by the municipality where the boat is registered, hence the name. Fishing permits are issued to fishing boats by the municipality where they intend to fish. Commercial fisheries include capture fishing operations using vessels of 3 GT and above. Commercial fishing vessels are required to secure a commercial fishing boat license from the Bureau of Fisheries and Aquatic Resources (BFAR) before they can operate. Until recently commercial fishing vessels were only allowed to operate in waters beyond 7 km from the shoreline. With the implementation of the Local Government Code in 1992, coastal waters within 15 km from the shoreline are now considered municipal waters and commercial fishing is not allowed within this area.

Fig. 2 illustrates the total marine, commercial and municipal fisheries catches in the Philippines from 1950 to 1994. Total marine landings showed

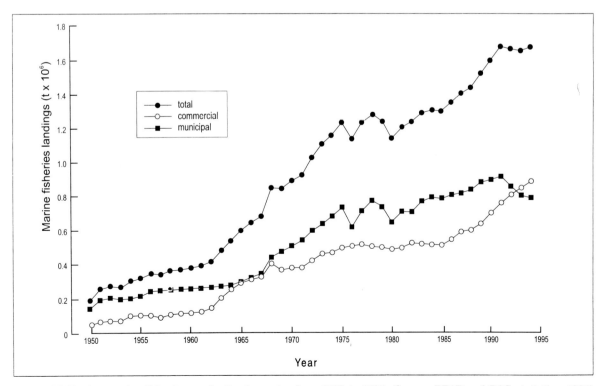

Fig. 2. Philippine marine fisheries production by sector from 1950 to 1994. (Source: BFAR and BAS statistics, 1994).

66

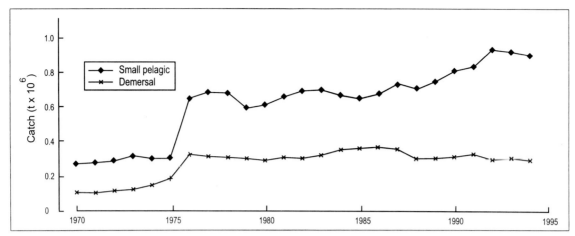

Fig. 3. Demersal and small pelagic fisheries in the Philippines from 1970 to 1994. (Source: BFAR and BAS statistics).

accelerated growth during the periods 1962-1975 and 1985-1991, but leveled off to around 1.65 million t in the early 1990s. This was brought about by the decline in municipal fisheries landings being compensated by the commercial fisheries sector, which may be indicative of increasing resource depletion and competition in nearshore fishing grounds.

In 1994 the marine landings of 1.67 million t consisted of 17% (277 000 t) demersal fishes, 53% (885 000 t) small pelagics, 18% (305 000 t) tunas and 12% (203 000 t) miscellaneous species or groups. Fig. 3 illustrates the trend in demersal and small pelagic fish production in the Philippines from 1970 to 1994. Note that the catch of demersals has leveled off since 1976. The catch of small pelagics, on the other hand, was almost the same from 1976 to 1988, increased rapidly from 1988 to 1992, then declined from 1992 to 1994. These catch trends in the face of continuously increasing fishing effort reflect assessments indicating overfishing of demersal and small pelagic fish stocks.

The trawl is commonly used by the commercial sector to exploit demersals while gillnets, hook and lines and 'baby' trawls are most commonly used by municipal fishers in catching demersal species or groups. Silvestre et al. (1986) provide a detailed review of trawl and demersal fisheries development in the Philippines and the consequent decline of demersal biomass in the country's coastal waters. Numerous trawl surveys have been conducted in Philippine waters (see Appendix III) which document this decline and which potentially offer numerous insights for improved fisheries management if standardized and analyzed more exhaustively.

The commercial sector commonly used bagnets, purse seines and ringnets for catching small pelagics while municipal fishers dominantly use gillnets, beach seines and round haul seines. Roundscads, sardines, anchovies, mackerels, big-eye scads, round herrings and fusiliers dominate small pelagic catches in the Philippines.

Six species of tuna dominate Philippine landings, i.e., yellowfin tuna (*Thunnus albacares*), big-eye tuna (*Thunnus obesus*), skipjack tuna (*Katsuwonus pelamis*), eastern little tuna (*Euthynnus affinis*), frigate tuna (*Auxis thazard*) and bullet tuna (*Auxis rochei*). The most common gear used by the commercial sector in catching tuna are purse seines and ringnets, while municipal fishers mainly use handlines. All these gears are operated jointly with fish aggregating devices known as 'payao'. Tuna production increased from about 9 300 t in 1970 to 125 000 in 1976 with the introduction of 'payao' in 1975. Wider use of 'payao' has contributed to the high level of tuna catches in the country.

The available information on fishing effort is limited. Table 1 gives a summary of the number of commercial fishing vessels in the Philippines, by gear type, and gross tonnage from 1967 to 1987. Note that the trawl, bagnet and purse seine were the most widely used gear during this period; also a shift to larger vessels (i.e., from 3-5 GT to 10-<50 GT) is evident. Similar vessel information prior to 1967 is unavailable, while information after 1987 is difficult to extract due to changes in methods of fishery statistics collection resulting from the transfer of such functions from BFAR to the Bureau of Agricultural Statistics (BAS). Information pertaining to municipal fisheries is far more limited, although periodic information may be available from national censuses occasionally conducted in the past (Dalzell et al. 1987).

The potential yield from the marine fishery resources of the Philippines has been studied extensively (Munro 1986; Silvestre et al. 1986; Dalzell and Ganaden 1987). Estimates of maximum sustainable yield (MSY) from conventional fishery resources vary widely between 1.1 and 3.7 million t. The higher estimates are based on overly optimistic yield-per-unit-area figures and do not consider productivity decline with depth. The scientific consensus since the 1980s is that MSY from conventional resources is

Table 1. Total number of commercial fishing vessels in the Philippines by gear type and tonnage from 1967 to 1987. (Source: BFAR statistics). See text.

Category	1967	1968	1969	1970	1971	1972	1973	1974	1975	1976	1977	1978	1979	1980	1981	1982	1983	1984	1985	1986	1987
Total number of vessels	2 361	2 225	2 273	2 284	2 180	2 222	2 513	2 286	2 543	2 571	2 269	2 133	2 464	2 407	2 349	2 580	2 634	2 666	2 484	2 283	2 516
Total gross tonnage	81 268	81 950	84 117	89 688	90 550	99 554	113 325	103 325	101 209	103 219	103 078	70 012	91 999	77 346	99 471	105 122	103 297	97 672	90 453	83 506	87 309
Powered vessels	2 124	2 033	2 040	2 061	2 142	2 169	2 455	2 202	2 488	2 525	2 218	2 123	2 421	2 366	2 310	2 473	2 514	2 621	2 343	2 168	2 331
Nonpowered vessels	225	174	182	179	311	53	54	82	55	46	51	10	43	41	39	107	120	45	141	17	16
Number by gear																					
Bagnet	1 002	883	796	858	743	650	791	584	713	656	504	642	641	624	552	603	578	652	602	565	618
Beach seine	45	44	41	49	35	44	41	51	77	75	103	46	63	77	58	59	59	55	70	60	37
Gillnet	7	5	15	15	15	12	29	27	23	23	19	16	18	23	6	9	3	7	5	2	3
Hook and line (handline)	111	88	97	88	83	94	81	76	63	73	67	50	57	97	109	61	93	120	104	91	77
Longline	9	14	9	9	5	10	3	5	5	5	8	5	22	26	70	61	62	62	55	41	62
Drive-in-net (muro-ami)	37	25	24	26	37	39	67	37	35	36	34	5	41	7	45	39	43	37	37	34	29
Purse seine (ringnet)	197	202	253	245	265	320	470	280	313	342	280	328	408	412	450	516	404	318	306	280	296
Pushnet					2	7	1	33	55	72	62	92	90	57	47	78	105	92	668	54	50
Otter trawl	593	653	667	653	652	690	794	767	763	786	684	769	877	848	739	829	932	884	763	702	769
Round haul seine	108	85	69	76	61	53	50	38	27	20	31	23	38	33	26	34	45	37	46	25	28
Others	240	208	251	221	275	303	212	388	469	483	477	157	143	158	247	291	315	402	448	428	547
Not reported	12	18	51	44	7		4	2					66	45			84				
Number by tonnage																					
3 to less than 5	832	711	677	666	664	261	268	205	285	246	234	214	219	205	165	144	149	186	103	93	94
5 to less than 10						358	426	450	571	627	587	575	611	606	533	584	618	674	539	477	494
10 to less than 20	534	476	469	478	443	466	529	480	559	532	432	515	583	653	636	730	672	686	659	633	715
20 to less than 50	411	421	414	426	422	444	534	468	497	531	470	421	503	467	483	508	567	611	602	562	629
50 to less than 100	416	437	480	478	440	458	495	427	395	397	328	291	311	278	277	275	277	265	251	232	230
100 to less than 200						195	203	195	178	179	154	86	141	137	145	156	158	138	136	118	119
200 to less than 500	156	162	182	192	211	40	58	61	58	59	62	13	47	17	51	55	57	48	43	47	44
500 and over	12	18	51	44							2	4	6	3	20	21	16	13	10	96	6
Not reported												4								98	169

68

1.650±0.250 million t, i.e., 600±100 thousand t of demersals, 800±100 thousand t of coastal pelagics and 250±50 000 t of tunas or oceanic pelagics (BFAR 1995).

Fig. 4 illustrates the results of the assessment of demersal fisheries. The figure indicates the following: MSY for exploited demersal resources (i.e., excluding offshore hard bottoms around Palawan, southern Sulu Sea area, and the central part of the country's Pacific coast) is about 340 000-390 000 t, while the maximum economic yield (MEY) ranges from 280 000 to 300 000 t worth of fish and invertebrates. Subsequent refinements of this assessment yielded similar results (Silvestre and Pauly 1987). Note that the MSY of 340 000-390 000 t, when combined with the MSY estimates for unexploited or lightly fished hard bottom areas of 200 000 t, come very close to the consensus demersal MSY of 600±100 000 t noted above.

Assessment of the exploitation status of currently fished demersal stocks indicates that these resources are biologically and economically overfished. Studies

indicate that the biomass of currently fished stocks has declined to about 30% of its original levels in the late 1940s (Silvestre et al. 1986). It can also be noted in Fig. 4 that fishing effort could be reduced by 3/5 without substantially reducing demersal yields. The annual rent dissipation from overfishing of the demersal stocks is about US$130 million per year. Reduction of fishing effort and reallocation of those displaced to lightly fished, hard-bottom areas is required.

Fig. 4 also illustrates the results of the assessment of small pelagic fisheries. The data indicates the following: MSY for small pelagics of about 550 000 t; MEY of 250 000 t for fish and invertebrates in the exploited fishing grounds, and; a maximum economic rent (MER) of US$290 million worth of fish and invertebrates (Dalzell et al. 1987). Subsequent refinements of this assessment have yielded similar results (Trinidad et al. 1993). The MSY figure of 550 000 t, when combined with MSY estimates of 250 000 t for lightly fished small pelagic resources or fishing grounds

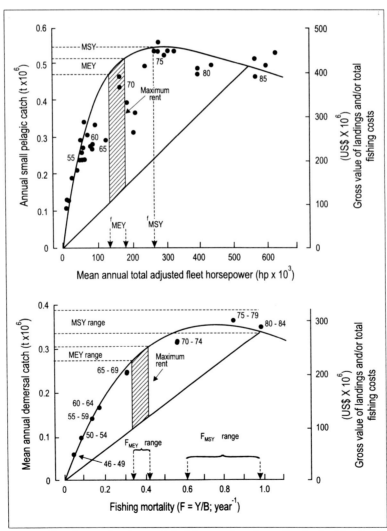

Fig. 4. Surplus production model of demersal and small pelagic fisheries in the Philippines providing estimates of MSY and economic rent for exploited areas/resources. (Sources: Silvestre and Pauly 1986; Dalzell et al. 1987).

Table 2. Coastal transect indicating activities and issues relevant to integrated coastal zone and coastal fisheries management in the Philippines.

Major zones	Terrestrial			Coastal		Marine	
	Upland (>18% slope)	Midland (8-18% slope)	Lowland (0-<8% slope)	Intertidal (1 km inland from HHWL-30 m depth)	Nearshore (30 m-200 m depth)	Offshore (>200 m depth-EEZ)	Deepsea (beyond EEZ)
Main resource uses/activities	Logging Mining Upland agriculture	Logging Farming Mining Dams Livestock production	Agriculture Urban development Industrial development Tourism Human settlement Freshwater fisheries Freshwater aquaculture	Aquaculture Municipal fisheries Mangrove forestry Tourism Ports/marine transport Human settlements Salt production Sand/gravel mining	Municipal fisheries Commercial fisheries Marine transport	Commercial fisheries Marine transport	Fisheries Marine transport
Main environmental issues/impacts on the coastal zone	Siltation/sedimentation Flooding Heavy metal pollution Agrochemical loading	Siltation Agrochemical loading Flooding Increased salinity Organic loading	Agrochemical loading Sewage disposal Industrial wastes Solid wastes Siltation Overfishing	Agrochemical loading Overfishing Mangrove depletion Coral reef degradation Oil spills Sewage disposal Habitat conversion Organic loading Siltation Red tides Reduced biodiversity	Overfishing Solid wastes Oil spills	Overfishing Oil spills Poaching	Oil spills Overfishing

70

(i.e., in waters off Palawan, parts of the country's Pacific coasts and some parts of Mindanao), virtually equals the consensus small pelagic MSY of 800±100 000 t noted above. Thus, available information also indicates that the small pelagics are biologically and economically overfished. The consequent rent dissipation is about US$290 million annually. Fishing effort is particularly high in nearshore coastal areas, especially the traditional fishing grounds. Effort level in the mid-1980s was already more than twice the level necessary to harvest MSY (Fig. 4). The obvious need for small pelagic fisheries is to reduce effort and reallocate the excess to lightly fished areas. There are indications that this expansion has occurred since the work of Dalzell et al. (1987).

The assessment of tuna indicates that the fishing pressure on the stocks in Philippine waters is high (Dalzell and Corpuz 1990; BFAR 1995). The magnitude of the catches and the concentration of fishing effort within a small surface area indicate a very high local fishing mortality. There is a need to expand to other fishing areas (further offshore) off the South China Sea and the Pacific coast of the country. Such a move may increase the size of tuna catches in Philippine waters, which largely do not meet export market requirements. Commercial fishing operations are known to have expanded to Indonesia, Micronesia and Papua New Guinea through joint venture arrangements from private sector initiatives.

The oceanic large pelagics such as marlin, swordfish and sailfish are not fully exploited at present. From a landing of 17 000 t-25 000 t in the 1980s, the large pelagic landings have declined to 9 000-15 000 t in the 1990s.

Nonconventional resources such as oceanic squid and deep-sea shrimp occur in Philippine waters. At present there is no established fishery for these resources and there is very little information to assess resource potential.

Management Issues and Opportunities

An excessive fishing effort level is evident from the various country-wide site-specific fisheries assessments conducted in the Philippines. Species composition changes reflective of growth, recruitment and ecosystem overfishing have occurred in many areas (Silvestre et al. 1986; Pauly et al. 1989;Cinco et al. 1994). Economic overfishing is also quite evident. There is a need to improve fisheries management, in general, and to effect effort reduction, in particular. It should be noted, moreover, that distributional inequity and conflict between municipal and commercial fisheries is an issue in many areas, particularly in nearshore, traditional fishing grounds. For instance, in the San Miguel Bay fisheries in the early 1980s, 50% of pure profits went to 95 trawlers

owned by 35 households, and the rest went to 2 288 municipal gear units owned by 2 030 households (Smith et al. 1983). Thus, effort reduction should be sensitive to considerations of distributional equity.

The decline of fishery resources in the Philippines, particularly of demersals and small pelagics, is presumably a combined effect of excessive fishing effort and coastal environmental degradation. The quantitative link of resource decline to habitat degradation, however, is difficult to document. Habitat degradation is more complex and serious in highly populated coastal areas. Habitat degradation due to pollution is mostly site specific. Table 2 illustrates a typical coastal transect, indicating activities and issues which require attention for effective coastal fisheries management in the Philippines. These issues, taken within the larger framework of integrated coastal zone management (ICZM), require increased attention and support in the Philippines. ICZM schemes have been implemented in 12 selected coastal areas in the country under the ADB-funded Philippine Fisheries Sector Program (1991-1997). Participatory schemes that develop the knowledge and capability of stakeholders and local governments (consistent with the 1992 Local Government Code) to manage their resources have been initiated successfully, requiring wider replication and continued support.

Improvement of information inputs into the fisheries management decision-making process requires immediate action in the Philippines. The statistical gaps resulting from the transfer of fisheries statistics collection from BFAR to BAS after 1987 need urgent attention. In this context we note the availability of numerous trawl surveys conducted in the Philippines (Appendix III). The work of Silvestre et al. (1986) and Silvestre and Pauly (1987) illustrate some of the insights which can be derived from these underutilized data. Retrospective analyses of these surveys using recently developed techniques such as population, community analysis (McManus 1986, 1989 and this vol.) and ecosystem analysis (Christensen and Pauly 1993, 1996) and along the lines suggested in Pauly and Martosubroto (1996) can provide numerous insights which can improve fisheries management in the Philippines. The availability of sound, scientific information will facilitate the elevation of resource allocation decisions into objective rather than emotional debates.

Acknowledgments

We wish to thank the staff of the Pelagic Resources Section, BFAR, particularly Ana Marie Coronel, Criselda Malit, Jenneth Manzano, Clarita Ulanday and Val Manlulu for their assistance in locating, preparing and encoding statistical information used in this paper. Thanks are also due to Salud Ganaden, Luz Regis and Leony Mijares for their assistance and suggestions in the preparation of the paper, and Geronimo Silvestre of ICLARM for his technical guidance.

References

Alcala, A.C. and E.D. Gomez. 1985. Fish yields of coral reefs in Central Philippines. Proc. 5th Int. Coral Reef Symp. 5: 521-524.

BFAR (Bureau of Fisheries and Aquatic Resources). 1950-1987. Fisheries statistics of the Philippines.

BFAR. 1995. On the allocation of fishing area for exclusive use by the municipal fisheries sector: a policy brief. Bureau of Fish. and Aquatic Res., Quezon City, Philippines.

Christensen, V. and D. Pauly (eds.) 1993. Trophic models of aquatic ecosystems. ICLARM Conf. Proc. 26, 390 p.

Christensen, V. and D. Pauly. 1996. Ecological modeling for all. Naga, ICLARM Q. 19(2): 25-26.

Cinco, E.A., J.C. Diaz, R. Gatchalian and G.T. Silvestre. 1994. Results of the San Miguel Bay trawl survey. In G. Silvestre, C. Luna and J. Padilla (eds.) Multidisciplinary assessment of the fisheries in San Miguel Bay, Philippines (1991-1993). ICLARM Tech. Rep. 47 (CD-ROM).

Dacay, B.U. 1992. The state of the seaweed industry in the Philippines, p. 23-27. In C. Calumpong and E. Mines (eds.). Proceedings of the Second Philippine-U.S. Phycology Symposium/Workshop, 6-10 January 1992, Silliman University Press, Dumaguete City, Philippines. 376 p.

Dalzell, P. and P. Corpuz. 1990. The present status of small pelagic fisheries in the Philippines, p. 25-51. In Pagdilao, C.R. and C.D. Garcia (eds.) Philippine tuna and small pelagic fisheries: status and prospects for development. Proc. of the Sem. Workshop, 27-29 July 1988. Zamboanga State College of Marine Science and Technology, Zamboanga City, Philippines. Philippine Council for Aquatic and Marine Research and Development Book Ser. No. 7.

Dalzell, P. and R.A. Ganaden. 1987. A review of the fisheries for small pelagic fishes in Philippine waters. BFAR Tech. Pap. Ser. 10(1), 58 p. ICLARM, Manila, Philippines and Bureau of Fisheries and Aquatic Resources, Quezon City, Philippines.

Dalzell, P., P. Corpuz, R. Ganaden and D. Pauly. 1987. Estimation of maximum sustainable yield and maximum economic rent from the Philippine small pelagic fisheries. BFAR Tech. Pap. Sci. 10(3): 23 p.

Fortes, M.D. 1990. Seagrasses: A resource unknown in the ASEAN region. ICLARM Educ. Ser. 5, 46 p.

Gomez, E.D. 1991. Coral reef ecosystems and resources in the Philippines. Canopy Int. 16(5): 1-12.

Gomez, E.D., P.M. Aliño, H.T. Yap and W.Y. Licuanan. 1994. A review of the status of Philippine reefs. Mar. Pollut. Bull. 29(1-3): 62-68.

McManus, J.W. 1986. Depth zonation in a demersal fishery in the Samar Sea, Philippines, p. 483-486. In J. Maclean, L. Dizon and L. Hosillos (eds.) The First Asian Fisheries Forum. Asian Fisheries Society, Manila.

McManus, J.W. 1989. Zonation among demersal fishes of Southeast Asia: the southwest shelf of Indonesia. p. 1011-1022. In O. Magoon, H. Converse, D. Miner, L. Thomas Tobin and D. Cark (eds.) Coastal Zone '89. Proceedings of the Sixth Symposium on Coastal and Ocean Management. American Society of Civil Engineers, New York.

Munro, J.L. 1986. Marine fishery resources in the Philippines: catches and potential, p. 19-45. In D. Pauly, J. Saeger and G. Silvestre (eds.) Resources management and socio-economics of Philippine marine fisheries. Dep. Mar. Fish. Tech. Rep. 10: 217.

Pauly, D., G.T. Silvestre and I.R. Smith. 1989. On development, fisheries and dynamite: a brief review of tropical fisheries management. Nat. Resour. Modelling 3(3): 307-329.

Pauly, D. and P. Martosubroto, Editors. 1996. Baseline studies of biodiversity: the fish resources of Western Indonesia. ICLARM Stud. Rev. 23, 312.

Primavera, J.H. 1991. Intensive prawn farming in the Philippines: ecological, sound and economic implications. Ambio 20: 28-31.

Sebastian, A.R. 1951. Oceanographic research in the Philippines. Philipp. J. Fish. (1): 147-153.

Silvestre, G. and D. Pauly. 1986. Estimate of yield and economic rent from Philippine demersal stocks 1946-1984. Paper presented at the IOC/WESTPAC Symposium on Marine Science in the Western Pacific: the Indo-Pacific Convergence, 1-6 December 1986, Townsville, Australia.

Silvestre, G.T., R.B. Regalado and D. Pauly. 1986. Status of Philippine demersal stocks-inferences from underutilized catch rate data, p. 47-96. In D. Pauly, J. Saeger and G. Silvestre (eds.) Resources, management and socioeconomics of Philippine marine fisheries. Tech. Rep. Dep. Mar. Fish. Tech. Rep. 10: 217.

Silvestre, G.T. and D. Pauly. 1987. Estimate of yield and economic rent from Philippine demersal stocks (1946-1984) using vessel horsepower as an index of fishing effort. Univ. Philipp. Visayas. Fish. J. 3(1-2): 11-24.

Smith, I.R., D. Pauly and A.N. Mines. 1983. Small-scale fishing of San Miguel Bay, Philippines: Options for management and research. ICLARM Tech. Rep. 11. 80 p.

Soegiarto, A. 1985. Oceanographic assessment of East Asian seas. In Environment and resources of the Pacific, p. 173-184. UNEP Reg. Seas Rep. Stud. 69: 244.

Trinidad, A.C., R.S. Pomeroy, P.V. Corpuz and M. Agüero. 1993. Bioeconomics of the Philippine small pelagic fishery. ICLARM Tech. Rep. 38, 74 p.

Wyrtki, K. 1961. Physical oceanography of the southeast Asian waters: scientific results of marine investigations of the South China Sea and Gulf of Thailand 1951-1961. Naga Rep. Vol. 2. Neyenesch Printers, California. 195 p.

Yap, H.T. and E.D. Gomez. 1985. Coral reef degradation and pollution in the East Asian Seas, p. 185-207. UNEP Reg. Seas Rep. Stud. 69: 294.

The Coastal Fisheries of Sri Lanka: Resources, Exploitation and Management

R. MALDENIYA

National Aquatic Resources Agency
Crow Island, Mattakkuliya, Sri Lanka

MALDENIYA, R. 1997. The coastal fisheries of Sri Lanka: resources, exploitation and management, p.72-84. *In* G. Silvestre and D. Pauly (eds.) Status and management of tropical coastal fisheries in Asia. ICLARM Conf. Proc. 53, 208 p.

Abstract

Fisheries are an important source of animal protein, employment, income and foreign exchange earnings for Sri Lanka, whose Exclusive Economic Zone (EEZ) is over 20 times larger than its land area. This paper reviews the marine environment and the results of surveys of the marine resources of Sri Lanka with emphasis on trawl surveys, including actual catches, potential yield, catch composition and the technology in use. Coastal fisheries in Sri Lanka today are in a stage of stagnation due to overexploitation, environmental degradation and breakdown of traditional sea tenure arrangements. The government has formulated a five-year plan with community-based approaches to develop the sector in a sustainable manner.

Introduction

The fisheries of Sri Lanka are a primary source of animal protein, employment and foreign exchange for the country. Fisheries contributed 65% of the animal protein consumed in Sri Lanka in 1995, provided full time and part time employment to around 115 000 persons (Anon. 1992) and contributed 2.3% to the Gross Domestic Product (GDP) of the country in 1995 (Anon. 1995), while exports of fish and aquatic products were valued at Rs.2 893 million[1] in foreign exchange.

Marine fisheries account for 90-95% of the total landings, their backbone being the coastal fisheries which contribute 70-80% of the total marine landings. The offshore fisheries are still in a developing stage.

In this paper, the status of Sri Lanka's marine fisheries is reviewed in terms of their resources and stage of development, exploitation patterns and the management issues they raise. Emphasis is given to the coastal fisheries and bottom trawl surveys - both historic and recent - that were conducted around Sri Lanka to explore and monitor the coastal demersal resources.

Marine Environment

Sri Lanka (formerly Ceylon) is an island in the Indian Ocean situated close to the southeastern corner of the Indian subcontinent. India and Sri Lanka are separated by the shallow Palk Strait and the Gulf of Mannar. The island, which has a land area of 65 610 km², is surrounded by highly productive ecosystems such as mangrove forests, seagrass beds, salt marshes, and coral reefs. The island's Exclusive Economic Zone (EEZ), declared in 1978, at 536 000 km² is over 20 times larger than its land area and consists of a territorial sea extending up to 12 nm, a contiguous zone extending from 12 to 24 nm and an EEZ proper extending from 24 to 200 nm (Fig. 1). The continental shelf of about 26 000 km² is narrow, with an average width of around 22 km. To the north and northwest the shelf widens to an extensive shallow bank where it forms the Gulf of Mannar, Palk Bay and Pedro Bank. With the exception of a few restricted areas in the Gulf of Mannar, off Pedro Bank and southeast of Hambantota, the edge of the continental shelf falls off rapidly from 20-65 m to 1 500-3 500 m.

[1] In 1997: US$1 = Rs.58.40.

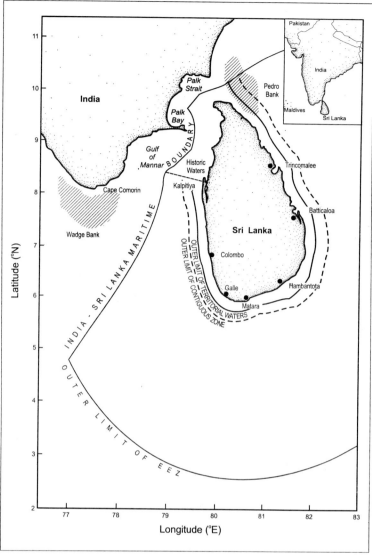

Fig. 1. Location of Sri Lanka and its EEZ.

The bottom sediment in the shelf area is generally rocky, particularly between Colombo and Batticaloa. However, the bottom of the northern shelf, particularly the Palk Strait, is predominantly muddy or muddy sand and hence trawlable (Fig. 2.).

Sri Lanka is subject to the influence of a monsoonal wind system. The southwest monsoon occurs from May to September causing heavy rains, mainly in the western part of the country. The northeast monsoon occurs from December to February, with the heaviest rainfall in the north and east and on the northeastern slope of the central hills. The sea surface currents are influenced by the monsoons and are often strong at the beginning and end of the southwest monsoon, and during the northeast monsoon, as shown by hydrographic observations made on board the *R.V. Dr. Fridtjof Nansen* in the three surveys made from 1978 to 1980. During April-June and August-September, the currents are governed by the southwest monsoon and the general oceanic circulation is from west to east with current velocities of up to 2-3 knots near

the shelf edge. During the northeast monsoon, the circulation is reversed and current velocities are only 1-2 knots. The depth of the thermocline also varies with the monsoon reaching 100-125 m on the west coast during the northeast monsoon and 40-60 m during the southwest monsoon. On the east coast, south of Pedro Bank, the depth of the thermocline is at 50-70 m from the end of the southeast monsoon to the start of the northeast monsoon, and 20-40 m at the start of the southwest monsoon. Upwellings are not observed. The tides around Sri Lanka are predominantly semidiurnal and microtidal, with the highest amplitudes in the Colombo area and the lowest around Jaffna and Trincomalee (Dassanayake 1994).

The better known fringing coral reefs in the country are located around Hikkaduwa, Galle, Batticaloa, Trincomalee, Mannar and around the Jaffna peninsula (Fig. 2); most show signs of degradation. The mangroves of Sri Lanka, never abundant due to the low tidal ranges, were estimated to cover about 12 000 ha in 1986 but have declined since. However, a man-

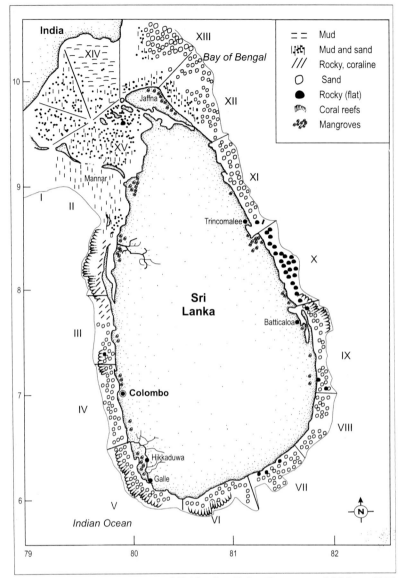

Fig. 2. Bottom structure around Sri Lanka. (After Pearson and Malpas 1926 and other sources).

grove park has been established in Negombo to promote research, education and public awareness of mangrove ecosystems, while several nongovernment organizations (NGOs) have launched mangrove replanting projects. Seagrass beds are composed of seed bearing, rooted marine plants that require clear shallow waters sheltered from high wave energy. They provide habitats and shelter for numerous species of fish and endangered animals, i.e., *dugong* and green sea turtle. Twelve species of seagrass have been recorded around Sri Lanka but the area covered by them has not been precisely estimated (Dassanayake 1994). Seaweed resources also occur around Sri Lanka but their types, extent and potential have not been adequately investigated. Some 260 species belonging to the Chlorophyceae (green), Rhodophyceae (red) and Phaeophyceae (brown algae) have been reported but only 20 of them are of com-

mercial importance in different parts of the world (Durairatnam 1961). Two species, *Gracilaria verrucosa* and *G. edulis,* are exploited in Sri Lanka (Jayasuriya 1989). *Sargassum* is another abundant species found in large quantities along the northern coast. A potential yield of 800 t·year^{-1} has been estimated for the southwestern and southern areas of the island (Durairatnam 1966).

Capture Fisheries Sector

The fishing industry provides full time employment to around 100 000 persons, part-time employment to around 10 000, while 5 000 persons work in related industries (Anon. 1992). Marine fisheries in Sri Lanka have traditionally been small-scale, producing largely for internal consumption. Foreign exchange is earned through export of tuna, shrimp, lobsters and nonconventional resources such as *bêche-de-mer* (sea cucumbers) and seaweed. Fishing effort is concentrated on a nearshore band of brackishwater and inshore waters. Traditional fishing gear made of natural fibers are deployed, with or without crafts/boats. Formerly the primary gear was the beach seine, the backbone of the small-scale industry, though hand lines, bottom-set gillnets, bottom longlines, set nets, set bags and stakes were also used. Fishing was also carried out with pole and line from outrigger canoes, the only craft used for fishing operations beyond immediate inshore areas (Table 1).

From the 1950s on, motorization gradually began to replace the traditional crafts operating in the outer range of the continental shelf (large outrigger canoes and *vallams*). This was followed by the replacement of log rafts (*teppam and kattumaram*), with 5.2-6.2 m fiberglass reinforced plastic (FRP), and outboard motor (OBM) boats, which expanded the coverage of the fishing grounds for small pelagics and demersals. However, there is still a high proportion of nonmechanized crafts in coastal fisheries. Newly designed boats of 3.5 GT, with lengths of 8-10 m and inboard engines have started fishing for large pelagics

without venturing far from shore and performing only day trips.

Prior to 1980, offshore fishing was dominated by bottom trawling on Wadge Bank, a fishery which closed in late 1978 as a result of the demarcation of the EEZ boundary between India and Sri Lanka. In the early 1980s 'Abu Dhabi boats' were built by the Northwest Coast Fisheries Development Project. These have an overall length of 34 ft, a 60 hp engine and are equipped with an insulated fish hold and crew accommodations allowing trips lasting several days. They were followed by even larger boats and, at present, 84% of the fishing boats have an overall length of between 34 and 36 ft (10 and 12 m) (Table 2).

From 1950 to 1975 the contribution of tuna to the total catch increased from 3.4 to 26% while that of rock fish or demersals declined from 24 to 11%, mainly due to shifts in effort. In 1990 clupeids accounted for 21% of total catch, while 19% was contributed by tuna. In the early 1990s 60% of total inshore landings originated from gillnets followed by bottom trawling (15%) and trolling (5%). In the offshore areas gillnets accounted for 80%, longlines 15% and troll lines 4% of landings (Joseph 1993). Fish production by fishing craft during 1979-1987 is given in Table 3.

Table 1. Types of fishing craft operating around Sri Lanka.

Craft type	Local name	Length (m)	Remarks	Operative offshore distance (km)
Indigenous crafts				
Log rafts	*Teppam*	3-5	Non-motorized/	1-2
	Kattuman	4-7	motorized OBM[a]	1-3
Outrigger canoes	*Oru, thony*	3-5	Sail/hand driven/mechanized OBM[a]	1-3
		10-12	Hand driven	1-15
Plan beach seine	*Paru, padu*	10-12	Hand driven	1-3
Dugout beach seine crafts	*Karavalai vallam*	10-12	Sail/hand driven/mechanized OBM[a]	1-3
Dugout with outrigger	*Vallam*	3-6	OBM[a], 8-25 hp	1-5
				1-10
Introduced crafts				
Fiberglass reinforced boat	*FRP[b] boat*	5.2-6.2	Inboard engine 39-54 hp	1-20
3.5-4.5 t boats	*3.5 boats*	8-10.3	Inboard engine 54-75 hp	> 20
11 t boats	*Abu Dhabi*	10.4-18	-	> 20

[a] Outboard motor.

[b] Fiberglass reinforced plastic.

Table 2. Age and size profile of the offshore fishing fleet of Sri Lanka. (Source: Joseph et al. 1995).

Year of construction	Length overall (feet)					Total	%
	28-32	34-36	38	40-45	> 50		
1981-83	-	2	-	-	-	02	1.6
1984-86	1	2	-	-	-	03	2.5
1987-89	2	3	-	-	-	05	4.1
1990-92	4	29	1	-	1	35	28.7
1993-95	2	67	4	4	-	77	63.1
Percentage	7.4	84.4	4.1	3.3	0.8	-	100.0

Table 3. The Sri Lankan coastal fleets and their catches, 1979-1987.

	Year								
	1979	1980	1981	1982	1983	1984	1985	1986	1987
1. Inboard motorized craft									
Catch (t)	50 105	54 825	56 454	60 379	57 375	46 625	47 862	49 249	50 960
No. of operating crafts	3 109	2 305	2 209	3 347	2 861	2 781	2 727	2 766	2 657
Catch/effort (t·craft^{-1}·year^{-1})	16.2	23.8	25.5	18.0	20.0	16.7	17.5	17.8	19.2
2. Outboard motorized craft									
Catch (t)	43 848	57 432	65 512	66 727	70 539	48 660	49 950	47 684	49 341
No. of operating crafts	9 723	8 020	8 865	9 745	10 086	10 800	11 515	10 340	10 543
Catch/effort (t·craft^{-1}·year^{-1})	4.2	7.2	7.4	6.8	7.0	4.5	4.3	4.2	4.7
3. Traditional artisanal craft									
Catch (t)	54 598	53 007	53 109	55 426	56 135	41 454	42 454	47 333	48 977
No. of operating crafts	15 330	15 721	12 855	14 101	14 312	14 404	13 303	13 412	13 865
Catch/effort (t·craft^{-1}·year^{-1})	3.6	3.4	4.1	3.9	3.9	2.9	3.2	3.5	3.5
Total output	148 851	165 264	175 075	182 532	181 049	136 642	140 266	144 266	149 210
Artisanal fisheries (%)	37	32	30	30	31	30	30	33	33

Fishery Resources and Potential

Small/Medium Pelagics

Canagaratnan and Medcof (1956) identified 53 species in the beach seine catches: 14 species belong to the family Clupeidae, 11 to the Carangidae and 9 to the Leiognathidae. Similarly, Joseph (1974) identified about 60 species in purse seine catches around Sri Lanka, while Dayaratne and Sivakumaran (1994), identified 55 species during bioeconomic surveys of the small pelagic fisheries along the west coast of Sri Lanka.

Gillnets (Karunasinghe 1990) and beach seines are the main gear used in the exploitation of small pelagics in Sri Lanka. Of the total number of crafts registered, around 65% of the traditional canoes and most of the FRP boats are engaged in small pelagic fisheries (Dayaratne 1994), either throughout the year or seasonally. A total small pelagic production of about 50 000 t was reported in 1994, down from about 76 000 t in 1992. Small pelagic species commonly harvested from inshore waters include sardines (Clupeidae), anchovies (Engraulidae), and scads (Carangidae) (Pajot 1977). The small and medium sized pelagics that are found outside the immediate inshore areas include Indian mackerel (*Rastrelliger kanagurta*), flying fish (Exocoetidae), halfbeaks (Hemiramphidae), gar fish (Belonidae), and ribbon fish (Trichiuridae).

Other species such as the Spanish mackerel (*Scomberomorus commerson*), frigate tuna (*Auxis thazard*), 'kawakawa' (*Euthynnus affinis*) and bullet tuna (*Auxis rochei*), usually categorized as 'large pelagics', are found far offshore and are not further discussed here (but see Maldeniya and Joseph 1987;

Maldeniya and Suraweera 1991). Experiments conducted by FAO/BOBP with inshore Fish Aggregation Devices (FADs) in the mid 1980s resulted in large catches of dolphin fish (*Coryphaena hippurus*) and rainbow runner (*Elagatis bipinnulata*) which usually do not contribute to inshore catches. This may suggest a potential for development (Hallier 1993).

Based on data from the *R.V. Dr. Fridtjof Nansen*, which carried out acoustic surveys of fish resources around the coastal waters of Sri Lanka from 1978 to 1980, and the work of Saetersdal and de Bruin (1978), the estimated average density of neritic fish is about 100 t per nm^2, which leads to a potential yield estimate of 250 000 t·year^{-1}. Of this, 170 000 t·year^{-1} may consist of small pelagics (Table 4 provides other estimates of resource potential).

Demersal fish and invertebrates

Except for a few areas, small-scale demersal fishing in Sri Lanka is carried out only on a seasonal basis, mainly from September to April. In 1991 demersal fish landings were over 27 000 t, encompassing a wide variety of species.

Based on a survey conducted from 1920 to 1923 by the *R.V. Lilla*, Pearson and Malpas (1926) listed 215 species in 55 families. Subsequent studies of demersal fish did not make detailed species lists and, in many cases, sampling was conducted with selective gear such as bottom longlines, handlines, bottom-set gillnets and traps.

Hinriksson (1980) noted that the distribution of demersal fish varies in accordance with the depth and bottom type (see also Berg 1971). Small, short-lived species such as pony fish (Leiognathidae) predominate over smooth bottom areas in the north and northwest,

Table 4. Fisheries resource potential in Sri Lanka. (Source: Sivasubramaniam 1995).

Resource	Potential yield (t·year^{-1})	Approach (remarks)	Author
Demersals	60 000	Exploratory trawl fishery	Tiews (1966)
Demersals	52 000	Primary production	Jones and Bannerji (1973)
Demersals	80 000	Acoustic survey	Blindheim and Foyn (1980)
Demersals	74 000	Acoustic and swept area	Sivasubramaniam (1983)
Pelagics	90 000	Primary production	Jones and Bannerji (1973)
Pelagics	170 000	Acoustics (inshore and offshore)	Blindheim and Foyn (1980)
Large pelagics	27 000	Production trend and survey catches (only for offshore area)	Sivasubramaniam (1977)

while larger, long-lived species such as snappers (Lutjanidae), emperors (Lethrinidae), groupers (Serranidae) and elasmobranchs (sharks and rays) dominate where the bottom is rough and uneven.

Based on the above-mentioned acoustic surveys, Saetersdal and de Bruin (1978) estimated a potential yield of demersal fish from the continental shelf area of 75 000-80 000 t·year^{-1}, 30 000 t·year^{-1} in the Palk Bay/Palk Strait area and 45 000 t·year^{-1} from the rest of the shelf.

Additionally, deep-sea fish such as *Chlorophthalmus bicornis*, *C. agassizi*, *Cubiceps* sp., and lanternfish (family Myctophidae) may represent a resource at 200-300 m depth along the shelf edge (Demidenko 1972; Saetersdal and de Bruin 1978; Dalpadado and Gjøsaeter 1993).

A total of 32 species of penaeid shrimps have been recorded from Sri Lanka (de Bruin 1970). Of these, four species are of commercial importance, i.e., *Penaeus indicus*, *P. monodon*, *P. semisulcatus* and *P. merguiensis*. *P. indicus* constitute about 50-70% of the total annual shrimp catch. In 1989 total shrimp production in Sri Lanka was 5 300 t, the bulk being from capture fisheries. In 1994 the 2 300 t of penaeid shrimp exported represented over 50% of the value of fishery exports from Sri Lanka.

Six species of spiny lobster occur in Sri Lankan coastal waters down to 50 m, but most are confined to 30 m. Preferred depth ranges vary among species: 1-5 m for *Panulirus penicillatus;* 1-20 m for *P. homarus*; 11-20 m for *P. versicolor* and *P. polyphaus*; and 11-30 m for *P. longipes* and *P. ornatus*.

Lobster resources are heavily exploited to meet the demand for export and tourist hotels. In 1989, export of lobsters was 663 t, 80% of which consisted of *P. homarus*, valued at Rs.278 million.

Bêche-de-mer or sea cucumbers (Holothurians) are harvested from larger, high salinity lagoons in the northwest, north and northeast and represent an important export product. Out of 70 species recorded from Sri Lanka, 13 species are consumed in many

parts of the world. The dominant species, *Holothuria scabra*, is selectively harvested (Joseph and Moiyadeen 1990).

There has been little effort directed towards the exploitation of cephalopods on a commercial scale but incidental catches are made by purse seines, scoop nets, shrimp trawlers, and hand lines. Perera (1975) identified only two species of cuttlefish and six species of squid in commercial catches. However, cephalopods represent a relatively high fraction of the prey of yellowfin tuna when these feed in coastal waters. Four species of cuttle fish and eight species of squid were recovered from their stomach contents (Maldeniya 1992).

Trawl fisheries and surveys

The first attempt to trawl in Sri Lanka occurred in 1902 when a private party deployed a steam trawler on the west coast off Colombo. Later an English steam trawler, the *Violet*, continued the operations off Colombo and extended them to the east coast (Malpas 1926). No details are available of these ventures but it appears that they encountered fierce resistance from artisanal fishers who attacked the vessels and damaged the gear. The survey carried out by the Ceylon Government trawler the *R.V. Lilla* in 1920-1923 indicated that the grounds off the southwest coast were small and rough and that catches were too small to justify the deployment of large steam trawlers. Therefore, large trawlers have not operated off the southwest coast except for the *Halpha* in 1952. However, trawling was done in this area by small boats such as paddled log rafts working in pairs and dragging a trawl known in Sinhalese as *katumaran dela* (Pearson and Malpas 1926). Sri Lanka had a well established commercial trawler fishery on the Wadge Bank, which the survey of the *R.V. Lilla* had identified as supporting high fish concentrations along with Pedro Bank and the Palk Strait (Malpas 1926). This fishery lasted from 1920 to 1975 with sporadic trawling on Pedro Bank during the same period. Following the demarcation of these waters and of the India-Sri Lanka sea boundaries in 1976, the whole of the Wadge Bank and one-third of Pedro Bank became part of the Indian EEZ and no further commercial trawling was undertaken by Sri Lanka-based trawlers. This paper, therefore, does not discuss the rich data set generated by these historic fisheries now lost to Sri Lanka.

Trawling was initially done with large mesh nets which did not retain small fish and invertebrates. The first attempt at trawling with a small mesh net was made by the *Halpha* in 1950 off Colombo and in the north, northwest and northeast of Sri Lanka. From 1953 to 1957 a series of trials with 45 ft trawlers, the *North Star* and the *Canadian*, were conducted on Pedro Bank and other areas around Sri Lanka. A summary of operations by the former vessel is given in Table 5. Trawling is now well established in some parts of Sri Lanka. The major surveys carried out in Sri Lankan waters are summarized in Table 6 (see also Appendix III). The majority of these surveys covered coastal waters down to about 60 m, while the surveys of the *R.V. Optimis* and the *R.V. Dr. Fridtjof Nansen* extended up to 300 m and 340 m, respectively, in some areas. Surveys in different areas are summarized in Tables 7 and 8. However, as the trawler size, engines and gear used were different, their catch rates cannot be compared directly to evaluate the stock status at different times.

During surveys of the *R.V. Lilla* (1920-1923) catch rates ranged from 1 to 120 kg·hour^{-1} (Fig. 3), with the highest densities occurring in the northern strata (Table 9). The catch rates of the the *Canadian* (1955-1956) ranged from 2 to 28 kg·hour^{-1} and were systematically lower than those of the *R.V. Lilla* even in the same strata. The gear was probably inappropriate. The catch rates of the *R.V. Dr. Fridtjof Nansen* (1978-1980) were higher than those reported by any of the previous surveys. Sivasubramaniam and Maldeniya (1985) analyzed the bulk of the available data on trawl surveys in Sri Lanka and noted that snappers (Family Lutjanidae), especially *Lutjanus malabaricus*, *L. rivulatus* and *L. lutjanus*, had dominated the catch of the *R.V. Lilla*, while emperors (Lethrinidae), especially *Lethrinus nebulosus* and *L. miniatus*, were very abundant in the catches of the *R.V. Dr. Fridtjof Nansen*. Hinriksson (1980) noted that shallow water demersal catches in the northern part of the island were dominated by pony fish (Leiognathidae), sharks and skates. In deeper waters, the surveys of the *R.V. Optimist* (1972) and the *R.V. Dr. Fridjoft Nansen* on the continental slope off the northeast coast revealed the occurrence of deep sea fish concentrations (Sivasubramaniam and Maldeniya 1985).

Fisheries Management in Sri Lanka

The administrative and policy making organization for fisheries in Sri Lanka is the Ministry of Fisheries and Aquatic Development (Fig. 4). The Department of Fisheries and Aquatic Resource (DFAR) is the line agency responsible for the functions of management, development and enforcement of the provisions of the fisheries and other related ordinances and acts. The National Aquatic Resource Agency (NARA) is the research arm of the Ministry

Table 5. Summary record of otter trawler *North Star* (Source: Medcof 1963).

Area	Period	Depth (m)	No. of hauls	Haul duration (hours)	Catch/effort (kg·hour^{-1})
Colombo West	1953	10-33	10	11.75	62.5
	1954	-	1	5	5.0
		-	1	5	2.3
	1955	-	6	35.3	7.2
		-	4	27.5	4.7
N.W. Talaimanaar	1953	6-7	3	2.5	121.8
N.E. Mullaitivu	1953	2-6	8	17	65.0
		9-22	16	38.2	47.7
		11-42	11	27.3	32.6
Trincomalee	1953	3-10	7	6	0.8
	1954	12-17	4	4.7	6.6
	1954	12-35	8	18.8	1.7
Point Pedro North	1953	7-22	5	4	64.8
Alampil North	1954	7-20	4	5	4.5
		11-25	6	11.6	14.3
Kayts North	1954	4-8	11	15.8	10.3
		3-9	10	15	8.9

Table 6. Summary of information on demersal trawl surveys conducted in Sri Lanka.

Years (months)	Vessel	Trawl gear	Depth range (m)	Towing speed (knots)
1920-1923	R.V. Lilla 38 m LOA 500 hp	Bottom trawl head rope 21 m Foot rope 30 m 13 cm tapering to 2.5 cm codend	mainly 4-35; few location 73.5-95	4
1955 (Jun) - 1956 (Oct)	R.V. Canadian 14 m LOA 80 hp	Bottom trawl Yankee 351 9.5 cm codend	-	1-2
1963 (Nov) - 1967 (Apr)	R.V. Canadian 14 m LOA 80 hp	Shrimp trawl 3.5 cm codend	2-80	-
1967 (May/Jun)	R.V. Myliddy 30 m LOA 240 hp	Granton trawl 8.0 cm codend, shrimp trawl 4.0 cm codend	1.5-3.0 for prawn 3-4 for fish	-
1971 (Mar) - 1992 (Dec)	R.V. Optimist	-	30-340	-
1975 (Nov) - 1978 (Jun)	R.V. Hurulla 11m LOA 96 hp	3.2-3.6 cm codend	3-28	6
1978 (Aug-Sep) 1979 (Apr-Jun) 1980 (Jan-Feb)	R.V. Dr. Fridtjof Nansen 45 m LOA	2.0, 3.0, 4.0 cm codend	-	-

and undertakes and coordinates research activities relating to fisheries and aquatic resources. The training of fisheries personnel is undertaken by the National Institute of Fisheries Training (NIFT). The Ceylon Fisheries Corporation (CFC) buys and sells fish in competition with private traders and in this process tries to regulate fish prices to help the producer and the consumer. The Ceylon Fishery Harbours Corporation (CFHC) is responsible for the maintenance of fishery harbors and related onshore facilities. The Cey-Nor-Foundation (CNF) is a state-owned company engaged in the building of boats and in the manufacture of fishing gear and nets. The primary responsibilities of the Coast Conservation Department (CCD) are policy formulation, planning and research, issuance of permits regulating coastal development activities and the construction and maintenance of shoreline protection works within designated coastal zones.

In accordance with national development policies, the Ministry of Fisheries and Aquatic Resources Development formulated a five-year plan for fisheries development (1996-2000), with the following objectives:
1. promoting economic growth through the optimal production of fish to improve the nutritional status of the population and to increase foreign exchange earnings;
2. reducing poverty by increasing gainful employment and income opportunities; and

3. enhancing resources and environment protection through improved resource management and conservation measures.

Usually a fishery can be expected to pass through three phases: developing, stable and declining. During the early part of the developing phase in Sri Lanka growth was slow due to the limitations of the traditional fishing systems, notably their being confined to inshore waters. With the introduction of modern, motorized craft and improved fishing methods the development of the fishing industry accelerated in the 1950s. It then began to stagnate, leading to the stable phase presently prevailing throughout most of the country's coastal fisheries. The stable phase exhibits declining catch rate with a leveling off in fishery production. This stage is characterized by endemic poverty among small-scale fishing communities, which pushes excess labor into coastal fisheries that are largely open access. This has led to conflicts among competing groups of fishers, which are worsened by subsidies, well-meaning incentives schemes and lack of enforcement of existing regulations.

Ironically fisheries management itself is not a new concept in Sri Lanka. Traditional sea tenure arrangements existed to regulate the deployment of fishing effort. Since the later part of the previous century these arrangements have been gradually replaced by

80

Table 7. Comparison of catch rates (kg·hour⁻¹) obtained for different areas during different surveys. (Source: Sivasubramaniam 1985).

Area	Code[b]	R/V Lilla 1920-1923	R/V Canadian Jun-Oct 1955-1956	R/V Canadian Nov-Apr 1963-1967	R/V Myliddy May/Jun 1967	R/V Hurulla Nov 1975-Jan 1978	R/V Dr. Fridtjof Nansen[a] Aug 1978-Feb 1980
Gulf of Mannar-Pamban	-	-	-	-	-	-	420
Kalpitiya	II	-	-	386	-	-	-
Puttalam-Chilaw	III	12.5	6.2	172	-	-	266
Chilaw-Negombo		40.2	5.0	73	-	-	-
Negombo-Colombo		11.6	-	-	-	-	-
Colombo-Kalutara		33.9	-	-	-	00	345
Kalutara-Bentota	IV	4.9	-	-	-	-	-
Bentota-Galle		20.5	5.3	-	-	-	167
Galle - Matara	V	12.5	25.6	-	-	-	-
Matara - Tangalla		9.4	-	-	-	-	-
Tangalla - Hambantota	VI	12.9	-	-	-	-	146
Hambantota - Yala		20.1	-	-	-	-	352
Yala - Amaduwa	VII	4.4	-	-	-	-	-
Amaduwa - Mumana		31.2	-	-	-	-	246
Mumana - Komari	VIII	3.9	-	-	-	-	-
Komari - Kallar		0.9	-	-	-	-	-
Kallar - Batticaloa	IX	13.4	-	60	60	-	93
Batticaloa - Valaichchenai		15.6	4.7	-	-	-	-
Valaichchenai- Panichanknai		29.9	6.6	-	-	-	-
Panichanknai - Trincomalee	X	4.4	7.5	-	-	-	215
Trincomalee - Pudawakattuwa		4.9	2.2	-	140	123	-
Pudawakattuwa - Pulmodai		27.6	27.9	-	-	-	-
Pulmodai - Mullativu	XI	22.3	13.3	115.6	-	-	227
Mullativu - Thalayaddi		19.6	11.5	181	-	-	240
Thalayaddi - Point Pedro	XII	6.2/62	-	49	250	71	-
Point Pedro-Jaffna	XIII	37.5	5.8	130	400	-	-
Island off Jaffna	XIV	16.1	-	-	-	-	-
		7.1	-	138	-	64	-
		12.9	-	-	-	-	-
Jaffna-Mannar	XV	35.7	6.8	-	200	-	-
	-	03.5	2.3	472.8	-	191	-
	-	106.2	-	-	-	-	-
Pedro Banks	-	62	-	-	365	-	240

[a] The catch/effort units of R.V. Fridtjof Nansen is kg/100 hooks and, hence, these catch rates cannot be directly compared with those obtained by trawling.
[b] Refer to Fig. 3.

government intervention. Ordinances enacted from 1889 to the 1930s provide legislation for the management of area-specific fisheries. A Department of Fisheries was created in 1929 to replace the Director of the Colombo Museum who had previously acted as the Government's Marine Biologist. The Fisheries Ordinance No. 24 enacted in 1940 gave a cabinet minister the power to regulate most of the fisheries activities in Sri Lanka. These changes with the introduction of new technology led to the decline of traditional management schemes and to increasing heterogeneity within the fishing community. Conflicts among user groups increased as did uncontrolled fishing and the use of destructive fishing methods. Environ-

Table 8. Average catch rates reported by trawlers in different areas of Sri Lanka. (Source: Medcof 1963).

Region and fishing vessel	Catch/effort (kg·hour⁻¹)
N.E. Pedro Bank Region	
Lilla (1920-1923)	84
Bulbul and Tongkol	130
Raglan Castle	185
Halpha	0
North Star	36
Canadian	12
N.W. Palk Strait Region	
Lilla (1920-1923)	130
Halpha	183
Northstar (1953)	101
Canadian (1954)	12
Regular hauls	26
Special hauls	2
Towing alone	5
Tandem towing with North Star	5-11
S.W. Galle to Chilaw	
Lilla (1920-1923)	21
Halpha	46
North Star (1953)	107
Katumaram dela	9

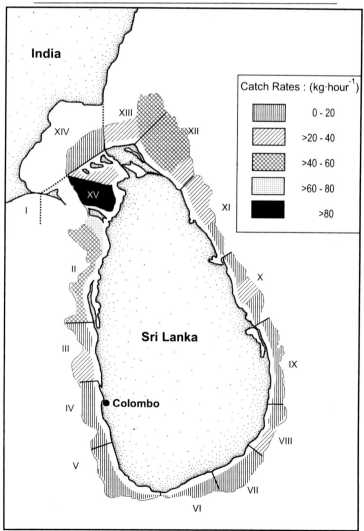

Fig. 3. Mean catch rates of demersal fish by area (down to 200 m) by 50 to 60 ft trawlers of approximately 100 hp. (After Sivasubramaniam and Maldeniya 1985).

mental problems such as pollution are affecting the sustainability of coastal fishery resources (see Table 10).

As the provision of the Fisheries Ordinance of 1940 appeared to be inadequate to meet these challenges, a new Fisheries and Aquatic Resources Act was enacted in 1995. This emphasizes the management of fisheries and their sustainable development, with due recognition of the need for conservation measures.

The DFAR has the tasks of implementing this Act as is reflected in its structure (Fig. 5).

Because of the problems facing coastal fisheries and the failure of top-down approaches, attention has been given in recent years to community-based approaches - in effect the re-establishment of some aspects of traditional management schemes. The National government and the communities are to share the

Table 9. Catch rates in different depth ranges from the *R.V. Lilla* cruises (kg·hour⁻¹). (Source: Sivasubramaniam 1985).

Depth range (m)	Stratum							
	IVa	Va	VIIa	XIa	XIb	XIc	XIIc	XIII
20 - 29	8.9	-	0.9	3.6	4.1	9.2	61.0	29.5
30 - 39	11.5	-	6.0	-	-	-	41.8	46.4
40 - 49	-	-	-	7.0	15.6	24.5	54.8	-
50 - 59	-	-	-	-	-	-	52.3	34.1
60 - 69	-	33.4	29.6	-	-	-	-	120.0
70 - 80	-	13.5	27.4	-	-	-	-	28.6

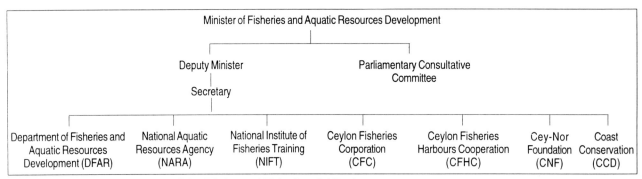

Fig. 4. Organizational structure of the Ministry of Fisheries, Sri Lanka.

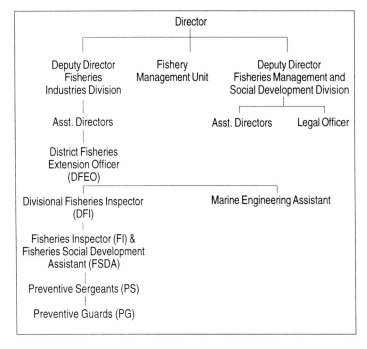

Fig. 5. Organizational structure of the Department of Fisheries and Aquatic Resources.

Table 10. A generic coastal transect in Sri Lanka illustrating impacts on the coastal zone.

Major zones	Terrestrial			Coastal		Marine
	Upland (>18% slope)	Midland (8-18% slope)	Lowland (0-<8% slope)	Tidal zone (1 km inland from HHWL-30 m depth)	Coastal (30 m-200 m depth)	Offshore (>200 m depth-EEZ)
Main resource uses/activities	Deforestation Irrigation Agriculture Animal husbandry	Agriculture Irrigation Deforestation Urbanization Animal husbandry	Tourism Industry Urbanization Land reclamation Mining Aquaculture Saltpans	Aquaculture Land reclamation Tourism Artisanal fishery	Artisanal fishery Coral mining Marine transport	Commercial fishing Marine transport
Main environmental issues/impacts on the coastal zone	Soil erosion Flooding Sedimentation Agrochemical pollution	Soil erosion Flooding Sedimentation Water pollution	Waste and industrial pollution Agrochemical pollution Sedimentation Soil erosion Flooding	Reduce biodiversity Habitat destruction and degradation Overfishing	Overfishing Reduced biodiversity Oil spills	Oil spills

authority for fisheries management through schemes that involve fisheries cooperative societies established at the village level (there were over 750 such societies with a membership of about 90 000 in 1995). Community-based approaches through these societies and similar self-help groups may be more successful than the earlier centralized approach as the resource users themselves are involved in the decision-making. Only through schemes involving some form of limited entry will it be possible to prevent Sri Lankan coastal fisheries from slipping into a declining phase.

References

Anon. 1992 Fisheries for Sri Lanka. Statistical Division, Department of Fisheries and Aquatic Resources.

Anon. 1995. Fisheries statistics, 1995. Data Management Unit. Department of Fisheries and Aquatic Resources.

Berg, S.E. 1971. Investigation on the bottom conditions and possibility for prawn and fish trawling in the north and east of Ceylon. Bull. Fish. Res. Stn. Ceylon, 22 (1&2): 53-88 p.

Blindheim, J. and L. Foyn. 1980. A survey of the coastal fish resources of Sri Lanka. Report No. 3, January-February 1980. Reports on the survey of the *R.V. Dr. Fridtjof Nansen*. Fisheries Research Station, Colombo and Institute of Marine Research, Bergen. 78 p.

Canagaratnan, P. and J.C. Medcof. 1956. Ceylon's beach seine fishery. Fisheries Research Station, Department of Fisheries, Ceylon, 4(4): 1-32 p.

Dalpadado, P. and J. Gjøsaeter. 1993. Lantern fishes (Myctophidae) in Sri Lankan waters. Asian Fish. Sci. 6(2): 161-168.

Dassanayake, H. 1994. Sri Lanka, p. 209-235. *In* S. Holmgren (ed). An environmental assessment of the Bay of Bengal region. Swedish Centre for Coastal Development and Management of Aquatic Resources and the Bay of Bengal Programme (BOBP/REP/67).

Dayaratne, P. 1994. Information needed for strategic planning of fishery management. Sri Lanka/FAO National Workshop on Development of Community-based Fishery Management Systems. Colombo, Sri Lanka. SRL/94/2. 41-51 p.

Dayaratne, P. and K.P. Sivakumaran. 1994. Biosocioeconomics of fishing for small pelagics along the southwestern coast of Sri Lanka. Bay of Bengal Programme. BOBP/WP/96. 38 p.

de Bruin, G.H.P. 1970. Small mesh trawling. Bull. Fish. Res. Stn., Ceylon. 21(1): 35-38.

Demidenko, U. 1972. Information about the results of the joint Soviet-Lankan fishery investigation carried out in waters adjacent to Ceylon island. Manuscript report. Fisheries Research Station, Colombo.

Durairatnam, M. 1961. Contribution to the study of the marine algae of Ceylon. Bull. Fish. Res. Stn. (Ceylon) No. 10: 1181.

Durairatnam, M. 1966. Quantitative survey of *Sargassum* along the southwest coast of Ceylon, Bull. Fish. Res. Stn. (Ceylon) 19(1/2) No. 182.

Hallier, J.P. 1994. Purse seine fishery on floating objects: what kind of fishing effort? what kind of abundance indices? p. 192-198. *In* Proceedings of the Fifth Expert Consultation on Indian Ocean Tunas, Mahé, Seychelles, 4-8 October 1993. FAO-UNDP-Indo-Pacific Tuna Development and Management Programme, Colombo, Sri Lanka. FAO-IPTP-TWS/92/2/25.

Hinriksson, T.G. 1980. Survey for demersal fish resources off the northwest, north and northeast coasts of Sri Lanka. Food and Agriculture Organization of the United Nations, Rome. DP/SR/72/051.

Jayasuriya, P.M.A. 1989. Preliminary observation of the culture of *Gracilaria edulis* using spore-setting techniques. Seminar on *Gracilaria* production and utilization in the Bay of Bengal Region, Songkhla, Thailand. Bay of Bengal Programme of the Food and Agriculture Organization (FAO). BOBP/REP/45.

Joseph, B.D.L. 1974. Preliminary report on experimental fishing with purse seine and lampara nets for small pelagic fish varieties around Sri Lanka. Bull. Fish. Res. Stn., Sri Lanka. 25(1/2): 1-3.

84

Joseph, L. and N.M. Moiyadeen. 1990. Ecology of commercially exploitable holothuroids on the northwest coast (Puttalam District). National Aquatic Resources Agency. Terminal Report - Natural Resources Energy and Science Authority, 166 p. (Unpublished).

Joseph, L. 1993. Coastal fisheries and brackishwater aquaculture in Sri Lanka. Coastal Resource Management Project of the University of Rhode Island, The Government of Sri Lanka, United States Agency for International Development.

Joseph, L., P. Dayaratne and R. Maldeniya. 1995. A profile of the offshore fishing fleet of Sri Lanka. Sri Lanka Fisheries Sector Development Project, Asian Development Bank. ADB/NRS/3 (Unpublished).

Karunasinghe, W.P.N. 1990. Some aspects of biology and fishery of trenched sardine *Amblygaster sirm* (Pisces: Clupeidae) from the coastal waters around Negombo, Sri Lanka. M. Phil. Thesis, University of Kelaniya, Sri Lanka, 102 p.

Maldeniya, R. and L. Joseph. 1987. On the distribution and biology of yellowfin tuna (*T. albacares*) from the western and southern coastal waters of Sri Lanka, p. 23-32. In Collective volume of working documents presented at the expert consultation on stock assessment of tunas in the Indian Ocean held in Colombo, Sri Lanka, 4-8 December 1986. FAO-UNDP-IPTP-TWS/86/18.

Maldeniya, R. and S.L. Suraweera. 1991. Exploratory fishing for large pelagic species in Sri Lanka. Bay of Bengal Programme. BOBP/REP/47. 53 p.

Maldeniya, R. 1992. Food and feeding habits of yellowfin tuna (*Thunnus albacares*) in Sri Lankan waters. M. Phil. Thesis. University of Bergen, Norway, 55 p.

Malpas, A.H. 1926. Marine biological survey of the littoral waters of Ceylon. Bull. Ceyl. Fish. 2: 13-165 p.

Medcof, J.C. 1963. Partial survey and critique of Ceylon's marine fisheries, 1953-55. Bull. Fish. Res. Stn. Ceylon 16(2): 29-118.

Pajot, G. 1977. Exploratory fishing for live bait and commercially important small pelagic species. UNDP-Sri Lanka Fishery Development Project. Technical Report No. 3. SRL/72/05 (Unpublished).

Pearson, J. and A.H. Malpas. 1926. A preliminary report on the possibilities of commercial trawling in the seas around Ceylon, Bull. Ceyl. Fish. 2: 1-12.

Perera, N.M.P.J. 1975. Taxonomic study of the cephalopods, particularly the Teuthoidea (squids) and Sepioidae (cuttle fish) in the waters around Sri Lanka. Bull. Fish. Res. Stn., Sri Lanka 26(1): 45-60.

Saetersdal, G.S. and G.H.P. de Bruin. 1978. A survey of the coastal fish resources of Sri Lanka, August-September 1978, Institute of Marine Research, Bergen, 88 p.

Sivasubramaniam, K. 1977. Experimental fishery survey for skipjack and other tunas by pole and line and driftnet fishery method in Sri Lanka. UNDP/Sri Lanka Fishery Development Project. SRL/72/051 (Unpublished).

Sivasubramaniam, K. 1983. A fresh look at Sri Lanka's demersal resources. Bay of Bengal News, 10: 14-17.

Sivasubramaniam, K. 1985. Sri Lanka's fisheries resources development and management in the past. Presented at the workshop on 25th anniversary of the Ministry of Fisheries and Aquatic Resources Development. 11 p. (Unpublished).

Sivasubramaniam, K. and R. Maldeniya. 1985. The demersal fisheries of Sri Lanka. Bay of Bengal Programme Work. Pap. 41, 41 p.

The Marine Fisheries of Thailand, with Emphasis on the Gulf of Thailand Trawl Fishery

M. EIAMSA-ARD
and
S. AMORNCHAIROJKUL
Southern Marine Fisheries Development Center
Department of Fisheries
Wichianchom Road
Muang District, Songkhla, Thailand

EIAMSA-ARD, M. and S. AMORNCHAIROJKUL. 1997. The marine fisheries of Thailand, with emphasis on the Gulf of Thailand trawl fishery, p. 85-95. *In* G.T. Silvestre and D. Pauly (eds.) Status and management of tropical coastal fisheries in Asia. ICLARM Conf. Proc. 53, 208 p.

Abstract

The Gulf of Thailand and Andaman Sea are the two major fishing areas of Thailand, jointly covering 394 000 km². The fisheries are diverse in size and type of operation, gear and techniques. In 1991, total fishery production reached about 2.5 million t, valued at 26 billion baht (US$1 040 million). Of this, the marine fisheries catches contributed about 82%. Catches in the Gulf consist of pelagic fishes (40%), demersal fishes (8%), shrimps (5%), squids (6%) and other groups including 'trash fish' (41%).

The demersal fisheries have developed rapidly since trawling was introduced in the late 1960s. Registered trawlers increased from 3 206 units in 1970 to 15 037 in 1989. Demersal catches grew from 51 000 t in 1978 to 149 000 t in 1992. In 1992, demersal catches of 149 000 t and 58 000 t were recorded from the Gulf of Thailand and the Andaman Sea coast, respectively. However, demersal trawl surveys conducted since 1961 found that the average catch/effort decreased from 298 kg·hour[-1] in 1961 to 39 kg·hour[-1] in 1981; the decline continuing in later years.

This paper reviews the status of Thailand's coastal environment and fishery resources. It emphasizes demersal fisheries and trawl surveys in the Gulf of Thailand.

Introduction

Marine fisheries are important in Thailand. However, the overexploitation of coastal resources, such as coastal demersal and small pelagic fishes, mangrove forests and coral reefs, and the deterioration of coastal water quality, are critical problems.

The Thai fisheries are diverse in size and type of operations. In 1991, total fishery production reached about 2.5 million t, worth 26 billion baht (US$1 040 million). Marine fisheries catches accounted for about 82% of the total.

This contribution reviews the status of the country's coastal environment and fishery resources, emphasizing demersal trawl surveys in the Gulf of Thailand (Fig. 1).

Coastal Environment

The Gulf of Thailand covers approximately 350 000 km². Four large rivers discharge their waters into the inner gulf. Mean water depth is 45 m; maximum is 80 m. This basin is enclosed by underwater ridges that cause an inadequate interchange of water between the gulf and the South China Sea, thus justifying its treatment as a large marine ecosystem (Piyakarnchana 1989) or as an ecologically distinct subset of the South China Sea (Pauly and Christensen 1993). The bottom of the inner gulf consists of loose mud, that of the central basin of soft mud, and that of the outer basin of soft or sandy mud (Shepard et al. 1949; Naval Hydrographic Department 1995).

Water temperature ranges from 28.4 to 30°C; surface salinity in the outer part of the gulf, from 31.4 to 32.7‰; and oxygen concentration at the surface, from 4.5 to 4.6 ml·l[-1] (Naval Hydrographic Department 1995). The Gulf of Thailand functions as a two-layered shallow estuary with lower-salinity surface water flowing out of the Gulf, while high-salinity, colder water enters from the South China Sea over a 67 m sill. Monsoon winds and tidal currents create complex circulation patterns, including localized upwellings and

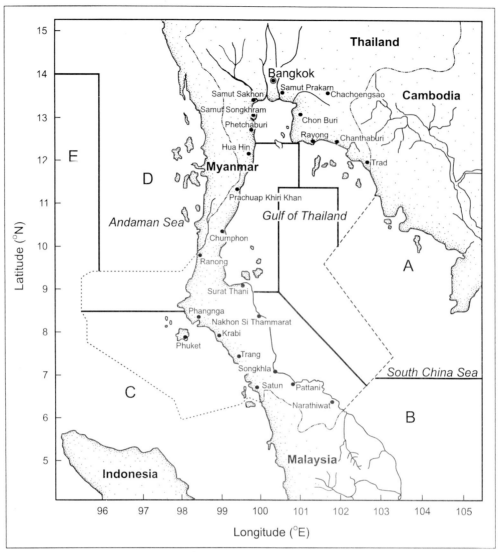

Fig 1. Fishing areas of Thailand, and locations mentioned in the text.

downwellings. Mean residual flow in the inner gulf is clockwise during the northeast monsoon and counterclockwise during the southwest monsoon.

In the Gulf of Thailand, as elsewhere, primary production is high in coastal areas near river mouths, and decreases with depth. Average primary production is 2.49 gC·m⁻²·d⁻¹ in the inner gulf and 2.96 gC·m⁻²·d⁻¹ off the western coast.

The concentration of phosphate in the inner gulf ranged from 1.02 to 1.59 mg-at N l⁻¹ from 1984 to 1989. Nitrate concentration ranged from 9.15 mg-at N l⁻¹ in 1984 to 24.86 mg-at N l⁻¹ in 1989 (Suvapepun 1991); Secchi disk transparency varies from 14 to 17 m.

In general, red tide occurrences reported in the inner gulf since 1958 have tended to be harmless. However, their frequency has increased of late: there were 7, 12 and 19 occurrences in 1991, 1992 and 1993, respectively. These events were mostly caused by

Noctiluca, generally from July to September in the eastern part of the inner gulf, and from December to February in the western part (Lirdwitayaprasit et al. 1995).

Along the gulf coast, most cases of marine food poisoning are caused by *Vibrio parahaemolyticus,* while cases of diarrhea are caused by *V. cholearae, Salmonella* and *Shigella. Vibrio parahaemolyticus* tends to occur in higher concentration in sediment and mollusc samples than in the surrounding waters. This contrasts with coliform bacteria, which tend to occur more frequently in surface waters than near the bottom.

Water circulation along the Andaman coast of Thailand is dominated by tidal currents, and alongshore flows. During the northeast monsoon, the nearshore surface waters generally move northward, with speeds of 2-4 cm·s⁻¹; during the southwest monsoon, these waters flow southward at a speed of

5-8 cm·s⁻¹. The hydrography of the region has been extensively studied. The northern stretch from Ranong to Phuket province is influenced by deep sea up-welling resulting in high salinity (32.9-33.4 ‰), while the southern stretch (Phuket to Satun Province) is influenced mainly by surface run-off, resulting in lower salinity (32.6-32.8 ‰). Dissolved oxygen, pH and temperature ranges are 5.5-6.4 mg·l⁻¹, 8.06-8.15 and 27.6-29.3°C, respectively. Nitrate and phosphate range from 0.12-3.40 mg·l⁻¹ and 0.08-0.87 mg·l⁻¹, respectively.

Over 300 major coral reefs have been identified along the Thai coast (both the gulf and Andaman Sea), covering approximately 12 000 km². The condition of coral reefs in Thailand ranges from very good to very poor. Over 60% of major reefs are in either poor or fair condition; less than 36%, good or very good. The provinces that still have significant areas of coral reefs in good and very good condition are Trad, Phangnga and Trang. The provinces where coral reef deterioration is most severe, due to human activities, are Chon Buri, Rayong, Surat Thani, Phuket and Satun.

Mangrove forests, the nursery area for many species of coastal fishes, are found along the muddy coastline and in the estuaries of rivers. Their distribution ranges from Trad in the eastern part of the gulf to Prachuap Kuirikhan in the west. Large areas are covered by mangroves on the southern region of the eastern coast from Chumphon to Pattani and, on the west coast, from Satun northward to Ranong. In 1991, the mangroves of Thailand were estimated to cover 1 734 000 ha, or 0.33 % of the country's land area. Loss of cover was approximately 6 500 ha·year⁻¹, attributable to legal and illegal logging, mining, human settlements, industrial development, ports and harbors, dredging, road construction and conversion into shrimp ponds.

Seven species of seagrass are found along the Gulf of Thailand, with *Halodule pinifolia* dominant along the east coast and *Enhalus acoroides* on the coarse sand of the west coast, at depth of 1-2 m (Sudara 1989). Along the Andaman Sea coast, nine species of seagrass range from Phuket to Satun Province. The largest seagrass beds are located in the shallow waters of Haad-Chaomai to Muk Island and Talibong Island, Trang Province (Phuket Marine Biological Center 1988).

Capture Fisheries Sector

Marine fisheries, which dominate Thai fisheries production, are conveniently divided into three categories: small, medium and large scale. Small-scale fisheries use specialized traditional gears such as shrimp gill net, crab gill net and mullet gill net, operated by local fishers in nearshore areas (Table 1). The medium-scale fisheries generally operate fishing

Table 1. Major fishing gears used in Thai waters.

Small-scale sector	Medium-scale sector
Pomfret gillnet	Otterboard trawl
Shrimp gillnet	Pair trawl
Mullet gillnet	Beam trawl
Mackerel gill net	Push net
Snapper gillnet	Anchovy purse seine
Threadfin gillnet	Thai purse seine
Sardinella gillnet	Chinese purse seine
Wolf-herring gillnet	Luring purse seine
Sand whiting gillnet	Mackerel encircling gill net
Marine striped catfish gillnet	King mackerel gill net
Crab and other gillnet	Bamboo stake trap
Squid and other cast net	
Acetes and other scoop net	
Shell rake with powered boat	
Hook and line, long line	
Squid hook and line	
Set bag net	

boats less than 18 m and remain in Thai waters, using gears such as trawls, purse seines and mackerel gill nets. The large-scale fisheries are equipped with modern, highly efficient gear, and operate mostly in international waters. Table 2 gives the number of registered fishing boats by major fishing gear from 1970 to 1989. Trawlers (otterboard and pair trawls) and push nets dominated the registered fishing boats in Thailand.

Catch and effort data are recorded in various forms. The information includes catches of economically important species, by fishing gear and fishing area. Analyses of catch, effort and biological data are carried out, the last with emphasis on length-frequency data of commercially important groups such as anchovy, shrimps, squids, scads and jacks. The data are collected through four different approaches: (1) marine fisheries census; (2) fishing logbook survey; (3) fishing community survey; and (4) research vessel survey.

The first marine fisheries census was initiated in 1967, and is done every 10 years. The census includes data on the number and nature of fish production factors such as fishing households, fishing boats and crew.

The fishing logbook survey emphasizes catch and effort data, particularly on medium-scale fisheries such as otterboard trawls, pair trawls, push nets, Chinese purse seines, Thai purse seines, luring purse seines, anchovy purse seines and king mackerel gill nets. Data collection, which began in 1969, involves random selection from a list of registered fishing units. Fishing logbooks were designed separately for each fishery to accommodate their different target species. The fishing areas in Thai waters were also identified for the survey: 5 subareas in the Gulf of Thailand; 5 along the Andaman Sea coast; and 2 in the South China Sea.

The fishing units are classified according to the type of gear and length of boat: (1) otterboard trawl, divided into 4 size classes (boats of less than 14 m, 14-18 m, 18-25 m and over 25 m); (2) pair trawl, in 3 size classes (less than 14 m, 14-18 m and over 18 m); and (3) Thai purse seines, in 2 size classes (less than 14 m and over 14 m). Other fishing units monitored include beam trawls, push nets, Chinese purse seines, anchovy purse seines, luring purse seines, King mackerel encircling gill nets and bamboo stake traps.

The fishing community surveys are used to collect catch (by species and gear/vessel type) and effort data from small-scale fisheries. These are conducted at selected fishing ports. Also, length-frequency data are collected monthly for key species, along with related catch and effort data by fishing area.

Research and monitoring trawl surveys are conducted with a standard bottom trawl net, at depths from 10 to 60 m. Data are studied in terms of changes in species composition and stock density and for estimating fishing impacts and future yields. Moreover, research surveys using purse seines and gill nets in the Exclusive Economic Zone (EEZ) are also being carried out to evaluate deeper fishing grounds for possible expansion of fishing activities.

Prior to World War II, fishing in Thai waters occurred mostly in inshore areas, using sailing vessels with inboard engines. In the 1930s, purse seines made of cotton twine were introduced from China, to implement a technique known as Chinese purse seine, which employs a sailboat with two small row boats for setting the net targeting pelagic fishes. Subsequent modification of this technique, leading to Thai purse seine, involved the transition to motorized vessels. Traditional gear such as the hand and troll lines, lift nets, cast nets, purse seines, nylon gill nets and bamboo stake traps were also used.

The development of trawling started in the 1950s. Between 1952 and 1960, several trials were conducted by private companies, using pair trawlers. However, these attempts failed due to lack of technical expertise and low price for the catch, given the lack of familiarity of local consumers with demersal fish. In the 1960s, otterboard trawling was introduced in Thailand by a German fisheries development project (Tiews 1962), and quickly became popular, first in Samut Prakan Province, then in the entire gulf coast (see below). However, subsequent developments soon led to excess capacity, a rapid decline of catch/effort and massive changes in catch composition (Tiews et al. 1967; Beddington and May 1982), including an increase of the proportion of small-size classes of commercial fishes, landed as 'trash fish' (Boonyubol and Pramokchutima 1984).

The Department of Fisheries (DOF) is composed of 28 divisions, with various responsibilities, notably:

1. implement various acts/laws, such as the Fisheries Act B.E. 2490 (1947) and Wildlife Conservation and Protection Act B.E. 2535 (1992);

2. conduct research in every field of fisheries, with emphasis on income generation; and

3. explore fishing grounds outside of Thai waters and promote fisheries cooperation with other states.

Of the 28 divisions, eight deal with marine fisheries and environment: (1) Fishery Planning and Policy Division; (2) Marine Fisheries Division; (3) Oceanic Fisheries Division; (4) Phuket Marine Biological Center; (5) Fishery Environment Division; (6) Fishery Resources Conservation Division; (7) Fishery Economics Division; and (8) Training Division.

The Fishery Planning and Policy Division plans and formulates fisheries development policy in conformity with the National Economic and Social Development Plan and government policy. This division prepares the department's annual budget.

The Marine Fisheries Division (MFD) is responsible for the study and restoration of fisheries resources, for marine environment, development of fishing gear, and professional development in the fisheries sector. MFD is divided into five 'centers': (1) Bangkok Marine Fisheries Development Center; (2) Eastern Marine Fisheries Development Center; (3) Central Marine Fisheries Development Center; (4) Southern Marine Fisheries Development Center; and (5) Andaman Sea Marine Fisheries Development Center.

The Oceanic Fisheries Division is responsible for exploration and analysis, development of fishing gear and techniques, and fishing grounds in foreign waters and deeper waters off Thailand. The Phuket Marine Biological Center is responsible for experimental research in marine biology and ecology, and for monitoring pollution impacts on fisheries.

The Fishery Environment Division is responsible for fishery environment planning and monitoring pollution impacts on fisheries. The Fishery Resources Conservation Division takes charge of enforcement of fisheries acts, ministerial decrees and regulations; surveillance of fisheries; and registration of vessels and fishing gear. The Fishery Economics Division is responsible for economic research, national and international fish market analyses, and fishery statistics. The Training Division plans and organizes courses for department personnel, fishing crews and other persons interested in fishery resources conservation.

Fishery Resources and Exploitation

In 1992, catches from the Gulf of Thailand contributed approximately 75% of total marine catches of the country; those from the Andaman Sea coast accounted for the rest. Total catches in the Gulf of Thailand in 1992 were 154 000 t for the small-scale fisheries, and 1 760 000 t for the medium-scale

89

Table 2. Registered fishing boats in Thailand, 1970-1989. (Source: Department of Fisheries).

Year	Otterboard trawl	Pair trawl	Push net	Mackerel encircling net	Purse seine	Anchovy purse seine	King mackerel gillnet
1970	2 452	440	314	-	-	-	-
1971	2 967	520	554	-	-	-	-
1972	3 783	698	1 327	-	-	-	-
1973	4 967	818	1 628	-	-	-	-
1974	4 415	848	1 213	-	-	-	-
1975	4 106	846	1 075	-	-	-	-
1976	4 371	832	844	-	-	-	-
1977	5 378	904	1 177	-	-	-	-
1978	5 599	854	1 426	-	-	-	-
1979	7 566	1 172	1 923	-	-	-	-
1980	9 181	1 240	2 257	307	747	34	296
1981	6 510	1 008	1 214	258	801	32	327
1982	10 067	1 406	1 899	238	784	56	281
1983	8 123	1 266	1 236	144	749	97	264
1984	8 096	1 166	960	245	806	155	265
1985	7 103	1 218	759	227	836	197	269
1986	6 322	1 084	664	203	853	143	329
1987	6 133	1 164	624	223	1 057	117	365
1988	5 807	1 130	531	146	1 255	199	461
1989	10 940	2 193	1 904	114	1 095	348	282

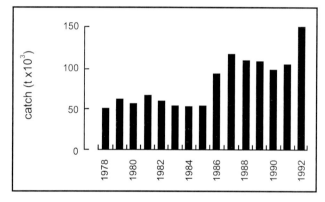

Fig. 2. Annual catch of demersal fish by trawlers in the Gulf of Thailand (1978-1992).

fisheries. The landed catch in the gulf consisted of 37% pelagic fish, 8% demersal fish, 6% squids, 5% shrimp, 10% miscellaneous fishes, and 34% 'trash fish'.

Demersal fish are caught mainly by otterboard trawlers, pair trawlers, beam trawlers and push netters. In the gulf, the number of registered trawlers increased from 3 206 units in 1970 to 15 037 in 1989 (Table 2). The annual catch of demersal fish increased from 51 000 t in 1978 to 149 000 t in 1992 (Fig.2); 78% of the catch originated from the gulf and the rest from the Andaman Sea coast.

The dominant genera in demersal catches are *Nemipterus, Saurida* and *Priacanthus*. Trawl surveys conducted in 1981 indicated that the catch consisted of squids (19%), Priacanthidae (11%), Nemipteridae (11%), Synodontidae, i.e., *Saurida* (7%), and other fish and invertebrates (54%). The 'trash

fish' included the following important groups: juvenile commercial fish (32%), small squids (2%), juvenile pelagic fish (2%) and early stages of various invertebrates (0.2%).

The demersals in Thai waters were overfished by the mid-1970s (Pauly 1979; Pope 1979), a situation that has not changed since (Pauly 1987, 1988; Pauly and Chua 1988; Fig. 4).

Several factors may have contributed to this, notably: increasing human population; decreased access to foreign fishing grounds by Thai trawlers, due to the declaration of EEZ by neighboring countries; development of processing techniques for turning low-price demersal fish into human food; and increasing number of animal feed plants that utilize trash fish.

Important small pelagic fish in Thailand are the mackerels (*Rastrelliger* spp.), scads (*Decapterus* spp.), sardines (*Sardinella* spp.) and anchovies (*Stolephorus* spp.). King mackerel (*Scomberomorus* spp.) and tuna are large pelagic fish not covered in this contribution.

In the past, Indo-Pacific mackerel (*R. neglectus*), or *pla tu,* was the most popular fish for Thai consumers. Hence, this species was extensively studied (see, e.g., Hongskul 1974), indicating that *pla tu* exhibit two spawning peaks, from February to March and from June to August. Following hatching, the larvae are transported into inshore waters; the young fish then migrate northward along the west coast, toward the inner gulf, from which they gradually move southward to spawn in the waters off Prachuap Kirikhan and Chumporn Provinces. The size at first maturity is about 17.5 cm.

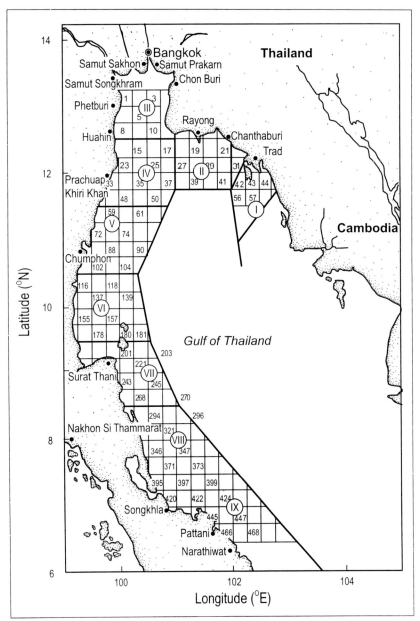

Fig. 3. Areas (I-IX) used for stratification of trawl surveys in 1966-1973 in the Gulf of Thailand. The numbers refer to successive stations.

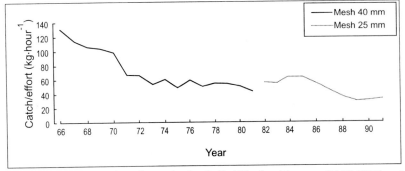

Fig. 4. Mean demersal catch rate in the Gulf of Thailand by year (1966-1991) and mesh size of trawl used in deriving catch rate.

The fishers use stationary gears such as bamboo stake trap, Chinese purse seine, Thai purse seine and luring purse seine. Catches of Indo-Pacific mackerel were about 40 000 t in 1971, or 47% of pelagic fish catches. However, following the development of improved pelagic fishing gears and techniques, the major contributors to the small pelagic catches in 1977 were sardines and roundscads, with 205 000 t and 131 000 t, or 41% and 26% of the total pelagic catches, respectively. These high percentages have declined since, due to the development of an offshore fishery for large pelagics, especially tuna.

The cephalopods in Thai waters consist of 10 families, 17 genera and over 30 species. The more important species for the squid fishery are *Loligo chinensis*, *L. duvauceli*, *L. singhalensis*, *L. edulis*, *Loliolus sumatrensis* and *Sepioteuthis lessoniana*; the cuttlefishes; *Sepia pharaonis*, *S. aculeata*, *S. recurvirostra*, *S. lycides*, *S. brevimana* and *Sepiella inermis* and the octopus; *Octopus membranaceous*, *O. dollfusi* and *Cistopus indicus* (Chottiyaputta 1993). The squids *L. chinensis*, *L. duvauceli*, *Sepioteuthis lessoniana* and *Loliolus sumatrensis* were abundant both in the gulf and the Andaman Sea; Supongpan (1988) assessed the first two species in the gulf. *L. edulis* and *L. singhalensis* are rare in the gulf, but more abundant in the deeper parts of the Andaman Sea coast. Cuttlefish, *S. pharaonis*, *S. aculeata*, *S. recurviorostra*, *S. brevimana* and *Sepiella inermis* occur both in the gulf and the Andaman Sea. *S. lycidas* is common along the Andaman Sea coast, but in the gulf occurred only south of 10°N. Octopus, *O. membranaceous*, *O. dalifusi* and *Cistopus indicus* are widely distributed in the gulf and the Andaman Sea. From 1985 to 1988, cephalopod catches consisted of squid (56%), cuttlefish (37%) and octopus (7%).

Trawl Surveys in Thailand

In 1961, under bilateral agreements between the Federal Republic of Germany and the Government of Thailand, several trawling surveys were conducted in the Gulf of Thailand, using different types of trawl (see Appendix III, this vol). The results showed that otterboard trawling was suitable for the Gulf of Thailand (Tiews 1965, 1972). Hence, the DOF and the Manila-based Asian Development Bank encouraged investments in the otterboard trawl fishery, and the number of trawlers and the landings of demersal fishes increased rapidly through the 1960s.

Stock monitoring was achieved via regular trawl surveys, which started in 1963, using the research vessel *R.V. Pramong 2* with two-sheet German trawl, 40 mm meshes in the codend. In 1977, the Marine Fisheries Division had acquired a new research vessel *R.V. Pramong 9*, and from 1978 to 1990, the survey program was conducted using both research vessels.

The first trawl surveys conducted in the gulf, from 1961 to 1965, did not involve any stratification scheme. From 1966 to 1973, the survey design was improved by dividing the gulf into nine survey areas (I-IX in Fig. 3), covered with 80 one-hour hauls, from shallow to deeper waters, i.e., a maximum of 720 hauls.

From 1974 to 1976, in response to budgetary limitations due to the 1973 oil crisis, the number of stations per area was decreased from 80 to 60, with an annual survey thus covering up to 540 hauls.

In 1977, a second trawler, *R.V. Pramong 9*, was added to *R.V. Pramong 2*. From 1977 to 1990, the two vessels covered an average of 67 stations per area. This generated a large amount of catch/effort and biological data, including length-frequency data, still being analyzed (see Table 3 for a set of intermediate results).

In 1991, the Marine Fisheries Division was divided into four centers (see above), each responsible for a research vessel and for monitoring a section of the Thai coast (Table 4).

Table 5 summarizes the key features of trawl surveys in the Gulf of Thailand from 1966 to 1995 (see also Appendix III, this vol.).

Based on the results of trawl surveys in the Gulf of Thailand, average catch of trawl with 40-mm mesh has declined from 132 kg·hour^{-1} in 1966 to 52 kg·hour^{-1} in 1973. In 1981, an average catch rate of 39 kg·hour^{-1} was obtained for the same mesh size. Likewise, catch rates of 53 kg·hour^{-1} in 1982 to 28 kg·hour^{-1} in 1981 were obtained using trawl net codends with 25-mm mesh (Fig. 4). Catch rates have been decreasing steadily in the entire gulf, as shown by catch rates in specific survey areas (Fig. 5).

Management Issues

In general, the coastal resources are in relatively good condition. However, in some areas, problems are evident because of multiple uses of the resources. The synoptic transect of key activities and issues relevant to coastal fisheries management in Thailand is shown in Table 6.

Problems continue to beset the Thai fisheries, particularly the demersal trawl fishery of the Gulf of Thailand. The issues include: (1) excessive fishing; (2) destruction of juvenile fish by shrimp fishing; and (3) small mesh sizes in codend of trawls.

From the coast to 50 m depth, the amount of fishing largely exceeds that required to extract maximum sustainable yield (MSY) - i.e., the stocks are overfished (Fig. 6) and the fleet is overcapitalized (Panayotou and Jetanavanich 1987).

A large number of juveniles of commercially valuable species are caught along with penaeid shrimps, the

Table 3. Estimated potential yields and some biological parameters of selected commercial species from the Gulf of Thailand.

Species	Potential yield[a] (10^3 t·year^{-1})	Length-weight relationship[b]		Growth parameters	
		a	b	$L\infty(cm)$[c]	K (year^{-1})
Demersal groups	893	-	-	-	-
Nemipterus spp.	24	-	-	-	-
N. hexodon	-	0.0182	2.91	28.6	1.73
N. mesoprion	-	0.018	2.93	19.5	2.15
Saurida spp.	12	-	-	-	-
S. elongata	-	0.00653	3.05	46.1	0.94
S. undosquamis	-	-	-	32.5 M	1.60
	-	-	-	42.3 F	1.02
Priacanthus spp.	17	-	-	-	-
P. tayenus	-	0.0182	2.95	27.7	2.11
Squids	36	-	-	-	-
Loligo duvauceli	-	0.374	2.00	26.6	0.86
Shrimps	22	-	-	-	-
Penaeus merguiensis	-	0.015	3.00	19.5	1.15
P. semisulcatus	-	0.00551	3.15	23.8	1.41
Pelagic fishes					
Indo-Pacific mackerel	-	-	-	-	-
Rastrelliger brachysoma	105	0.00614	3.21	20.0	3.83
Indian mackerel	-	-	-	-	-
R. kanagurta	53	-	-	24.6	2.53
Sardine	150	-	-	-	-
Sardinella gibbosa	-	0.00626	3.10	18.7	3.36
Longtail tuna	45	-	-	-	-
Thunnus tonggol	-	0.021	2.98	-	-
Little tuna	41	-	-	-	-
Euthynnus affinis	-	0.015	3.02	-	-
Auxis thazard	-	0.020	2.99	-	-
Carangid	50	-	-	-	-
Selaroides leptolepis	-	0.00438	3.30	19.2	1.54
Round scad	110	-	-	-	-
Decapterus maruadsi	-	-	-	23.1	1.32

[a] Estimated by application of an exponential surplus production model (Fox 1970).
[b] Of the form $W = aL^b$; total length in cm and weight in grams.
[c] Total length.

Table 4. Distribution of responsibilities for trawl-monitoring surveys along the Thai coast.

Center	Research trawler	Area (no. of stations)	Area (km²)
Bangkok Marine Fisheries Development Center	*Pramong 4*	III-IV (14)	20 850
Southern Marine Fisheries Development Center	*Pramong 9*	VII-VIII-IX (25)	39 372
Eastern Marine Fisheries Development Center	*Pramong 15*	I-II (12)	11 668
Central Marine Fisheries Development Center	*Pramong 1*	V-VI (17)	29 594

Table 5. Key statistics/features of otter trawl surveys in the Gulf of Thailand at 10-50 m depth[+].

Year[a]	No. of hauls[b]	Year	No. of hauls
1966	712	1981*	159
1967*	713	1982	211
1968*	719	1983	328
1969*	720	1984	172
1970*	718	1985*	228
1971*	720	1986*	260
1972*	720	1987	-
1973*	718	1988*	125
1974*	540	1989*	179
1975	480	1990*	21
1976	261	1991*	21
1977	579	1992*	21
1978	442	1993*	22
1979	235	1994*	24
1980*	245	1995*	23

[+] Vessels used: *R/V Pramong 2* (1966-1989) and *R/V Pramong 9* (1978-1995) in areas I-IX (1966-1989) and areas VII-IX (1990-1995).

[a] Years with * are those for which raw data are still available; all with L/F data, except for 1966-1969.

[b] Towing speed: 2.5 knots; duration: 1 hour.

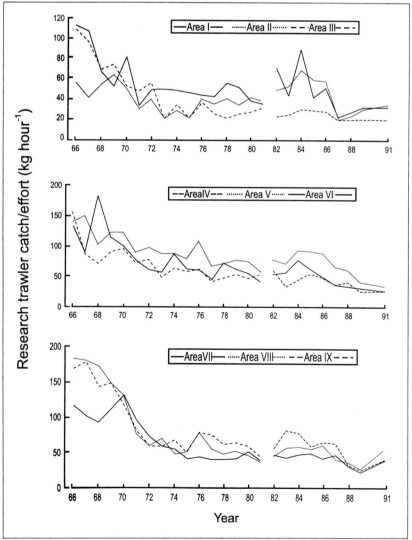

Fig. 5. Mean catch rate by year (1966-1991) and areas of the Gulf of Thailand.

94

Table 6. Synoptic transect of key activities and issues relevant to coastal fisheries management in Thailand.

| Major zones | Terrestrial | | | Coastal | | Marine |
	Upland (>18% slope)	Midland (8-18% slope)	Lowland (0-<8% slope)	Intertidal zone (1 km inland from HHWL-30 m depth)	Nearshore (>30 m-200 m depth)	Offshore (>200 m depth-EEZ)
Main resource uses/activities	Tourism Logging	Rubber plantation Mining	Agriculture Human settlements Industries	Mangrove forestry Aquaculture Land reclamation Tin mining Tourism Artisanal fishing Ports	Tourism Fishing Marine transport	Industrial fishing
Main environmental issues/impacts on the coastal zone	Siltation	Endangered fauna Siltation	Oganic loading Sewage pollution	Agriculture runoff Improper agricultural practices Agrochemical loading Water pollution Organic loading Coral reef destruction	Mangrove destruction/ conversion Water pollution Oil spills Inequitable distribution of tourism benefits Overfishing	Change in catch composition Overfishing

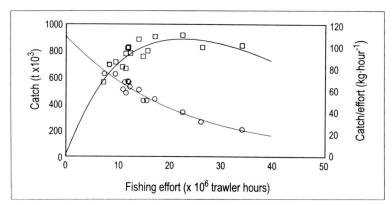

Fig. 6. Relationship between catch and effort of demersal resources in the Gulf of Thailand.

target of trawlers operating nearshore (Eiamsa-ard and Shamniyom 1979,1980; Eiamsa-ard 1981; Charnprasertporn 1982). These small fish have no value for human consumption, so they are either dumped or used for fishmeal or duck feed. These losses are added to those generated by the collection of juvenile shrimps in coastal waters using push net and small shrimp trawls.

The mesh sizes in the codends of commercial trawls are mostly 20-25 mm. Fish are thus caught at sizes that generate much less than optimum yield per recruit (Meemeskul 1979, 1988; Sinoda et al. 1979). Also, such small-meshed nets have a high drag, thus in-

creasing the amount of fuel required for trawling and decreasing the efficiency of capture of larger fish. Initiatives to address these issues are sorely needed.

Acknowledgments

We would like to thank ICLARM and ADB for organizing, and inviting us to, the workshop at which this contribution was presented. We also thank the director and our colleagues at the Southern Marine Fisheries Development Center for their valuable suggestions during the preparation of this paper.

References

Beddington, J.R. and R.M. May. 1982. The harvesting of interacting species in a natural system. Sci. Am. 247(5): 42-49.

Boonyubol, M. and S. Pramokchutima. 1984. Trawl fisheries in the Gulf of Thailand. ICLARM Trans. 4, 12 p.

Charnprasertporn, T. 1982. An analysis of demersal fish catches taken from otterboard trawling surveys in the Gulf of Thailand, 1980. Demersal Fish. Rep. No.1, Mar. Fish. Div., Department of Fisheries, Bangkok. (In Thai.)

Chottiyaputta, C. 1993. Cephalopod resources of Thailand. Recent advances in Cephalopod fisheries biology. p. 69-78.

Eiamsa-ard, M. 1981. An analysis of demersal fish catches taken from otterboard trawling survey in the Gulf of Thailand, 1979. Demersal Fish. Rep. No.1, Mar. Fish. Div., Department of Fisheries, Bangkok. (In Thai.)

Eiamsa-ard, M. and D. Shamniyom. 1979. An analysis of demersal fish catches taken from otterboard trawling surveys in the Gulf of Thailand, 1977. Demersal Fish. Rep. No.1, Mar. Fish. Div., Department of Fisheries, Bangkok. otterboard trawling surveys in the Gulf of Thailand, 1978. Demersal Fish. Rep. No. 2, Mar. Fish. Div., Department of Fisheries, Bangkok. (In Thai.)

Fox, W.W., Jr. 1970. An exponential surplus-yield model for optimizing exploited fish populations. Trans. Am. Fish. Soc. 99: 80-88.

Hongskul, V. 1974. Population dynamics of *pla tu Rastrelliger neglectus* (van Kampen). Proc. Indo-Pac. Fish Counc. 12(2): 297-342.

Lirdwitayaprasit, T., T. Vicharangsan and N. Sawetwong. 1994. Occurrences of red tide phenomenon in the inner Gulf of Thailand during 1991-1993. *In* A. Snidvongs, W. Utoomprukporn and M. Hunspreugs (eds.) Proceedings of the NRCT-JSPS Joint Seminar on Marine Science Bangkok, Thailand. Chulalongkorn University.

Meemeskul, Y. 1979. Optimum mesh size for the trawl fishery in the Gulf of Thailand. Indo-Pac. Fish. Counc., RRD/II/79/INF13, 20 p.

Meemeskul, Y. 1988. Effects of a partial increase of the mesh size in the multispecies and multifleet demersal fisheries in the Gulf of Thailand, p. 493-506. *In* S.C. Venema, J.M. Christensen and D. Pauly (eds.) Contributions to tropical fisheries biology. Papers by the participants of FAO/DANIDA follow-up training courses on fish stock assessment in the tropics. Hirtshals, Denmark, 5-30 May 1986, and Manila, Philippines, 12 January-6 February 1987. FAO Fish. Rep. (389).

Naval Hydrographic Department. 1995. Report on the analysis of hydrographic data in the central part of the Gulf of Thailand during 1982-1993, 269 p. (In Thai.)

Panayotou, T. and S. Jetanavanich. 1987. The economics and management of Thai fisheries. ICLARM Stud. Rev. 14, 82 p.

Pauly, D. 1979. Theory and management of tropical multispecies stocks: a review, with emphasis on the southeast Asian demersal fisheries. ICLARM Stud. Rev. 1, 35 p.

Pauly, D. 1987. Theory and practice of overfishing: a southeast Asian perspective, p. 146-163. *In* Papers presented at the Symposium on the Exploitation and Management of Marine Fishery Resources in Southeast Asia, 16-19 February 1987, Darwin, Australia. RAPA Rep. 1987/10.

Pauly, D. 1988. Fisheries research and the demersal fisheries of Southeast Asia, p. 329-348. *In* J.A. Gulland (ed.) Fish population dynamics. John Wiley and Sons, Chichester and New York.

Pauly, D. and T.E. Chua. 1988. The overfishing of marine resources: socioeconomic background in Southeast Asia. Ambio 17(3): 200-206.

Pauly, D. and V. Christensen. 1993. Stratified models of large marine ecosystems: a general approach and an application to the South China Sea, p. 148-174. *In* K. Sherman, L.M. Alexander and B.D. Gold (eds.) Large marine ecosystems: stress, mitigation and sustainability. AAAS Press, Washington, D.C.

Phuket Marine Biological Center. 1988. Project II: Living resources in coastal areas with emphasis on mangrove and coral reef ecosystems. Subproject on inventory and monitoring on coral reefs, seagrass and soft bottom communities in the Andaman Sea. ASEAN-Australia Cooperative Program on Marine Science. 244 p.

Piyakarnchana, T. 1989. Yield dynamics as an index of biomass shifts in the Gulf of Thailand, p. 95-142. *In* K. Sherman and L.M. Alexander (eds.) Biomass yields and geography of large marine ecosystems. AAAS Symposium 111. Westview Press Inc., Boulder.

Pope, J.G. 1979. Stock assessment in multispecies fisheries, with special reference to the trawl fishery in the Gulf of Thailand. South China Sea Fisheries Development and Coordinating Programme, SCS/DEV/79/19, Manila, 106 p.

Shepard, F.P., K.D. Emery and H.R. Gould. 1949. Distribution of sediments on east Asiatic continental shelf. Allan Hancock Foundation, Occas. Pap. 9. University of Southern California Press, Los Angeles.

Sinoda, M., S.M. Tan, Y. Watanabe and Y. Meemeskul. 1979. A method for estimating the best codend mesh size in the South China Sea area. Bull. Choshi Mar. Lab., Chiba University, 11: 65-80.

Sudara, S. and Team. 1989. ASEAN-Australia Cooperative Program on Marine Science. Project II: Living resource in coastal areas, with emphasis on mangrove and coral reef ecosystems. Site selection survey report: the Gulf of Thailand, a seagrass community. Department of Marine Science, Chulalongkorn University, Bangkok. 68 p.

Supongpan, M. 1988. Assessment of Indian squid (*Loligo duvauceli*) and mitre squid (*L. chinensis*) in the Gulf of Thailand, p. 25-41. *In* S.C. Venema, J.M. Christensen and D. Pauly (eds.) Contributions to tropical fisheries biology. Papers by the participants of FAO/DANIDA follow-up training courses on fish stock assessment in the tropics. Hirtshals, Denmark, 5-30 May 1986, and Manila, 12 January-6 February 1987. FAO Fish. Rep. (389), 519 p.

Suvapepun, S. 1991. Long-term ecological changes in the Gulf of Thailand. Mar. Pollut. Bull. 23: 213-217.

Tiews, K. 1962. Experimental trawl fishing in the Gulf of Thailand and its results regarding the possibilities of trawl fisheries development in Thailand. Veröff. Inst. Küst. Binnenfisch. (25): 53 p.

Tiews, K. 1965. Bottom fish resources investigations in the Gulf of Thailand and an outlook on further possibilities to develop the marine fisheries in Southeast Asia. Arch. FischWiss. 16(1): 67-108.

Tiews, K. 1972. Fishery development and management in Thailand. Arch. FischWiss. 24(1-3): 271-300.

Tiews, K., P. Sucondharman, A. Isarankura. 1967. On the changes in the abundance of demersal fish stocks in the Gulf of Thailand from 1963/1964 to 1966 as consequence of trawl fisheries development. Department of Fisheries, Bangkok. Mar. Fish. Lab. Contr. 8: 1-39.

Overview of the Coastal Fisheries of Vietnam

PHAM THUOC
and
NGUYEN LONG
Research Institute of Marine Products
170 Le Lai Street, Hai Phong, Vietnam

THUOC, P. and N. LONG. 1997. Overview of the coastal fisheries of Vietnam, p. 96-106. *In* G.T. Silvestre and D. Pauly (eds.) Status and management of tropical coastal fisheries in Asia. ICLARM Conf. Proc. 53, 208 p.

Abstract

This paper briefly reviews the status of the coastal fisheries resources of Vietnam, with emphasis on the exploitation of the stocks of demersal fishes and invertebrates. Vietnamese coastal waters offer little scope for expansion of fisheries, although the offshore waters may offer some potential for expansion. In coastal waters it is necessary to put emphasis on the rational exploitation and management of fish and invertebrate resources and to create sites where coastal fisheries are restricted, including 'no-take' marine protected areas.

Introduction

Vietnam has an elongated shape, and an extremely long coastline, such that its Exclusive Economic Zone, covering over 1 million km², is over three times larger than its land area. Given the relatively high primary productivity of the part of the South China Sea bordering Vietnam (Nguyen 1989), marine resources have a potential well beyond the role they traditionally played in the Vietnamese economy. Yet knowledge of these resources is poor and their long-term potential unrealized, mainly due to overexploitation of nearshore resources and underexploitation of offshore ones.

This paper summarizes information on the natural conditions of Vietnamese waters. It emphasizes the biological basis of demersal fisheries, i.e., the status and distribution of demersal fishes and invertebrates. Conclusions are drawn on possible directions for fishery development, and the rational management and conservation of these resources.

Marine Environment

Given its long coastline, the marine environment of Vietnam is characterized by a wide range of climatic and hydrological conditions, and of features (biological, economic, etc.) resulting from these primary factors.

The climate is dominated by the northeast and the southwest monsoons, from October to March and from April to September, respectively. Typhoons are frequent, especially from July to October.

The continental shelf is wide in the north and south, but narrow and steep in the central coast (Fig. 1). Numerous limestone islands are scattered along the coast, both in the south and north, especially in the Gulf of Tonkin. The coastline from northern to central Vietnam is predominantly sandy, stretching into complex lagoon systems in the area of Hué. Further south, the coastline becomes rugged.

Three regions can be identified based on their hydrological regime: the Gulf of Tonkin, central Vietnam and south Vietnam. The Gulf of Tonkin area has four distinctive seasons; its salinity and turbidity are strongly affected by inflow from the Red River. The other two regions have two seasons: 'summer' (southwest monsoon) and 'winter' (northeast monsoon).

The surface currents of Vietnam's 'Eastern Sea' (i.e., the South China Sea) are induced by the monsoon winds, and hence shift direction in the course of the year (Fig. 2).

Sea surface temperatures (SST) usually range from 21 to 26°C. These are generally high in the south and low in the north, especially in January-February, when air temperatures in the north drop to 15-16°C. In August, when air temperatures are highest, SSTs range from 28 to 29°C.

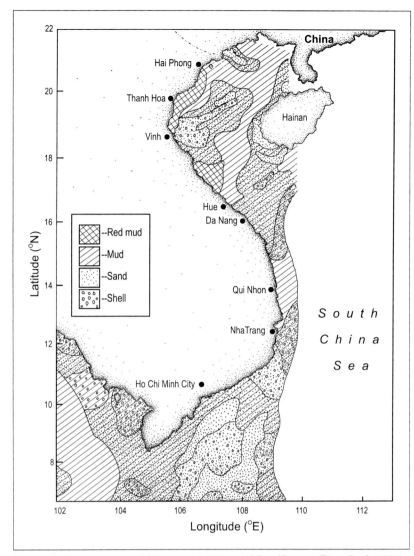

Fig. 1. Sediment types off Vietnam, 0-200 m. (After Nguyen Tien Canh 1984).

All coral reefs in North Vietnam are of the fringing type. Most are short and narrow or in the form of patches, and then to be shallow. In South Vietnam, where natural conditions are more favorable, coral reefs can be found along the coast from Da Nang to Binh Thuan Province, around islands and submerged banks in the southeast, and around most islands in the Gulf of Thailand.

Recent faunal studies of coral reef fish generated a consolidated list of only 346 species.

Some 20 sandy lagoons making up a total area of 50 000 ha are scattered along the coast of Vietnam, especially in the central region. These lagoons, varying from a few hundred to 21 000 ha (the largest, Tam Giang - Cau Hai, near Hué), also range widely in hydrological and biotic characteristics and resource utilization (capture fisheries and aquaculture), depending on freshwater influx and exchanges with the sea.

Mangrove ecosystems are found in tidal mudflats along most estuaries and seashores of Vietnam. The biological resources of these areas are diverse and abundant, with many aquatic species of high value. Also, mangroves serve as temporary feeding, overwintering and/or breeding ground for numerous species of migratory and other birds, and as permanent habitat for several mammal species.

Mangrove areas in Vietnam are becoming degraded; the areas of barren land are increasing; and saline water is moving farther and farther inland (Le Xuan Sinh 1994). This has led to the loss of nurseries for many species of shrimp, crab and fish, whose populations have subsequently declined. There are no effective measures to counteract mangrove deforestation, as this is driven by the lucrative and powerful shrimp aquaculture industry.

Fishery Resources and Exploitation

Demersal and Pelagic Fish, and Invertebrate Resources

The Vietnamese fisheries sector depends mainly on pelagic and demersal fish, which contribute 80 to 90% of fisheries yield. The rest is contributed by valuable invertebrates such as penaeid shrimp, crab, lobster, cuttlefish and squid.

Seasonal upwellings occur in various areas, especially in the central region, mainly from June to September (Fig. 2).

Primary production along the Vietnamese coast is high (Nguyen 1989); this supports high biomasses of zooplankton (Fig. 3) and, via detrital pathways, of zoobenthos as well (Fig. 4). A model of the Vietnamese shelf ecosystem presented by Pauly and Christensen (1993) which suggested that, in the late 1980s, 85% of fish production based was itself consumed by fish, and 4% by invertebrates, while fishery took about 11%, a low value compared with similar areas off Malaysia and Thailand.

Coral reefs have long been regarded in terms of biodiversity as the marine equivalent of tropical rainforests. Coral reefs are useful for fisheries, coastline protection and marine tourism. In Vietnam, coral reefs are distributed over a wide range of latitudes, with increasing coverage and structural and species diversity from north to south.

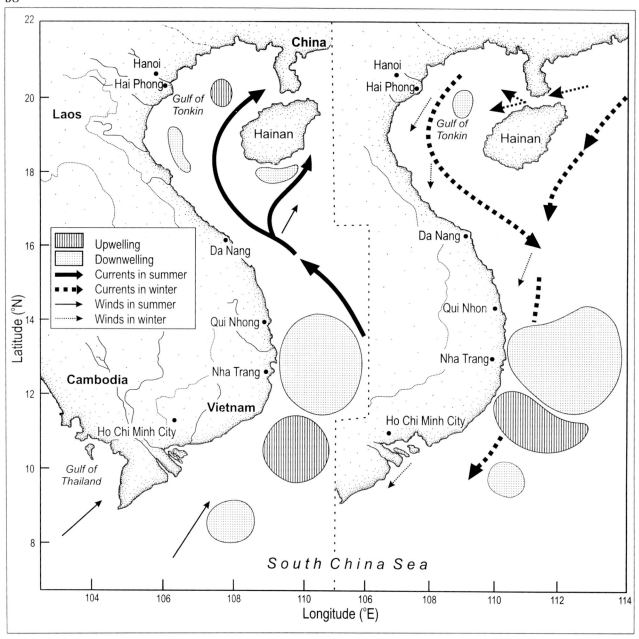

Fig. 2. Sea surface currents, upwelling and downwelling areas in summer (southwest monsoon; left) and winter (northeast monsoon; right). (After Nguyen Tien Canh 1984).

Overall, the total number of marine fish species reported exceeds 2 000, belonging to over 700 genera and nearly 200 families. Around 70% of these species are demersals with tropical affinities. Subtropical species are found mainly in the Gulf of Tonkin.

Some dominant species of demersal fish are *Lutjanus erythropterus* (snapper), *Priacanthus macracanthus* (red big-eye), *Saurida undosquamis* (true lizardfish), *Upeneus sulphureus* (yellow goatfish) and *Nemipterus zysron* (slender threadfin bream). Table 1 lists the percentage distribution of demersal species frequently encountered in demersal trawl catches in different areas of Vietnam.

Pelagic fish schools tend to be small and scattered. The major species of pelagic fish are *Sardinella jussieui* (spotted herring), *Sardinella aurita* (gilt sardine), *Pampus argenteus* (silver pomfret), *Decapterus maruadsi* (yellowtail roundscad), *Decapterus russelli* (roundscad), and various mackerel and tunas.

The demersal and pelagic fishes in shallow waters tend to be small, generally ranging from 10 to 30 cm. Most species are batch spawners with short life cycles of one to five years.

The fish spawn throughout the year, with seasonal peaks, and individual fecundity is high. Food types are varied and feeding intensity does not exhibit strong

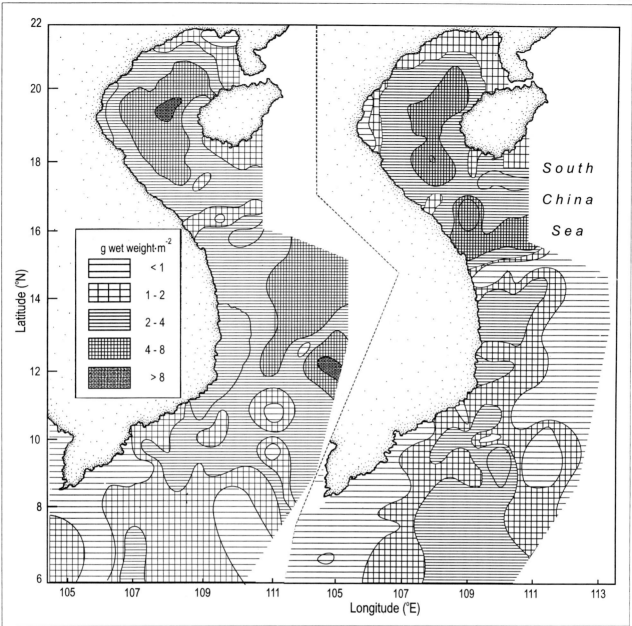

Fig. 3. Zooplankton distribution off Vietnam during the southeast (left) and northeast monsoon (right). (After Nguyen Tien Canh 1984).

seasonal fluctuations. Seasonal mitigations occur leading to fish biomass increases e.g., during the northeast monsoon, in the Gulf of Tonkin.

More than a 100 species of shrimp are found in Vietnam, and over half of these are of commercial interest. Shrimps are concentrated in shallow waters (< 30 m), especially in estuaries and river mouths. Vietnam's main shrimping grounds are along the coast of the Gulf of Tonkin and in the Mekong Delta.

Squat lobster (*Thenus orientalis*) is exploited down to a depth of 50 m. The important cephalopods are cuttlefish (*Sepia tigris*), squid (*Loligo edulis*) and octopus.

Table 2 summarizes the key features of the trawl surveys conducted in Vietnamese waters between 1977 and 1987. These surveys led to the establishment of some of the general patterns presented above (Fig. 5).

Status of Coastal Resources and Trawl Fisheries

In 1995, Vietnam's marine fisheries landed 929 000 t, of which over 80% came from the south and central-southern regions. Catches from the north (above 20°N) and central-north (North of Da Nang) amounted to only 20% (Table 3).

100

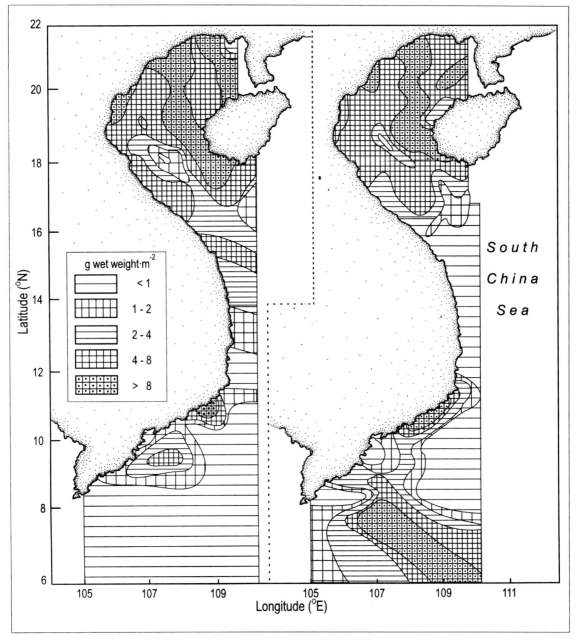

Fig. 4. Zoobenthos biomass off Vietnam during the southeast (left) and northeast monsoon (right). (After Nguyen Tien Canh 1984).

Standing stocks and some earlier rough estimates of potential yields of demersal and pelagic fish in various areas of Vietnam are presented in Table 4.

Overexploitation of coastal resources is clearly shown by the decline in catch/effort, related to the recent massive increase of fishing effort (Table 5). Thus, catch/effort in the Gulf of Tonkin dropped from 0.75 t·hp^{-1}·year^{-1} in 1980 to 0.66 t·hp^{-1}·year^{-1} in 1990. Similarly, in the central part of Vietnam, catch/effort declined from 1.06 t·hp^{-1}·year^{-1} in 1986 to 0.66 t·hp^{-1}·year^{-1} in 1991; in southern Vietnam, it declined from 2.05 t·hp^{-1}·year^{-1} in 1988 to 1.2 t·hp^{-1}·year^{-1} in 1995.

The general decline of the coastal resources of Vietnam is also reflected in individual species. Thus,

Nematalosa nasus, of which 1 000 t·year^{-1} were landed three decades ago, is now extremely rare. In the north, the catches of three Clupeidae - *Tenualosa reveesi*, *Clupanodon thrissa* and *Konosirus punctatus* - have plummeted over the last two decades. The catches of invertebrates such as lobsters (*Panulirus*), abalone *(Haliotis)*, scallop *(Chlamys)* and squid *(Loligo)* have also declined markedly.

Most of the fishing effort is exerted by relatively small vessels: 98% of all fishing craft in Vietnam either have no engine (33% of all craft) or have engines with less than 60 hp. In 1995 there were about 420 000 fishers in Vietnam, mostly engaged in small-scale, nearshore operations. The government encourages

Table 1. Contribution of major species to the catches of research trawlers in various areas of Vietnam (% weight).

Major species	Southwest (1979-1983)	Southeast (1979-1983)	Central Vietnam (1979-1983)	Tonkin Gulf 1974	1975
Pennahia argentata	1.2	-	3.4	-	6.0
Netuma thalassina	3.4	0.9	-	-	-
Decapterus kurroides	-	9.5	0.4	-	-
Decapterus maruadsi	-	9.6	0.8	12.4	11.7
Leiognathus equulus	18.2	-	-	-	-
Leiognathus splendens	7.7	1.6	-	-	-
Loligo edulis	-	-	-	4.1	1.4
Malakichthys wakiyae	-	-	29.3	-	-
Parargyrops edita	-	-	-	35.0	9.1
Priacanthus macracanthus	-	3.4	4.4	0.4	3.1
Promethichthys prometheus	-	-	8.5	-	-
Rastrelliger kanagurta	3.8	-	-	-	-
Saurida tumbil	1.2	4.0	7.2	3.1	2.6
Saurida undosquamis	-	29.4	2.1	1.0	1.6
Selaroides leptolepis	10.7	2.5	-	0.2	2.3
Trichiurus lepturus	-	-	22.2	-	-
Upeneus mollucensis	-	1.3	-	3.8	3.5
Others	56.2	44.1	28.4	41.6	67.1

Table 2. Summary information on demersal resources survey (bottom trawl) in the coastal waters of Vietnam, 1977-1987.

Month/year	Vessel	Area	Engine (hp)	No. of hauls	Depth range (m)
6/1977 - 6/1978	R.V. Bien Dong	Tonkin Gulf	-	111	20 - 100
9,10,11,12/1978	R.V. Bien Dong	Tonkin Gulf	1 500	33	20 - 60
9/1978 - 7/1980	R.V. Bien Dong	S.E. Vietnam	-	173	30 - 100
1,5,6,7,9,10,11/1979	R.V. Bien Dong	S.E. Vietnam	1 500	80	30 - 80
1,2,5,7,10,12/1979	Aelita	Centr. and S. Vietnam	800	1 482	20 - 80
1-4/1979	Kalper	-	3 800	507	40 - 300
1-7/1979	Elsk	-	1 000	504	20 - 100
4-6/1979	Yalta	-	1 350	578	20 - 150
6-7, 11-12/1979; 1-3, 8/1980	Nauka	-	1 350	808	20 - 150
12/1979; 1/1980; 1-3, 6-7/1981	Semen Volkov	-	1 000	803	20 - 100
1,2,3,4,5,7/1980	R.V. Bien Dong	S.E. Vietnam	1 500	63	30 - 80
7-9/1980; 1-3/1981	Marlin	-	800	362	20 - 80
11-12/1980; 1/1981	Vozrozdenhie	-	3 800	109	40 - 300
11-12/1981; 1-3/1982	Zavetinsk	-	800	644	20 - 80
7-10/1982	Milogradovo	-	2 300	246	30 - 150
9/1982	Trud	-	800	154	20 - 80
7-10/1983; 5-6/1987	Gerakl	-	2 300	347	30 - 150
12/1983; 1-3/1984	Antiya	-	1 000	137	20 - 100
1-3/1984	Gidrobiolog	-	1 000	33	20 - 100
10-12/1984	Uglekamensk	-	1 000	55	20 - 100
12/1984; 10-12/1987	Ochakov	-	2 300	162	30 - 150
11-12/1985; 01-02/1986	Omega	-	2 300	122	30 - 150
5-6/1986	Shantar	-	2 300	61	30 - 150
8-12/1987	Mux Dalnhi	-	2 300	246	30 - 150
2-3, 5-6/1987	Mux Tichi	South Vietnam	2 300	194	30 - 150
1-5/1988	Kizveter	-	2 300	205	30 - 150

102

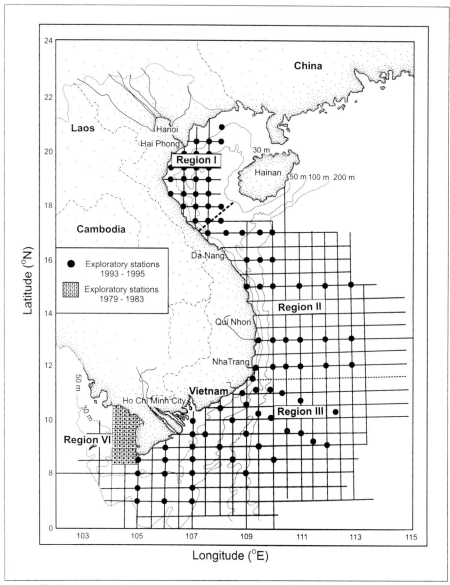

Fig. 5. Exploratory trawl survey stations off Vietnam, 1979-1983 and 1993-1995.

fishers to invest in larger boats capable of operating offshore, while some provinces do not permit construction of boats with engines of less than 33 hp.

Based on the number of units and production, major fishing gear includes trawls (fish and shrimp), gillnets, lift nets, longlines and handlines. Gillnets are widely used by fishers in the northern provinces while longlines and handlines predominate in the central region. Trawl fishing contributes almost 50% of fish production in the southern provinces of Vietnam.

The technical characteristics of various fishing gear and methods are described in Vinh and Long (1994). Trawling was a traditional form of fishing in Vietnam, conducted from pairs of sailing boats. In 1959 motorized pair trawlers of 80-90 hp were introduced, followed by single trawlers with motors of 250 hp. 1967 saw the deployment of even larger trawlers, with 800 - 1 000 hp engines and refrigerating systems,

allowing longer trips, and hence offshore fishing. The deployment of effort still remains unbalanced, with most applied to the nearshore resources, which are clearly overexploited, while offshore resources remain underutilized.

Management Issues and Opportunities

Overfishing, habitat degradation, industrial and domestic pollution and siltation are the major problems faced in the management of coastal resources in Vietnam (Table 6). These problems are collectively caused by various resource uses and activities. They are localized in highly urbanized or developed areas.

The management of coastal resources, especially the sustainable use of fishery resources, will require: (1) strengthening of fishery laws and regulations as well as their enforcement; (2) establishment of marine

Table 3. Fishery and aquaculture production of Vietnam (1981-1995).

Year	Marine fisheries (t)	Aquaculture (t)	Total (t)	Export value (US$ '000)
1981	419 740	180 380	600 120	19 000
1982	476 597	188 901	665 490	55 063
1983	519 384	204 530	723 914	64 000
1984	530 650	223 279	753 929	80 000
1985	550 000	230 400	780 400	90 000
1986	582 077	242 866	824 743	100 000
1987	624 445	251 000	875 445	138 900
1988	622 364	253 791	876 155	166 744
1989	651 525	283 327	934 852	174 196
1990	672 130	306 750	978 880	205 000
1991	730 420	335 910	1 066 330	252 000
1992	737 150	349 630	1 086 800	305 300
1993	793 324	372 845	1 166 169	368 604
1994	878 474	333 022	1 211 496	458 200
1995	928 860	415 280	1 344 140	550 100

Table 4. Estimates of standing stock and potential yields of demersal and pelagic resources in Vietnamese waters. (Thuoc 1985).

Area	Year	Group	Standing stock (t)	Potential yield (t·year[-1])
Tonkin Gulf	1981	Pelagics	390 000	156 000
	1977	Demersals	504 839	166 596
Central Areas	1976	Pelagics	500 000	200 000
	1985	Demersals	118 125	398 810
Southeast Areas	1981	Pelagics	524 000	210 000
	1985	Demersals	676 230	223 156
Southwest Areas	1973	Pelagics	316 000	126 000
	1985	Demersals	541 000	178 670
Total	-	Pelagics	1 730 000	692 000
	-	Demersals	1 840 619	607 404

protected areas; and (3) monitoring of fisheries stocks or resources.

The principal responsibility for fisheries in Vietnam lies with the Ministry of Fisheries and its associated departments, enterprises, research institutions and technical training schools. The main tasks of the ministry are to control and coordinate fisheries activities (including aquaculture) of the country and to oversee enterprises which are centrally owned.

The Ministry is organized into specialized departments: fisheries economics and planning, management, research and technology, finance, international cooperation, control, transport and infrastructure. It coordinates the operations of four national fisheries research institutes, three vocational schools and one technical training school.

Under the Ministry of Fisheries are the provincial, district and village Fisheries Authorities established within relevant People's Committees. They are responsible for fisheries and aquaculture activities in their own jurisdictions, setting up plans, collecting statistics and controlling local fisheries enterprises. These groups engage in fishing, aquaculture operations and/or export of fisheries products, when authorized by the central government.

The most recent fisheries-related legislation in Vietnam is the Ordinance on Fisheries Resources Protection promulgated in April 1989 and updated in 1996. This covers a wide range of issues, including fishing seasons and catch quotas, prevention of pollution in fishing grounds, stock assessment research, technology development, local and international investments in fisheries and aquaculture, and improvement of fishers' livelihood.

The realization of the aims of this legislation will require transcending conventional approaches to fisheries management, as these have failed to protect the stocks on which coastal fisheries depend. One of these nonconventional approaches is the setting up of marine protected areas (MPAs). There is currently no MPA in Vietnam. The only restricted areas are 12 fisheries control zones, from which only limited catches may be extracted, a small marine park around Hon Mun (Box 1), and Cat Ba National Park, which encompasses only 5 400 ha of coastal seascape. Other sites, notably the Tam Giang (Hué) and O Loan (Phy Yen) lagoons and tidal marshes near the Red River and Mekong Delta, have been proposed as possible MPAs.

Table 5. Effort and catch/effort in the marine fisheries of Vietnam (1976-1995).

Year	Number of boats	Total horsepower[a]	Catch/effort (1) (t/hp)	Number of fishers	Catch/effort (2) (t/person)
1976	34 833	543 431	1.10	-	-
1980	28 021	453 915	0.80	-	-
1981	29 584	453 871	0.91	-	-
1982	29 429	469 976	1.00	-	-
1983	29 117	475 832	1.09	-	-
1984	29 549	484 114	1.15	218 558	2.54
1985	29 323	494 507	1.27	220 770	2.84
1986	31 680	537 503	1.11	269 279	2.20
1987	35 406	597 022	1.07	291 441	2.20
1988	35 744	609 317	1.09	299 300	2.21
1989	37 035	660 021	1.00	269 467	2.15
1990	41 266	727 585	0.92	253 287	2.65
1991	43 940	824 438	0.87	275 035	2.60
1992	54 612	986 420	0.76	338 927	2.20
1993	61 805	1 291 550	0.65	363 486	2.19
1994	-	1 443 950	0.61	389 533	2.28
1995	-	1 500 000	0.62	420 000	2.21

[a] Note that 'horsepower' is translated from the French *cheval/vapeur,* or *c.v.,* which differs from the English measure of horsepower.

Table 6. Synoptic transect of key activities and issues relevant to coastal fisheries management in Vietnam.

Major zones	Terrestrial			Coastal		Marine	
	Upland (>18% slope)	Midland (8-18% slope)	Lowland (0<18% slope)	Interface[a] (1 km inland from HHWL-30 m depth)	Nearshore (30 m-200 m depth)	Offshore (>200 m depth)	Deepsea (beyond EEZ)
Main resource uses/activities	Logging Mining Agriculture Tourism	Logging Mining Agriculture Tourism	Urbanization Industries Agriculture	Mining Mangrove forestry Salt production Aquaculture Small-scale fisheries Tourism	Small-scale fisheries Commercial fishing Marine transport	Marine transport Fishing Mining	Marine transport
Main environmental issues/impacts on the coastal zone	Siltation Erosion Flooding	Chemical fertilizer Erosion Siltation Flooding Water pollution	Siltation Erosion Domestic pollution Industrial pollution	Reduced biodiversity Habitat degration Overexploitation	Reduced biodiversity Overfishing Oilspills		

[a] Including all coastal systems such as mangroves, saltmarshes and sand dunes.

However, due to the wide variety of economic and human pressures in different marine, insular and coastal environments that may be considered for setting up MPAs, the classification of protected areas and well-defined objectives and rules within these areas are essential to conserve living resources. Thus, while pristine coral reef sites with little human pressure may be set aside as strict 'no-take' MPAs, sites already utilized may be better off as fisheries-regulated zones or multiple-use areas managed in ways that minimize conflicts while somehow protecting the resources. Given the limited resources available for such schemes in Vietnam, protected areas need to be economically viable. Funding sources or incomes fore-gone must be taken into account when designing MPAs.

In any case, fisheries scientists should be actively involved in fisheries monitoring. They should focus on heavily exploited or declining species and on sensitive areas where fishing may result in overexploitation and/or environmental degradation. This includes bottom trawling in coastal water at depths of less than 30 m and controlling the collection of juvenile individuals (lobsters, groupers) for the export-oriented mariculture industry.

Regional and international cooperation is vital. This will enable Vietnamese scientists to build on experiences gained abroad in these areas of research.

Box 1. Lessons from Vietnam's first marine park[a].

The recent establishment of a marine park in Vietnam offers lessons for future marine reserves in the country as well as for other countries considering their establishment.

The waters around Hon Mun, a small (1.1 km²) island in the Nha Trang, and a famous beach-side tourist town in southern Vietnam, have recently been designated as Vietnam's first marine park. The park proposal was initiated after a survey by a World Wildlife Fund (WWF) funded team of marine scientists from the Institute of Oceanography. The purpose is to conserve its coral reef biodiversity while encouraging low-impact marine tourism and maintaining a sustainable fishery. However, there have been problems as well as successes.

"Marine Parks" - a Mistake

The term "marine park" led to overuse of the park before any controls and management capability could develop. The loss of all six moorings (to prevent anchor damage) in two weeks was one result. A better term would be "marine reserve" or "fisheries sanctuary" which emphasize protection and the value of the site in terms of fisheries. The naming of future marine protected areas (MPAs) should be done with caution and in line with their main objectives.

Over-publicity

On one hand, publicity hastened up the approval of the park by showing the government what there was to protect at Hon Mun. On the other hand, it created an influx of visitors to the island. Tourist boat operators were successfully selling their "Trips to Hun Mun Marine Park" even before the designation of the park, while fishers took their last chance before legislation took control.

The negative impacts of the publicity could have been prevented if the rules and the authority for implementation had been ready during the promotion of the park.

Concepts New and Foreign

MPAs are a new concept in Vietnam. There is no government agency with a clear mandate to deal with them, although an array of departments from fisheries science to military services are involved.

A greater involvement of Vietnamese scientists might have convinced the government of the importance and urgency of park protection. It is hoped that they will become more active in this area. Changes in the status of the coral reefs in and out of the park, before and after protection, should be monitored, documented and publicized regularly for evaluation and encouragement.

Lack of Public Participation

The park was developed without consultation with its resource users - tourist boat operators and fishers. The customary barrier between the government and users was too strong to remove. The exclusion of the users from the project has resulted in a total lack of understanding and support from the users who undoubtedly continue to exert pressure on the resources.

Meetings between the park authority (yet to be identified) and tourist boat operators and fishers, are needed to incorporate opinions from all park users for the establishment of park rules and management strategies.

Insufficient Funding

Insufficient funds are partly a reflection of the low priority given to marine conservation by both local and foreign agencies with financial potential. The situation was worsened by the waste of available funds. All the money spent on the moorings was wasted because no authority would take the responsibility of protecting them.

With the thriving tourism of Nha Trang, Hon Mun Marine Park is fully capable of generating its own operational funds. Tourist boat operators, tourists and fishers should contribute to the maintenance of the park. Fund-raising events such as photo contests can be encouraged. A well-managed protected area will attract sponsors.

Six other coral reef sites have been surveyed and shown to have high conservation priorities. The lessons from Hon Mun are useful for the planning of these sites to avoid "paper parks" with no protection or management such that their value for conservation, fisheries and tourism is eventually lost.

[a] Adapted from Cheung (1994).

Acknowledgements

We thank the workshop organizers for this opportunity to present this review of the state of Vietnam's fishery resources to an international forum.

References

Canh, N.T. 1989. An attempt to assess fish biomass and potential production in the sea of Vietnam from studies on plankton and benthos. Akademia Rolnicza W. Szczecinie No. 118.

Cheung, C. 1994. Lessons from Vietnam's first marine park. Naga, ICLARM Q. 17(4): 13.

Kyokuyo Hogel Co. Ltd. 1970a. Trawling survey in the South China Sea off the coast of the Republic of Vietnam. First Annual Report to FAO, March 1970. Tech. Rep. No. 1.

Kyokuyo Hogel Co. Ltd. 1970b. Offshore fishery development, the Republic of Vietnam trawling survey in the northern South China Sea. Annual Report prepared for FAO. Tech. Rep. No. 2, July 1970.

Kyokuyo Hogel Co. Ltd. 1970c. Offshore fishery development Rep. of Vietnam: trawling survey in the South China Sea and Gulf of Thailand. Semi-annual Report prepared for FAO. Tech. Rep. No. 3, December 1970.

Kyokuyo Co. Ltd. 1972. Offshore fishery development, the Republic of Vietnam trawling survey in the northern South China Sea. Annual Report prepared for FAO. Tech. Rep. No. 4, March 1972.

Kyokuyo Co. Ltd. 1972. Trawling survey in the South China Sea and the Gulf of Thailand (1969-1971). Final Report.

Le Xuan Sinh. 1994. Mangrove forest and shrimp culture in Ngoc Xien District, Minh Hai Province, Vietnam. Naga, ICLARM Q. 17(4): 15-16.

Marine Environment and Resources. 1991. Some hydrogeochemical characteristics and rational utilization of brackishwater ponds in the coast of North Vietnam. Vol. 1. 89 p.

MOF/NACA. 1995. Vietnam National Workshop on Environment and Aquaculture Development, 17-19 May 1994. NACA Environment and Aquaculture Development, Haiphong. Ser. No. 2. Ministry of Fisheries, Hanoi, Vietnam and Network of Aquaculture Centers in Asia-Pacific, Bangkok, Thailand. 507 p.

Nguyen, Tac An. 1989. Energy balance of the major tropical marine shelf of Vietnam. Biol. Morya/Soviet J. Mar. Biol. 15(2): 78-83.

Pauly, D. and V. Christensen. 1993. Stratified models of large marine ecosystems: a general approach and an application to the South China Sea, p. 149-174. In K. Sherman, L.M. Alexander and B.D. Gold (eds.) Large marine ecosystems: stress, mitigation and sustainability. AAAS Press, Washington, D.C.

Thuoc, P. 1986. Fish resources of Vietnamese seas: biological and fisheries characteristics, biomass assessment, and exploitation potentials. Akademia Polnicza W. Szczecinie. No. 106, 157 p.

Thuoc, P. 1986. Regional study and workshop on environmental assessment and management of aquaculture development. TCP/RAS/2253.

Vinh, C.T. and N. Long. 1994. Marine fish harvesting in Vietnam. INFOFISH Int. 1: 54-56, 58.

FAO Program Perspectives
in the Fisheries Sector, with Emphasis
on South and Southeast Asia

P. MARTOSUBROTO
Marine Resources Service
Fishery Resources Division
Food and Agriculture Organization of the United Nations
Via delle Terme di Caracalla, 00100 Rome, Italy

MARTOSUBROTO, P. 1997. FAO program perspectives in the fisheries sector, with emphasis on South and Southeast Asia, p. 107-111. *In* G.T. Silvestre and D. Pauly (eds.) Status and management of tropical coastal fisheries in Asia. ICLARM Conf. Proc. 53, 208 p.

Abstract

This paper provides an overview of the goals, objectives and strategies pursued under FAO's fisheries program for the 1996-1997 biennium. Priority activities under the program include: (1) promoting responsible fisheries at the global, regional and national levels; (2) increasing the contribution of responsible fisheries and aquaculture to world food supply and security; and (3) global monitoring and strategic analysis of fisheries. Activities of FAO projects (particularly the Bay of Bengal Program) and fisheries bodies (particularly the Asia-Pacific Fishery Commission) relevant to assessment and management of coastal fisheries in South and Southeast Asia are briefly discussed.

Introduction

The UN Food and Agriculture Organization (FAO) program of work is developed every biennium under the close scrutiny of the governing bodies representing its member states. In the early process of program development, the proposed sectoral programs developed by the respective departments (agriculture, forestry, fisheries) are first presented to and discussed in the corresponding sectoral committees (e.g., Committee on Fisheries, or COFI). Each sectoral committee then submits its report to the FAO Council. The report of the council on the program is then discussed at the FAO Conference for final endorsement.

The Program Committee, whose members consist of selected representatives of member states, scrutinizes the proposed program of the sectoral departments and takes into account the financial situation of FAO as discussed and reported by the Finance Committee. In the fisheries sector, program development starts with the discussion of the governing bodies in COFI's session, which convenes every two years. The present program, which covers the 1996-1997 biennium, was developed with inputs from the last (21st) session of COFI in March 1995.

The Fisheries Program of 1996-1997[1]

The long-term goal of the fisheries program is to promote and enhance sustainable food supplies from fisheries and aquaculture through improved planning, management and technology. In formulating goals and objectives, various considerations have to be taken into account, including current problems. These include the following:

1. Managers responsible for national policies for the management and development of fisheries face: increasing fishing and environmental pressure on fishery resources; high and increasing natural variability of fishery stocks; stagnating or decreasing institutional and financial resources to meet these challenges in the developing world; and inadequate information bases to work on.

2. The catching capacity of fishing fleets exceeds significantly the quantities that could be harvested in a sustainable manner, and much is wasted through discarding.

3. High-seas fisheries, including those of straddling stocks and highly migratory stocks, are insufficiently regulated, often overfished and a cause of high political tension at the international level.

---

[1] The programme for the next biennium (1988-1999) is being finalized and subject to the FAO Conference scheduled in November 1997.

Main Objectives and Strategies

While meeting these challenges, various objectives must also be met:

1. Improve the social and economic performance of fisheries and aquaculture through responsible management and development, enhanced trade and more responsible harvesting and utilization, recognizing the need to balance fishing, capital and labor at sustainable levels.

2. Rehabilitate the natural resource base and promote the protection of critical fishery habitats, through the design of improved fishery management scheme, promotion of applied fishery research and enhancement of international collaboration.

3. Increase the contribution of fisheries to food production and security, through enhancement of inland and coastal fisheries, development of aquaculture, reduction of waste and discards, and better use of small pelagic and underutilized species.

These objectives and the strategies developed for meeting them are then translated into priority action that takes into account the current financial situation of FAO.

Priority Areas and Activities

In 1996-1997, the efforts of the Fisheries Department are concentrated on the priority areas and activities outlined below.

PROMOTING RESPONSIBLE FISHERIES SECTOR MANAGEMENT AT THE GLOBAL, REGIONAL AND NATIONAL LEVELS

Priority activities:
- Finalize the guidelines for implementing the Code of Conduct for Responsible Fisheries.
- Identify, adapt and transfer improved strategies for sustainable development and management of capture fisheries, including inland fisheries, considering natural fluctuations in productivity, socioeconomic factors, food needs and competing demands from other sectors.
- Enhance the role of the FAO fishery bodies, through appropriate extrabudgetary support and increase in management powers, with focus on shared stocks, straddling stocks and other high-seas resources and trade.
- Transfer technologies to improve selectivity of fishing gear and optimize energy usage.
- Collect, analyze and distribute information on management approaches and systems, and their performance.

- Adapt and transfer methodologies and technologies to raise national capacity in management-oriented fishery research.
- Follow up UN initiatives such as the UN Convention on the Law of the Sea, UN Conference on Environment and Development, and UN Conference on Straddling Fish Stocks and Highly Migratory Fish Stocks.

INCREASING CONTRIBUTION OF RESPONSIBLE FISHERIES AND AQUACULTURE TO WORLD FOOD SUPPLIES AND FOOD SECURITY

Priority activities:
- Identify and transfer approaches to sustainable fisheries enhancement and aquaculture development.
- Integrate culture systems into more generalized rural and coastal contexts.
- Improve national capacity in fish utilization and marketing by promoting and coordinating the adaptation and transfer of methodologies, technologies and information.
- Promote better utilization of underexploited species (particularly small pelagics), reduction of waste and postharvest losses, and improved quality as a means for development and improved food security.

GLOBAL MONITORING AND STRATEGIC ANALYSIS OF FISHERIES

Priority activities:
- Conduct biennial review on the state of world fisheries and aquaculture.
- Collect, analyze, interpret and disseminate statistics on fisheries, including catches, commodities and fleets; and monitor world fisheries, with focus on the high seas, as well as information on the state of fish stocks.
- Compile and distribute data and information on trade in fish and fishery products.
- Initiate development of a global Geographic Information System on fisheries, their resources and their environment, and contribute to the FAO Digital Atlas on Agriculture, Forestry and Fisheries.

Support to FAO Member Countries

The assistance of FAO to member countries takes various forms such as: providing technical guidelines, policy advice, training and extension; strengthening regional fisheries bodies; and enhancing technical cooperation among developing countries and other forms of partnership programs. The assistance can be executed in the form of technical and investment projects to be funded by donor agencies or by FAO itself (through the Technical Cooperation Program -- TCP). The projects can be national, regional and global. Continued financial constraints faced by the United Nations System in recent years have, however, lowered the number of FAO projects. A single regional project remains in South and Southeast Asia, called the Regional Program for Coastal Fisheries Management in the Bay of Bengal, or, in short, the Bay of Bengal Project (BOBP). The BOBP covers seven countries: Bangladesh, India, Indonesia, Malaysia, Maldives, Sri Lanka and Thailand (Box 1). Another project, the Fishery Resources Training Project (GCP/INT/575/DEN), is global and relevant to member countries in tropical Asia.

Regional Fisheries Bodies

There are a number of regional fisheries commissions under FAO. In the South and Southeast Asian regions, two commissions are relevant. One is the Indian Ocean Fishery Commission (IOFC) covering the countries around the Indian Ocean, extending from South Asia to East Africa, and dealing with marine fisheries in the Indian Ocean. The other is the Asia-Pacific Fishery Commission (APFIC), covering both inland and marine fisheries, and embracing countries in Asia (including landlocked Nepal) and countries bordering the Western Pacific Ocean and South Pacific.

One of the recent developments under IOFC is the establishment of the Indian Ocean Tuna Commission, which is intended to take care of a large part of the

Box 1. The Bay of Bengal Program.

The Bay of Bengal Program (BOBP) of the United Nations Food and Agriculture Organization (FAO) is a multiagency regional fisheries program composed of seven countries bordering the Bay of Bengal: Bangladesh, India, Indonesia, Malaysia, Maldives, Sri Lanka and Thailand. The program started in 1979; it is funded by the United Nations Development Fund (New York), ODA (U.K.), DANIDA (Denmark), SIDA (Sweden) and the International Maritime Organization (Vienna).

In its first two phases (1979-1994), BOBP assisted in developing, demonstrating and promoting new techniques, technologies and development ideas to achieve its mission. The key activities during these phases were: (1) development of more efficient fishing technology; (2) aquaculture techniques; (3) postharvest fisheries; (4) assessment of fishery resources; (5) biosocioeconomics; (6) people's participation; (7) fisherfolk education; (8) facilitating credit; (9) fisheries extension services; and (10) dissemination of information.

The goal of the third phase (1995-2000) is to serve as catalyst for developing innovative participatory approaches and solutions to small-scale fisheries management. Specifically, during this phase the program aims to achieve: (1) sustainable exploitation of fisheries resources; (2) equitable access to and use of aquatic resources for livelihood security; (3) promotion of integrated coastal area management; and (4) promotion of resource users as resource managers and decentralization of fisheries management responsibility and authority.

The components of the third phase are: (1) building and increasing awareness and public opinion among stakeholders on the needs, benefits of and options for fisheries management; (2) building the capacities of local fishing communities and national fisheries agencies to plan and carry out integrated fisheries management; (3) human resources planning and development, emphasizing utilization of local capacity; and (4) providing technical assistance and other identical needs of member countries for implementation.

Moreover, the third phase is centered on the following general strategic principles: (1) reduce excess or surplus fishing capacity; (2) create awareness to instill in resource users a pride of ownership of the ideas and concept of management; (3) reexamine fishing technologies to educate fisherfolk on gears that deplete living resources and impair resource habitat; (4) promote action-oriented projects that produce measurable positive effects; (5) implement program activities that reflect the opinions of the public involved in fisheries and coastal communities; (6) for BOBP's national counterparts to assume greater responsibility and leadership in project implementation (national execution); (7) encourage government to allocate necessary financial resources for fisheries management; (8) coordinate national and regional information systems and exchanges; (9) integrate the environmental impacts of fishing practices and vice-versa into fisheries management decisions; (10) coordinate and consult in fisheries management decisions among national agencies with jurisdiction in the coastal zone; and (11) establish a regional forum or mechanism to reduce fishing conflicts between member countries and minimize arrests of fisherfolk caught illegally fishing in the national waters of BOBP member and nonmember countries.

The third phase of BOBP is funded by the governments of Denmark and Japan. In addition, member countries contribute funds for specialized project activities, such as information services, and make in-kind and cash contributions for project implementation.

work previously done by the Indo-Pacific Tuna Development and Management Program (IPTP), originally funded by the United Nations Development Program and later by the Japanese government. APFIC in the meantime has established two committees: one covers activities in aquaculture and inland fisheries; the other, in marine fisheries.

With the current decentralization of FAO, the role of the regional fisheries bodies has become more important.

Asia-Pacific Fishery Commission

FAO member states established the Indo-Pacific Fisheries Council in 1948. In 1976, the council became a commission, the Indo-Pacific Fishery Commission, to be able to take action directly related to fishery management and development in its area of responsibility.

The agreement to establish the commission was further amended during a reorganization in 1993. This was to meet emerging issues, especially the boom of coastal aquaculture (particularly shrimp culture) in Asia, and to put more emphasis on the marine fisheries of the South China Sea region. The commission again changed its name to the Asia-Pacific

Fishery Commission (APFIC), whose secretariat is based in Bangkok, with backstopping from the technical divisions of the Fisheries Department at FAO Headquarters, in Rome.

The main responsibility of APFIC is to promote the full and proper utilization of living aquatic resources through the development and management of appropriate fishing and culture operations and of related processing and marketing activities, in conformity with the objectives of the members. In pursuing its activity, APFIC is supported by the Aquaculture and Inland Fisheries Committee and the Committee on Marine Fisheries. In turn these two committees are supported by working parties and joint working parties (Fig. 1). The current members of APFIC are Australia, Bangladesh, Cambodia, the People's Republic of China, France, India, Indonesia, Japan, the Republic of Korea, Malaysia, Union of Myanmar, Nepal, New Zealand, Pakistan, the Philippines, Sri Lanka, Thailand, the United Kingdom, the United States of America and Vietnam.

APFIC is scheduled to convene its 25th Session on 15-24 October 1996 in Seoul. In conjunction with this session, an APFIC Symposium will also be convened with the theme "Environmental Aspects for Responsible Fisheries". In the meantime, it has been planned that the first APFIC Working Party on Marine

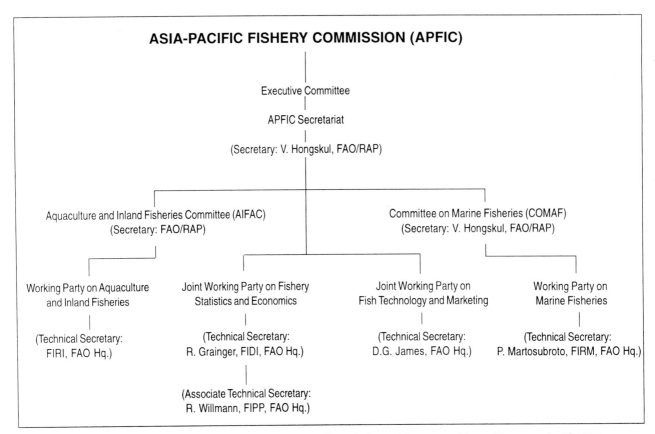

Fig. 1. Organizational structure of the Asia-Pacific Fishery Commission, with names of key officers as of mid-1996.

Fisheries will convene in early 1997, with a special agenda to discuss status, trends and issues of the small pelagic fisheries in the region.

FAO promotes the establishment of regional fisheries bodies with a mandate in certain aspects of fisheries that member states wish to have. For example, FAO was instrumental in establishing the Network of Aquaculture Centers in Asia and the Pacific, based in Bangkok, and INFOFISH, in Kuala Lumpur. FAO has worked closely with ICLARM in developing computer software relevant to fisheries management (FishBase, FiSAT; Gayanilo et al., this vol.), and with the Southeast Asian Fisheries Development Center in organizing various regional work-

shops that dealt with shared stocks, fishery information and statistics, and fisheries planning. FAO, within available means, will continue to support regional activities and keep the spirit of cooperation with existing regional fisheries bodies for the benefit of member states.

References

APFIC. 1995. Asia-Pacific Fishery Commission: Structure, Function and Directory of its Subsidiary Bodies. RAP Publ. 1995/1: 25 p.

FAO. 1995. The Director-General's Programme of Work and Budget for 1996-97. FAO Rome: C 95/3: 330 p.

Program Perspective: SEAFDEC-MFRDMD in Fisheries Resources Development and Management

M.I. MANSOR
and
M.N. MOHD-TAUPEK
Marine Fishery Resources Development
and Management Department
Southeast Asian Fisheries Development Center
Chendering 21080 Kuala Terengganu, Malaysia

MANSOR, M.I. and M.N. MOHD-TAUPEK. 1997. Program perspective: SEAFDEC-MFRDMD in fisheries resources development and management, p. 112-115. *In* G.T. Silvestre and D. Pauly (eds.) Status and management of tropical coastal fisheries in Asia. ICLARM Conf. Proc. 53, 208 p.

Abstract

This paper illustrates the active role of the Southeast Asian Fisheries Development Center-Marine Fishery Resources Development and Management Department (SEAFDEC-MFRDMD) in fishery resources development and management in Southeast Asia. It summarizes the various research activities of SEAFDEC-MFRDMD in its current three-year (1995-1997) program. MFRDMD studies in the Marine Fisheries Biology and Stock Assessment Division cover marine fisheries biology, fish and invertebrate stock assessment, and turtle ecology, conservation and management. The Oceanography and Resource Exploration Division has conducted researches on fisheries oceanography, remote sensing, resource exploration, fishing gear technology and resource rehabilitation.

Background of MFRDMD

The establishment of the Marine Fishery Resources Development and Management Department (MFRDMD) in Kuala Terengganu, Malaysia, as the fourth department of the Southeast Asian Fisheries Development Center (SEAFDEC) was agreed on at the 23rd SEAFDEC Council Meeting in December 1990. The general objective of MFRDMD, which started its activities in 1992, is to assist member countries of SEAFDEC in developing and managing the marine fishery resources in the Exclusive Economic Zone (EEZ) waters of Southeast Asia.

The functions of MFRDMD are to:
- act as a center for providing guidelines for the proper development and management of fishery resources in the EEZs of member countries;
- monitor the state of the marine fishery resources;
- provide the scientific basis for the proper development and management of the marine fishery resources of member countries;
- serve as a regional forum for cooperation and consultation on marine fishery resources research and management among member countries; and
- coordinate and implement programs to improve the capability of member countries in developing, managing and conserving fishery resources in their respective EEZs.

To fulfill its functions, MFRDMD:
- carries out regular research on biology, ecology and population dynamics of fish stocks;
- participates in monitoring surveys in the EEZs of member countries;
- undertakes programs for regional cooperation on research, management and exploitation of shared stocks;
- provides training on methodologies for marine fishery resources assessment and management;
- carries out research on methods to enhance and rehabilitate the productivity of marine fishery resources in the region;

- establishes a marine fishery resource information center to collect and disseminate information on the conservation and management of marine fishery resources; and
- disseminates information in the form of publications, documents and other media.

Overall, MFRDMD's research, training and information dissemination activities (as approved by the SEAFDEC Council) aim to promote sustainable development and management of fishery resources in Southeast Asia.

MFRDMD Activities

MFRDMD's research activities are conducted through two main divisions: (1) the Marine Fisheries Biology and Stock Assessment Division, and (2) the Oceanography and Resource Exploration Division. Both divisions are responsible for research at the national (Department of Fisheries, Malaysia) and regional (SEAFDEC) levels. Research under the Fisheries Biology and Stock Assessment Division covers: (a) marine fisheries biology; (b) fish and invertebrate stock assessment; and (c) turtle conservation, management and ecology. The Oceanography and Resource Exploration Division conducts research on: (a) fisheries oceanography; (b) remote sensing; (c) resource exploration; (d) fishing gear technology; and (e) resource rehabilitation and enhancement.

A brief account of MFRDMD research projects is given below. These are part of SEAFDEC's current three-year (1995-1997) program.

Marine Fisheries Biology in Southeast Asia

Substantial biological data on migratory fishes have been produced by research institutions in Southeast Asia. These need to be compiled regionally (and should include recent and updated information) to provide inputs for assessing regional/shared stocks. MFRDMD proposes to continuously update this compilation via occasional publications for fishery managers and scientists in Southeast Asia.

Distribution of Marine Fish Species in Southeast Asia

A complete list of marine fish species would be useful to fishery scientists studying fishes in Southeast Asia. Among other things, such list could be used as a reference for accurate identification, and to facilitate studies on population structure and dynamics. Thus, a checklist of the important marine fish species in SEAFDEC member countries was made from published materials as well as through personal communication with other scientists (especially taxonomists) in the region. The list includes species coding for the data information system of SEAFDEC/MFRDMD. This draft checklist of marine species is being reviewed by experts from Japan.

Regional Tuna-Tagging Program

Tunas are commercially important migratory fish whose stocks are believed to be extensively shared by Southeast Asian countries. Thus, excessive fishing pressure imposed on the tuna stocks in one country will have a negative effect on the fishery of another country. Collaborative tuna tagging would help identify migration routes within the South China Sea.

The tuna-tagging project was approved by the SEAFDEC Council in 1994. It aims to tag tunas off the coasts of Malaysia, Thailand, Brunei, Vietnam, the Philippines and Indonesia. Preliminary results from the Malaysian National Tuna-Tagging Program in 1992 and 1994 suggest that tuna migration in the region generally resembles the pattern shown by surface water currents in the South China Sea.

Tuna Resources and Statistics in Southeast Asia

International management of living resources is important to ensure that they are exploited on a sustainable basis. To this end, collection of comprehensive tuna-catch statistics and related information is included as part of SEAFDEC/MFRDMD main activities. The objective is to carry out a regional collaborative study to identify the various tuna resources in Southeast Asia, including, where possible, sampling at the major landing centers. Questionnaires, relating to tuna landings and size-frequency distribution, will be sent to liaison officers from countries bordering the South China Sea, i.e., Malaysia, Thailand, Singapore, the Philippines, Brunei Darussalam, Vietnam and Indonesia. The information obtained will be compiled, published annually and disseminated.

Results from the Indo-Pacific Tuna Development and Management Program (IPTP) sampling activities have indicated that interactions exist between the tuna fisheries among the SEAFDEC member countries. So far, however, the mutual impact of these fisheries has not been well established. It is hoped that close collaboration among the scientists of member countries will promote a better understanding of the tuna stocks and help develop a framework for rational management strategies.

114

Regional Squid-Tagging Program

Like tunas, squids are also one of the most commercially important pelagic species, and some of their stocks may be shared by Southeast Asian countries. Collaborative squid tagging would help identify migration patterns within the South China Sea, besides establishing closer cooperation between personnel of the various countries involved.

Fishery Oceanography

Information on oceanographic parameters of the South China Sea is scant and not well documented, and come mostly from the coastal waters. Most of the information is not generally available to researchers, but kept at individual research institutions of the SEAFDEC member countries.

The objectives of this project are to: (1) carry out oceanographic surveys using *M.V. SEAFDEC* in waters off the coasts of Malaysia, Thailand, the Philippines and Vietnam; (2) compile the oceanographic data from the countries bordering the South China Sea; and (3) apply remote-sensing techniques in mapping oceanographic parameters.

M.V. SEAFDEC was deployed to conduct collaborative resource and oceanographic surveys in the South China Sea involving a number of Malaysian and Thai researchers (box). In most cases, fish resource is assessed as part of the acoustic survey (using the FQ 70 scientific echosounder), while oceanographic parameters are measured simultaneously using standard methods.

Oceanographic data will be obtained daily from the NOAA receiver and processed to produce maps of sea-surface temperatures and to locate fronts and other productive areas.

Premonsoon and postmonsoon surveys were successfully completed by MFRDMD and Thailand Department of Fisheries researchers in September 1995 and April-May 1996, covering the waters of the Gulf of Thailand and East Coast Peninsular Malaysia, respectively. The next surveys were scheduled for July-August 1996 and April-May 1997 in the waters off Sarawak, Sabah and Brunei.

Marine Turtle Research and Conservation

The waters of the South China Sea, the Strait of Malacca and the Andaman Sea host a number of marine turtle species. However, in recent years, the turtle populations have drastically declined. Current information on the biology and ecology of the turtles is scant. Concerted efforts and serious commitments from the various relevant agencies and parties are necessary to ensure that marine turtles are conserved and will survive for future generations.

The project aims to: (1) compile and disseminate marine turtle statistics of this region; (2) carry out a collaborative program for tagging marine turtles in the region; and (3) analyze the marine-turtle tagging data for estimation of growth and the sexual maturity of some of the major species.

The first workshop on marine turtle research and conservation was conducted on 15-18 January 1996 at MFRDMD in Kuala Terengganu, with delegates from Malaysia, Thailand, the Philippines, Singapore, Brunei Darussalam and Japan.

Fishing Gear Selectivity

The trawl is the main fishing gear in most countries of this region. Despite its major contribution to marine fish landings, the gear is found to be destructive as it generates large quantities of by-catch, consisting mostly of small juveniles of commercial fish. Steps should be taken to reduce this by-catch.

The objective of the project is to study the selectivity and efficiency of selected fishing gear in order to eventually reduce their by-catch. A trouser trawl has been designed, constructed and tested in collaboration with the SEAFDEC Training Department. A comparative study on the 38-mm diamond- and square-mesh cod end has been carried out. Also, a workshop on 'Research in the Selectivity of Fishing Gear in Southeast Asia and Selective Shrimp Fishing' was jointly organized by SEAFDEC/MFRDMD, INFOFISH and FAO on 28-30 May 1995.

Fish Stock Assessment

The major marine fish stocks within the EEZ of SEAFDEC member countries need to be studied by region; the results obtained should be relayed back to the fishery managers of the countries concerned. The objectives of this project are to: (1) establish at MFRDMD a system for the computerization of fisheries data; (2) undertake compilation, review and publication of historical regional catch and effort data in the new, improved format; and (3) perform analyses on the status of fish stocks in the region, using available historical catch and effort data, combined with data obtained by the *M.V. SEAFDEC* acoustic survey.

Catch-effort data are being compiled based on questionnaires sent to SEAFDEC member countries. Also, acoustic surveys using *M.V. SEAFDEC* have been conducted in September 1995 and April/May 1996.

Box 1. SEAFDEC collaborative research projects (1995-1997).

By SEAFDEC / Training Department and Thai local institutions

- Distribution, Abundance and Species Composition of Macrobenthos
- Distribution, Abundance and Species Composition of Phytoplankton: (1) in the Gulf of Thailand and East Coast of Peninsular Malaysia, and (2) in the West Coast of Sabah, Sarawak and Brunei
- Distribution, Abundance and Composition of Zooplankton in the Gulf of Thailand and Malaysia
- Species Composition, Diversity and Biology of Economically Important Fishes
- Composition, Abundance and Distribution of Fish Larvae in Thai-Malaysian Waters
- Distribution of Dinoflagellate Cysts in the Surface Sediments of Asian Waters: (1) the Gulf of Thailand and the East Coast of Peninsular Malaysia, and (2) the West Coast of Sabah, Sarawak and Brunei Darussalam
- Fishery Oceanography of Southeast Asian Continental Shelf Waters: (1) Numeric Modeling of Shelf Circulation, (2) Pelagic Food-Chain Analysis, (3) Marine Pollution Studies, and (4) Remote Sensing and GIS Applications
- Biomass Estimation in Southeast Asian Waters
- Study on the Physical and Chemical Characteristics of Bottom Sediments in the Thai-Malaysian Waters

By SEAFDEC / MFRDMD and Malaysian local institutions

- Longitudinal Variability of Physical Parameters on the Western Side of South China Sea
- Trace Metals and Mineral Composition of Sediments of the Gulf of Thailand and South China Sea
- Chlorophyll *a* Content of the South China Sea and the Gulf of Thailand
- Photosynthetic Values, Light Intensity and Related Parameters of South China Sea and the Gulf of Thailand
- Microplankton of the South China Sea and the Gulf of Thailand
- Microzooplankton (including Dinoflagellate and Foraminiferans) of the South China Sea and the Gulf of Thailand
- Distribution, Abundance, Species Composition and Biological Studies of Economically Important Pelagic Fishes in the East Coast of Peninsular Malaysia
- Distribution of Nutrients in the South China Sea
- Distribution, Abundance and Species Composition of Phytoplankton in the South China Sea and the Gulf of Thailand
- Distribution, Abundance and Species Composition of Benthos in the Gulf of Thailand and South China Sea
- Density, Biomass, Species Composition and Distribution of Pelagic Fish Species

---

Workshops and Training Activities

A workshop on shared stocks involving the countries of Southeast Asia was organized to discuss the status of the fisheries (including catch-effort information, fishing gear, population dynamics and biological parameters of the shared stocks). Similar workshops will be held to analyze oceanographic and environmental data related to the shared stocks, and to discuss collaborative research on the marine environment and resource surveys among SEAFDEC member countries.

Conclusion

This paper documents the active role taken by SEAFDEC/MFRDMD in fisheries resource assessment and management in Southeast Asia. The department welcomes any comments concerning its research activities, and would be greatly interested to participate in any collaborative efforts for the benefit of fisheries in the region.

Acknowledgments

I thank the workshop organizers for the opportunity to present this paper and to participate in the meeting, the Director General of Fisheries (Malaysia) for permission to attend the workshop, the Chief of MFRDMD for comments on the manuscript, and SEAFDEC for funding support to attend the workshop.

Toward a Generic Trawl Survey Database Management System[1]

F.C. GAYANILO, JR.

International Center for Living Aquatic Resources Management
MCPO Box 2631, 0718 Makati City, Philippines

T. STRØMME

Nansen Programme, Institute of Marine Science
PO 1870, 5024 Bergen, Norway

and

D. PAULY

International Center for Living Aquatic Resources Management
MCPO Box 2631, Makati City 0718, Philippines
and
Fisheries Centre, University of British Columbia
Vancouver, B.C., V6T 1Z4, Canada

GAYANILO, F.C., JR., T. STRØMME and D. PAULY. 1997. Toward a generic trawl survey database management system, p. 116-132. *In* G. Silvestre and D. Pauly (eds.) Status and management of tropical coastal fisheries in Asia. ICLARM Conf. Proc. 53, 208 p.

Abstract

The first assessments of fisheries potential in tropical Asia were based largely on various demersal trawl surveys. In many countries, trawl surveys continue to be used for monitoring the status of stocks. This contribution presents the concept of a trawl survey database management system to organize, store, retrieve and exchange historical and contemporary trawl survey data. Some of the data analyses that can be made based on trawl survey data are also discussed. The prototype of a trawl survey database management system developed along these lines (called TrawlBase) is presented.

Introduction

Trawl surveys constitute an important fisheries-independent method for assessing and monitoring demersal stocks. To a large extent, data collected through this method, as well as analyses performed, are fairly standard in form (Pauly 1996). Whether surveys are random, systematic, stratified-random or encounter-response in design, the data they generate can be easily handled by a database management system (DBMS). A number of DBMSs can accommodate trawl survey data. These are either generic, such as the NAN-SIS software (Strømme 1992), or tailored to the needs of a given country (Vakily 1992).

Recent developments in computer technology — such as higher speed, large increases in memory, the Windows multitasking environment and standard methods for data exchange — open potentials for powerful analyses that in the past were reserved for scientists with specialized skills in computer programming.

For a DBMS to be effective and efficient, special attention should be given to how data are stored (Dates 1981). The design of the DBMS should guarantee that: (1) redundancy of stored data is avoided; (2) inconsistencies within the database are avoided; (3) data are or can be shared and exported to other forms and standards; (4) standards in data entry are enforced; (5)

---

[1] ICLARM Contribution No. 1380.

116

conflicting requirements are balanced among users; and (6) integrity of data is maintained through data validation and/or controls in data encoding.

This paper represents an attempt to identify the essential features of a generic DBMS for trawl surveys, given DBMS design considerations and computer technology developments. A special effort was made to include all the functionalities of NAN-SIS, plus other routines thought to be important, given the development of related systems, notably FishBase 97 (Froese and Pauly 1997), ReefBase (McManus and Ablan 1996) and FiSAT (Gayanilo et al. 1996).

Trawl-Survey Data

A description of equipment used in a survey is the first step in designing a trawl survey DBMS. This includes a general description of the vessel or boat used. The items usually included here are:

Vessel Particulars

- name of the vessel;
- total length;
- gross tonnage;
- name and horsepower of the main engine; and
- cruising speed.

Gear Particulars

The trawl gear is identified by:

- length of the headrope;
- circumference;
- length of footrope;
- headline width and height;
- codend mesh size;
- bottom gear;
- otter boards (type, size);
- bridles (length); and
- estimated mean door distance.

Survey Particulars

A cruise logbook is also accomplished before and after a trip, usually with the following items:

- cruise number (related to a project);
- positioning system (geographic positioning system, or GPS, radar, etc.)
- position accuracy;
- log system (surface referenced, bottom referenced);
- names and watch duties;
- head of fishing operation;
- trawl-monitoring system with list of sensors;
- sampling scheme (random, stratified random, adaptive);

- universal time convention, or UTC, for onset and offset of daylight;
- dates of sailing and return to port; and
- time and position of important events along cruise tracks (course change, conductivity-temperature-depth, or CTD, stations, etc.).

Station Particulars

At each trawling station, the following items are recorded:

- station number (or ID);
- date;
- purpose code (biomass estimation, experimental station, etc.);
- starting and ending time (UTC);
- starting and ending positions (latitude and longitude);
- starting and ending fishing depths;
- starting and ending bottom depths;
- starting and ending time of trawling;
- gear damage code;
- haul validity code; and
- quality control code.

In some applications, the ending positions are not recorded, i.e., they are replaced by the towing direction and speed. The final position can be computed using a simple geometric relationship (Box 1) and may be plotted (Box 2).

Box 1. Estimating final position, given initial position, speed and bearing.

Assuming a constant towing speed and that the effects of the surface current and wind speed are negligible, the following relationship may be used to estimate the final position of a trawler.

If we let

X_{orig} = starting latitude position
X_{final} = final latitude position
Y_{orig} = starting longitude position
Y_{final} = final longitude position
S_{towing} = towing speed
T_{start} = time trawling started
T_{end} = time trawling ended
T_{total} = total trawling time $(=T_{end} - T_{start})$
D = distance traveled $(=S_{towing} \cdot T_{total})$
d_{towing} = towing direction (in degrees)

then

$X_{final} = X_{orig} + D \sin d_{towing}$
$Y_{final} = Y_{orig} + D \cos d_{towing}$

Wind speed and direction have a negligible effect on the direction of the vessel during normal trawling operations. On the other hand, water currents do affect vessels and the geometry of the trawl gear. In

Box 2. Mapping trawl surveys[a]

Ideally, a trawl-data management system should be linked with a Geographic Information System (GIS). However, GIS often requires users to have undergone a long training and the resulting applications usually cannot be distributed for copyright reasons.

A geographic component is therefore proposed for TrawlBase, consisting of a preprogrammed set of maps, not requiring user inputs, which can be distributed without restriction. This is based on WINMAP, the successor of MAPPER (Coronado and Froese 1994), developed as a distributable component of FishBase (see text).

As presently designed, the mapping module of TrawlBase is designed to overlay station points off the coast of a selected country.

Generating the plot is done in three steps:
1. Read the position defining the stations (latitude, longitude; use means for stations defined by two positions) for a selected set of stations.
2. Convert the position to X-Y coordinates and save these as a formatted text file (*.TXT).
3. Use WINMAP to plot the country and stations (as solid circles).

The figure in this box presents a facsimile of a map of Sri Lanka and 20 hypothetical stations.

Further developments of this module may include making the points sensitive to clicking by a mouse, thus enabling users to access station-specific information, and performing geostatistical analyses of the station data (e.g., Kriging).

Acknowledgments

Many thanks to Mr. E. Garnace and Mr. N. Quirit for their valuable technical support in the development of the module.

References

Coronado, G.U. and R. Froese. 1994. MAPPER: a low-level geographic information system. ICLARM Software 9, 23p.

- A.B.A.K. Gunaratne[b] and F.C. Gayanilo, Jr.[c]

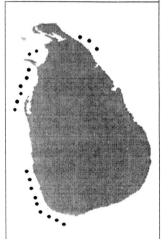

[a] Developed during the training course of Mr. Gunaratne at ICLARM, Manila, 2-29 August 1996. The trip was funded by the World Bank through the Agriculture Research Project, SMI Building, Colombo 3, Sri Lanka.
[b] National Aquatic Resources Agency, Crow Island, Colombo 15, Sri Lanka.
[c] ICLARM, e-mail: f.gayanilo@cgnet.com.

Box 3. Accounting for the effects of surface current when estimating the distance covered by a trawl.

Let W_d be the direction of the water current, W_s the speed of the surface current and other parameters as defined in Box 1, then the distance (D) from the original position is,

$$D = [(W_s)^2 + (S_{towing})^2 + 2 \cdot W_s \cdot S_{towing} \cdot \cos (d_{towing} - W_d)]^{0.5}$$

The distance covered (D) by the vessel can also be computed when X_{final}, X_{orig}, Y_{final} and Y_{orig} are given as follows:

$$D = 60 \cdot ([(X_{final} - X_{orig})^2 + (Y_{final} - Y_{orig})^2 \cdot \cos^2 ((X_{final} + X_{orig})/2)]^{0.5}$$

- reference number to CTD, plankton or grab stations, if taken;
- wind speed (Box 4);
- wind direction;
- current speed and direction;
- bottom type (Box 5);
- sea-surface temperature (SST);
- secchi disc reading;
- salinity;
- dissolved oxygen at bottom;
- trawling speed referred to sea bottom;
- distance towed referred to sea bottom;
- towing warp length (or wire out);
- towing course; and
- area of influence (i.e., area of the polygon surrounding a station, wherein all points of the polygon are closer to the station than to any other neighboring stations).

Catch Particulars

Following the entry of station-related information, the catch is recorded. This includes the following key information:

cases where both the surface current and its direction are given, the resulting distance is best computed using a simple vector analysis with the assumption that the frictional forces are negligible (Box 3).

Other information usually collected at the trawl station includes:

- name of taxon (preferably species name; genus or higher when specific identification is not possible);
- weight of catch (by species or higher groupings);
- numbers caught (by species or higher groupings);
- reference number or tag indicating whether a length sample was taken; and
- reference number or tag indicating whether biological samples or specimens were taken.

For target species, length-frequency samples are usually taken (and weighted), with the length frequencies (total length, fork length or standard length) usually grouped by constant class intervals. Ungrouped length records are sometimes collected, along with individual weights, for some target fish. In some surveys the sex, maturity stage and stomach contents of some of the fish caught are also recorded.

Box 4. Recording wind observations.

Wind speed and direction can be observed through the use of a wind vane or an electronic instrument. Rough estimates of wind speeds can be recorded using the Beaufort scale.

Beaufort scale, as used to record wind speed (Branson 1987):

Beaufort scale	Wind speed (knots)	Descriptive term	Ave. height of waves (m)	Max. height of waves (m)
0	<1	Calm	-	-
1	1 - 3	Light air	0.1	0.1
2	4 - 6	Light breeze	0.2	0.3
3	7 - 10	Gentle breeze	0.6	1.0
4	11 - 16	Moderate breeze	1.0	1.5
5	17 - 21	Fresh breeze	2.0	2.5
6	22 - 27	Strong breeze	3.0	4.0
7	28 - 33	Near gale	4.0	5.5
8	34 - 40	Gale	5.5	7.5
9	41 - 47	Strong gale	7.0	10.0
10	48 - 55	Storm	9.0	12.5
11	56 - 63	Violent storm	11.5	16.0
12	>64	Hurricane	>14.0	-

Box 5. Recording bottom types.

Bottom types can be assessed using a core sampler or a Petersen's grab sampler or by examining the underside of the doors of a trawl net. There are qualitative and quantitative methods of classifying bottom type. One useful classification of bottom types is the Wentworth grade scale.

Wentworth grade scale (English et al. 1994):

Name	Grade Limits (mm)
Boulder	>256
Cobble	256 - 64
Pebble	64 - 4
Granule	4 - 2
Very coarse sand	2 - 1
Coarse sand	1 - 0.5
Medium sand	0.5 - 0.25
Fine sand	0.25 - 0.125
Very fine sand	0.125 - 0.062
Silt	0.062 - 0.0039
Clay	<0.0039

A frequently used maturation scale consists of these stages: immature, developing, active, developed, gravid, ripe-running or spent (Munro and Thompson 1983).

Data Storage

When laying down the specifications for a trawl-survey database, the amount of data that will have to be recorded must be estimated.

To avoid the need for the user to enter vast numbers of characters (hence minimize encoding errors and conserve storage space), codes were utilized in the early 1970s for most elements of a trawl survey (vessel and gear type, cruise, station, and taxa caught). The use of codes is obsolete, with recent advances in computer technology and new interfaces involving the use of choice fields. However, a generic trawl-survey DBMS should support the use of codes in important routines to allow incorporation and transfer of older data sets. Codes are still used in modern computer programs to classify and sort data sets. However, this occurs 'behind the scene' and is of no direct concern to users of the software.

120

It is also important that the database is organized so it can work on subsets of data. A subset could be one survey or a group of surveys, forming what in NAN-SIS is called a 'project' and organized in separate files or group files. The grouping could make use of directories or filename suffixes, as in NAN-SIS. This will facilitate data maintenance and exchange. The analytical programs in the package must be able to pool 'projects', to enable comparative and time series analyses.

Fig. 1 summarizes the concepts presented above, which can be straightforwardly realized using commercially available relational database systems such as Foxbase, Visual FoxPro, Oracle, Visual dBase, PowerBuilder or Microsoft Access. The choice of the system is not very important and any of the commercially available database development systems can be used since the data are either alphanumeric or numeric. The relationship among tables is so simple that even first-generation database systems such as dBase II (once distributed by Ashton-Tate Co., USA)

can easily be configured to fit the main requirements. Moreover, most of the commercially available database systems include utilities to export and import data from other database systems, making problems of intersoftware communication a thing of the past (Box 6).

The choice of a database system to use is more a question of maintenance. The DBMS should be developed using a system that guarantees both the long-term availability of knowledgeable programmers and a commitment of the company that initially developed the database system to continue to support and upgrade it.

Data Analysis

Storing the data from a trawl survey is only the first step. To be useful, the data must be interpreted, analyzed or presented in some other form. The most useful analysis, and the one most commonly required, is the estimation of biomass or 'standing stock'.

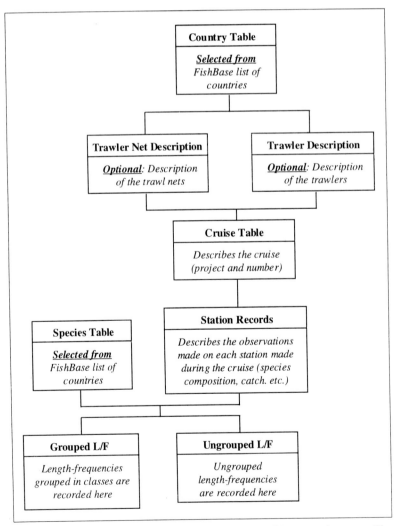

Fig. 1. Schematic representation of the data gathered in a trawl survey. The relationship is one-is-to-many for all the data.

T. Strømme[a]

---

[a] E-mail: tore@imr.no or in writing to T. Strømme, The Nansen Programme, Institute of Marine Science, PO 1870, 5024 Bergen, Norway.

Box 6. Towards a common fisheries resources survey information system.

In monitoring and assessing fisheries resources, surveys play an essential role. Various generic options for storing and processing survey data are available, but often institutions develop in-house applications tailored to specific needs. In the worst but not uncommon case, data storage is left to the responsibility of the individual scientists, leading to formats that make data exchange even within the institution complicated. Another problem is that well-developed systems may wither away when contracts with computer companies expire, maintenance funds go dry or scientists and/or computer experts quit the institution.

In the light of recent developments in computer systems and database applications, the creation of a Common Fisheries Data Format (CFDF) is proposed to accompany a toolbox of analytical applications that can be composed of a menu-driven shell system tailored for the specific needs of institutions.

By agreeing on a common format for data storage, exchange of data is facilitated and scientists and programmers are invited to contribute to the development of a common box of analytical tools which will be open and transparent to all. Rules for crediting a subprogram to an author/developer should be established and it should be adopted to a common toolbox that will be available on the Internet following a peer review procedure. It is envisaged that after the initial phase, such an arrangement will be self-perpetuating and reinforcing, and the ownership will belong to the scientific community at large, free of commercial interests.

If the CFDF meets acceptance, the Dr. Fridtjoft Nansen Project will port the main functions of the NAN-SIS software package to the new environment.

By transfering historic data to CFDF, global analyses in fisheries ecology will be facilitated as well. It is also suggested that this procedure be followed when building a common database and analytical programs for the Southern African BENEFIT Project.

Scientists, programmers or others who would support the idea of a common format for survey data are requested to contact the author in order to create an Internet forum for development of the idea.

Biomass Estimation

The biomass (B) of demersal fishes can be estimated using the "swept-area" method (Pauly 1984a, 1984b). It is defined by the equation:

$$B = \frac{\overline{C/f} \cdot A}{a \cdot X_1} \quad \dots 1)$$

where C/f is the mean catch per unit effort; A is the total area covered by the survey; X_1 is the escapement factor, i.e., the fraction of the fish in the path of the trawl that was actually retained; and 'a' is the area swept by the trawl in one unit of effort (usually measured in one hour). The swept area is defined by:

$$a = t \cdot v \cdot h \cdot X_2 \quad \dots 2)$$

where 'v' is the trawling speed; 'h' is the length of the trawl's headrope; 't' is the duration of trawling; and X_2 is the effective width of the trawl relative to the length of its headrope. Commonly used values of X_2 range from as low as 0.4 (SCSP 1978) to as high as 0.66 (Shindo 1973), with a value of 0.5 probably being the best compromise (Pauly 1980). Similarly, the value of X_1 in Equation 1, usually ranges from 0.5 (Pauly 1980) to 1.0 (Gulland 1979). Few studies have been made to actually estimate values of X_1 and X_2, even though they drastically affect the standing stock estimates. Pauly (1980) showed that $X_1 = 0.5$ and $X_2 = 0.5$ generated fishing mortalities compatible with those estimated for various species in the Gulf of Thailand.

To increase the precision of the biomass estimate, a survey area can be divided into several *strata*. For stratified random sampling, Equation 1 is re-expressed as:

$$B = \frac{\sum_{i=1}^{n} A_i}{X_1 \cdot n} \cdot \sum_{i=1}^{n} \left[\left(\overline{C/f} \right) / a \right] \quad \dots 3)$$

where 'n' is the number of strata considered in the analysis and 'i' is the stratum. The variance of B can be computed based on Equation 3 as follows:

$$Var(B) = \left(\frac{\sum A_i}{X_1} \right)^2 \cdot \frac{1}{n(n-1)} \cdot \sum_{i=1}^{n} \left[(\overline{C/f})_i - \left(\frac{1}{n} \cdot \sum_{i=1}^{n} (\overline{C/f})_i \right) \right]^2 \quad \dots 4)$$

Alternatives to Equation 4 exist (Box 7). One of these assumes the catch/effort data to be log-normally distributed. In this case, an arbitrary value of 1 is added to each catch/effort value (before taking logs) to allow for analysis of data sets with zero entries (as frequently occurs when analyzing catch/effort data pertaining to a single taxon rather than to the total catch). Also, the "D-distribution" (Aitchinson 1955) can be used, which explicitly accounts for the occurrence of zeroes and of rare but extremely high catches (Smith 1981; Pennington 1983).

As an alternative to Equation 4 and its variants, Monte-Carlo analysis can be used to estimate the variance of the biomass estimate from trawl-survey data.

The advantage of this method is that it allows consideration of the uncertainty in the value of all parameters used in estimating biomass, rather than considering only the variance of C/f (Box 8).

Box 7. Non-normal distributions and the estimation of biomass.

Estimating biomass when there are lots of small catches and a few very large ones can be done using a log-transformation of the catches, as follows (Som 1973; Cochran 1977; Pauly 1984b):

Step 1. Compute the catch per-hour $(c/f)_j$ per stratum (j). Add 1 to all catches if hauls with zero catch occurs.

Step 2. Take the natural logarithm (ln) of each value $\left(Xi,j = \ln((c/f)_j)\right)$ in Step 1.

Step 3. Compute the uncorrected geometric mean $\left((\dot{c}/f)_j\right)$

$$\left(\dot{c}/f\right)_j = \exp\left(\sum_{i=1}^{n} X_{i,j}\right);$$ where n is the number of hauls in a stratum.

Step 4.   Subtract the value of 1 previously added to obtain partially corrected mean

$$\left((\ddot{c}/f)_j = (\dot{c}/f)_j - 1\right).$$

Step 5. Compute the corrected mean catch per unit of effort $(\bar{c}/f)$

$$\left(\bar{c}/f\right)_j = \left(\ddot{c}/f\right)_j \cdot \exp\left(\mathrm{Var}\left(x_{i,j}\right)\right)^2 \text{ where Var}\left(X_{i,j}\right) = \sum X_{i,j}^{2}/(n-1)$$

Step 6. Estimate the variance of the corrected mean $\left(\mathrm{Var}((\bar{c}/f)_j)\right)$

$$\mathrm{Var}\left((\bar{c}/f)_j\right) \approx \left(\bar{c}/f\right)_j^{2} \cdot \exp\left(\mathrm{Var}(X_{i,j})\right) \cdot \left((\exp(\mathrm{Var}(X_{i,j}))) - 1\right)/n$$

Note: Step 1 to 6 should be done for each stratum separately.

Step 7. Compute the biomass (B$_j$) per stratum

$$B_j = q_j \cdot \left(\bar{c}/f\right)_j \quad \text{where } q_j = 2A/a_j \text{ (A and a}_j \text{ are as defined in Eq. 1)}$$

Step 8. Compute the overall biomass per stratum.

$$B_{total} = \sum B_j$$

Step 9. Compute the standard error of the overall biomass (B$_{total}$).

$$\mathrm{s.e.}\left(B_{total}\right) = \sqrt{\sum\left(\mathrm{Var}((\bar{c}/f)_j) \cdot q_j^{2}\right)}$$

The degrees of freedom (d.f) is the number of all the hauls for all strata minus the number of strata used in the analysis.

Box 8. A Monte-Carlo approach for estimating the variance of the biomass estimates.

The procedure outlined in Box 7 does not consider the uncertainties associated with the estimates of parameters such as the trawling speed, the escapement factor (X_1) and the effective width of the headrope (X_2). A Monte-Carlo approach can be used to obtain a picture of the confidence region of the biomass estimate.

Using a 'triangular' probability function given the lower limit (i.e., smallest acceptable value), the upper limit (the largest acceptable value), and the 'best' estimate (the value found most credible), and drawing several random values from each of the probability functions, a series of biomass estimates can be computed using the procedure in Box 7. The estimates obtained can be plotted in the form of a histogram from which an estimate of the variance of the biomass can be obtained.

The 'triangular' probability function may not necessarily be symmetrical (i.e., the 'best' estimate does not have to be the midpoint between the lower and upper limits). Theoretically, a log-normal distribution of the biomass estimates is expected when drawing a large number (e.g., 10 000 runs) of random samples.

Other Analyses

Trawl surveys are usually conducted in highly heterogenous environments, with distinct fish assemblages associated with a variety of identifiable habitats (McManus 1985, this volume., Bianchi 1996). Several classification techniques can be used to define such assemblages and their relationships to their habitat. Two-way indicator analysis (TWIA, Hill 1979) is widely used for this (McManus 1985, 1986; Bianchi 1991, 1996). Several software like PRIMER (Clarke and Warwick 1994), CANOCO (ter Braak 1991), TWINSPAN (Hill 1979) and other commercially available statistical packages can also execute the required routines. However, unlike biomass estimation, which can easily be programmed as part of a DBMS, the data required for community analysis are best exported to formats that can be read by specialized software.

Often, these independent software will contain other functions such as ordination methods, e.g., DCA (detrended correspondence analysis) and MDS (multidimensional scaling analysis), to further support the results of the community analyses. On the other hand, approaches designed to measure ecological stresses (Pielou 1977; Warwick 1986; McManus and Pauly 1990) which can use trawl survey data can be made a part of the DBMS as they require relatively little programming.

When length frequencies are available, they can be used to estimate growth and mortality parameters useful for predicting the behavior of the stock with changes in fishing pressure through variants of the models of Beverton and Holt (1957) and Thompson and Bell (1934), as incorporated in the FiSAT software (Gayanilo et al. 1996). Thus, exporting length-frequency data to FiSAT is a straightforward way for a trawl-survey DBMS to accommodate length-frequency data.

The Prototype

The following presents the prototype of a generic DBMS for trawl surveys called TrawlBase. The prototype is structured around the principles discussed above.

System Requirements

TrawlBase was developed for use with microcomputers running on Microsoft Windows 3.x (or Windows 95 and NT). It requires a minimum of 4 megabytes of free disk space. Although the software will run on 4 megabytes of memory (RAM), 16 megabytes is highly recommended.

Installation

A SETUP.EXE is available on the first disk (Disk No. 1) to install TrawlBase on a computer. It may also be installed on a network server, but this should have control functions to prevent two users from updating the database at the same time. SETUP.EXE creates a group in the Program Manager with an icon which can be used to activate the program.

Main Menu

The first screen display is the *Main Menu* containing five command buttons: (1) *Trawl Survey*, (2) *View Species List*, (3) *View Country List*, (4) *Import / Export Data* and (5) *Exit* (Fig. 2). The *Trawl Survey* command button will open the form that allows the user to enter and retrieve survey data and estimate biomasses. The *View Species List* and *View Country List* buttons open forms that display information imported from FishBase 97 (Froese and Pauly 1997). The *Import / Export Data* command button opens forms that allow users to export data to a LOTUS 1-2-3, Microsoft Excel or FiSAT data formats, and to import NAN-SIS data files.

Trawl Survey

This command button opens the main form of TrawlBase (Fig. 3), and the default view is the first record in the database. There are 12 command buttons on the lower portion of the working screen. The first of the two command buttons allow *Biomass* estimation, based on the approaches presented above. Immediately following the *Biomass* command buttons are two reserved buttons: *EcoStress*, for estimating indices of ecological stresses as briefly mentioned above; and *Plot*, for plotting the locations of the stations, by survey, in the form of a simple map. The next four command buttons are (from left to right) *Print Record*, *Add a Record*, *Delete Record* and *Find a Record*. Sample table outputs under the *Print Record* command — typical of outputs frequently required from trawl surveys — are given in Tables 1 to 5. The next four buttons are used to scroll through the database, and the last is to close the form and return to the *Main Menu*.

Data entry starts by clicking the *Add a Record* command button. The entry form is displayed, and the user will be asked to provide values for the fields. The name for the country where the data were collected is the first input. The country list imported from FishBase 97 (Froese and Pauly 1997) is used for this purpose and can be modified or added to. The command button at the right of the country name may be used to view key information on the country selected.

The gear code is the next input. If the code is not on the list, the program will automatically open the entry form for gear description. The command button located on the right of the gear name may also be used to add a gear record. The entry form to describe the gear (Fig. 4) includes data that describe the trawl. The data enable the software to estimate the biomasses based on the swept-are method described above.

The trawler used during the survey is the next input.The command button on the right of trawler's name opens the form to record data about the trawler (Fig. 5). The next input is the cruise number. The command button on the right of the cruise's description opens the form to record details concerning the cruise (Fig. 6). The station number is the next input, and the command button on the right of the sampling date opens the form to enter station-related data (Fig. 7a and 7b).

The next set of inputs are the catches. These include the name of the taxon, the weight of the catch, and the numbers caught. When the species name entered is not in the TrawlBase list (imported from FishBase 97), the program will search the database for synonyms (also from FishBase 97). It will display the valid name if found, or alert the user to a possible error. It will accept the entry, nonetheless, if the user chooses to proceed with the name entered (this allows

the entry of names of invertebrates, not included in FishBase).

When length frequencies are available, two buttons can be used to open the form designed for these data. For grouped length frequencies (as in NAN-SIS), the class interval and the smallest (lower) class limit are the first required inputs, along with the length type (Fig. 8). When available, sample weights may also be entered. Ungrouped length measurements these may be entered along with individual weight (in g), sex (male, female or unknown), as well as sexual maturity code (Fig. 9).

The length frequencies (grouped and ungrouped) can be readily exported to FiSAT (Gayanilo et al. 1996). FiSAT allows for growth, mortality and related parameters to be estimated from these data, and for yield-per-recruit and other analyses to be performed based on the parameters estimated.

View Species List

This command button opens the form containing species information imported from FishBase 97 (Fig. 10). This cannot be modified by users (but will be periodically updated using new releases of FishBase). A click in any of the fields opens a dialog box allowing users to search the database.

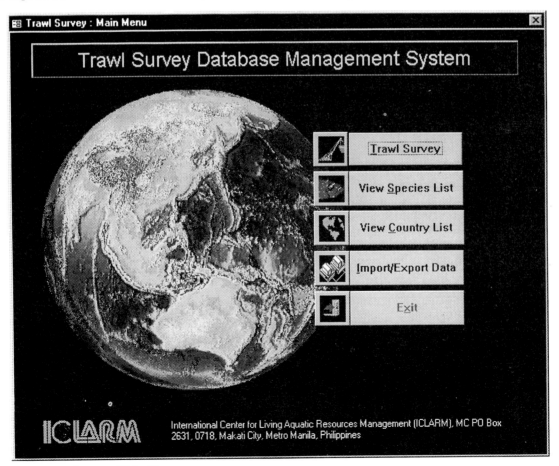

Fig. 2. Trawl survey *Main Menu.*

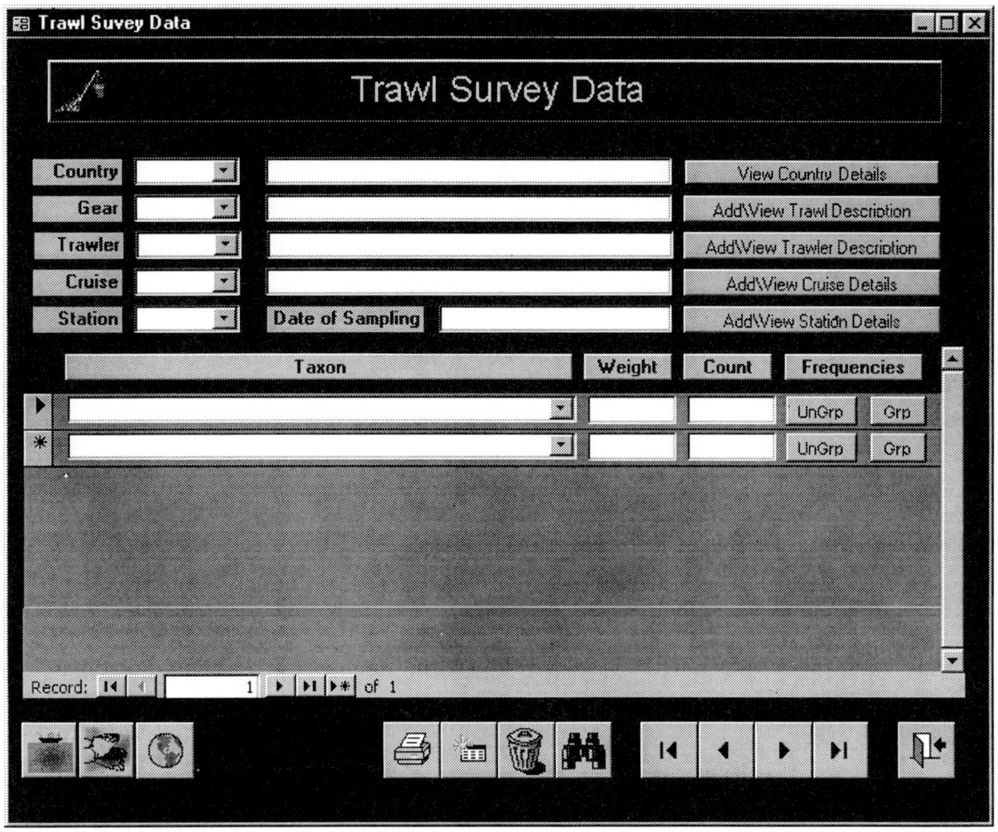

Fig. 3. Main form of TrawlBase.

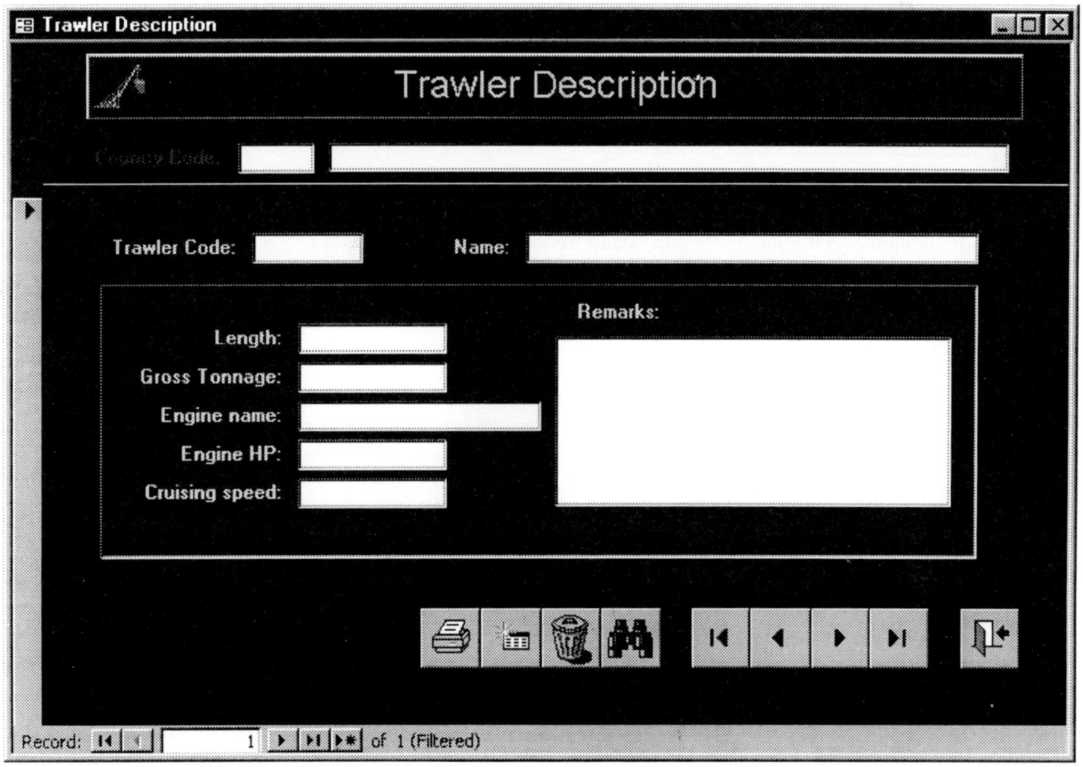

Fig. 4. Entry form to describe the trawl net.

126

Table 1. Table output summarizing the catches at one station.

Country   : _____
Trawler   : _____
Trawl net   : _____
Cruise   : _____
Station   : _____
Date of Sampling   : _____

Species name	Weight (kg)	Count (n)
Species 1		
Species 2		
Species 3		
....		
....		
....		
Species n		
Total		

Table 2. Summary output for grouped length-frequencies by country and project.

Country   : _____
Trawler   : _____
Trawl net   : _____
Cruise   : _____
Species name   : _____
Length unit   : _____

Length Class	Date 1	Date 2	Date 3	.	.	.	.	.	.	.	Date n	Total
Total												

Table 3. Table output summarizing the inputs for ungrouped frequencies by country and project.

Country   : _____
Trawler   : _____
Trawl net   : _____
Cruise   : _____
Species name   : _____

Species name	Length (cm)			Weight	Sex	Maturity stage
	Total	Fork	Standard			
Species 1						
Species 2						
Species 3						
....						
....						
Species n						

Total weight _____

Table 4. Table output of biomass analysis. The definition of the strata (here by depth in m) and the breakdown for the catch/effort data are user-defined.

Species name	Catch per unit effort (kg/hr)						% incidence	Mean density (t/m²)	Mean density by stratum (t/m²)				
	<30	30-90	90-300	300-900	900-3000	>3000			<50m	50-90m	100-199m	200-399m	≥400
Species 1													
Species 2													
Species 3													
....													
....													
....													
Species n													
Sum													

Table 5. Table output summarizing the catches for all stations by species and by cruise for a particular country.

Country : _____
Cruise : _____

Species/Station	01	02	03	04	05	06	07	08	09	.	.	.	.	n	Total
Species 1															
Species 2											.				
Species 3															
....															
....															
....															
Species n															
Total															

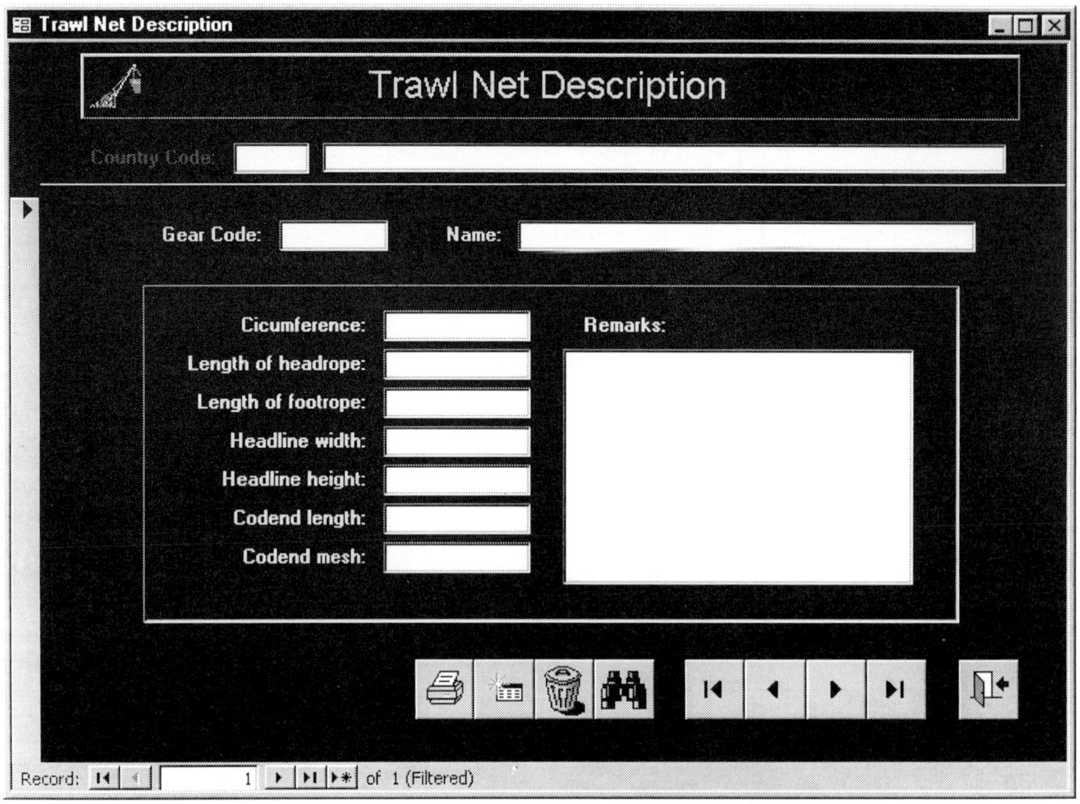

Fig. 5. Entry form to describe the trawler.

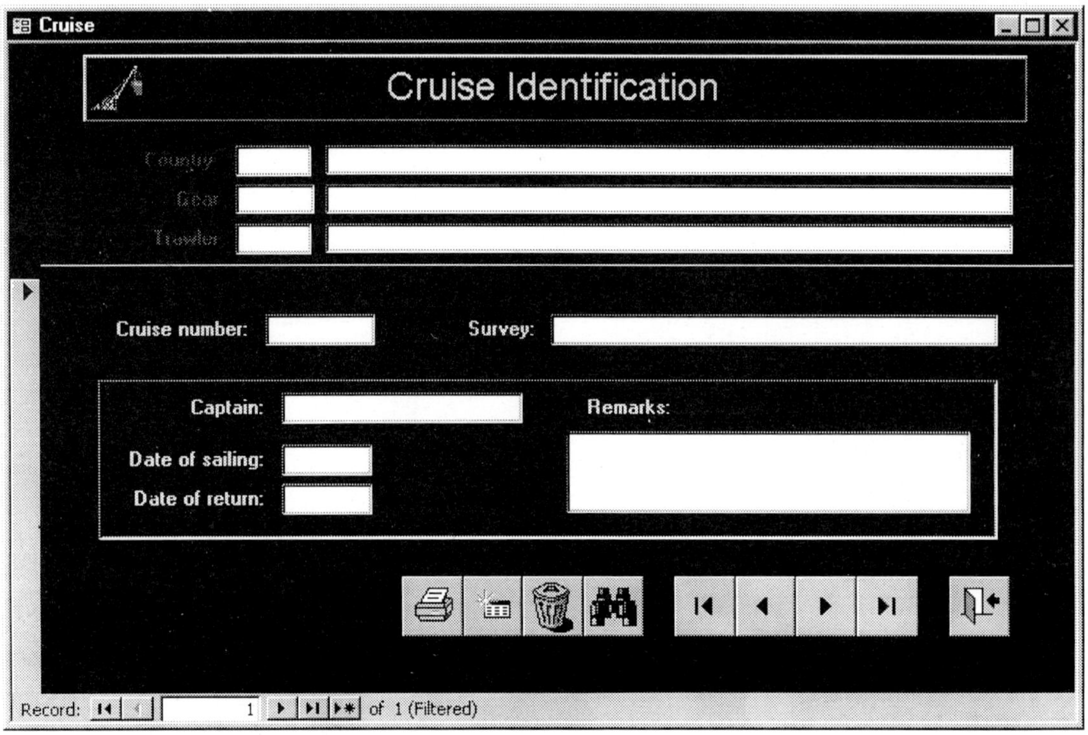

Fig.6. Entry form for the cruise's description.

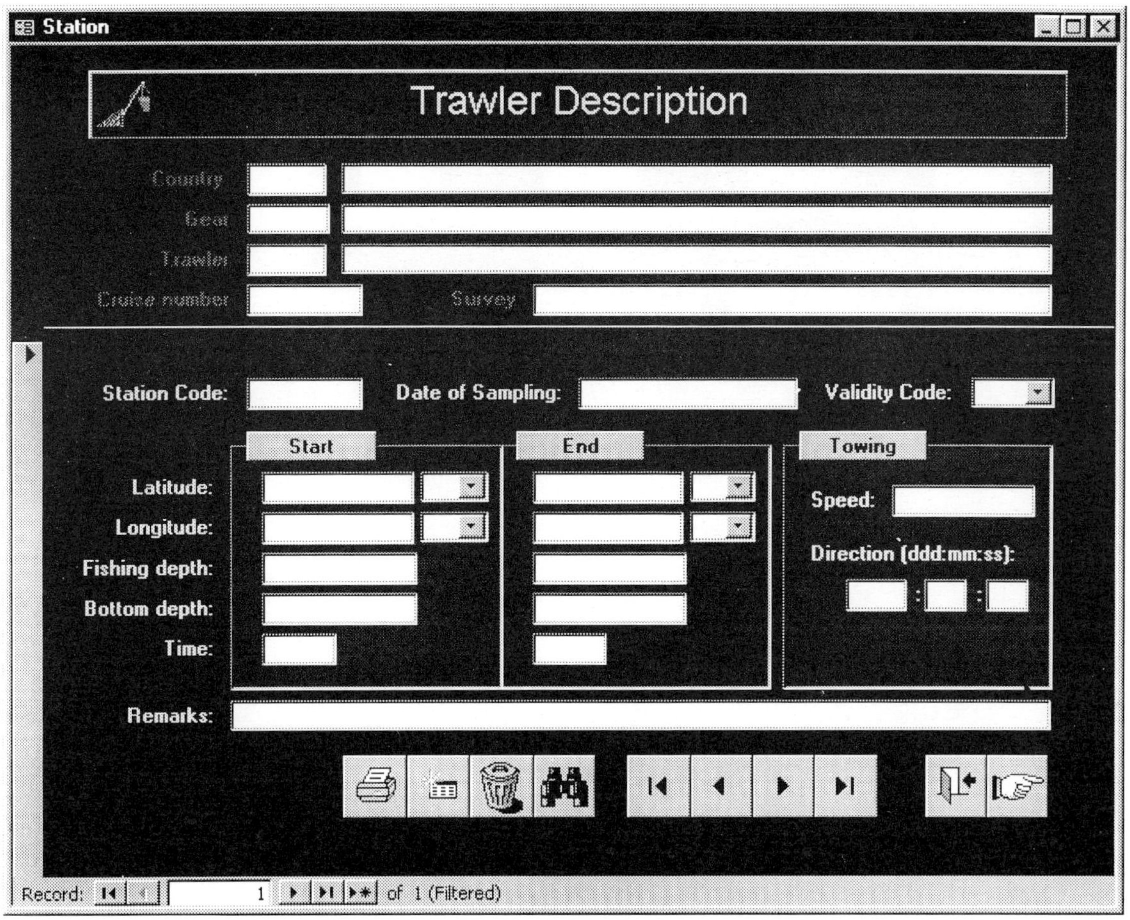

Fig. 7a. Entry form for station-related data.

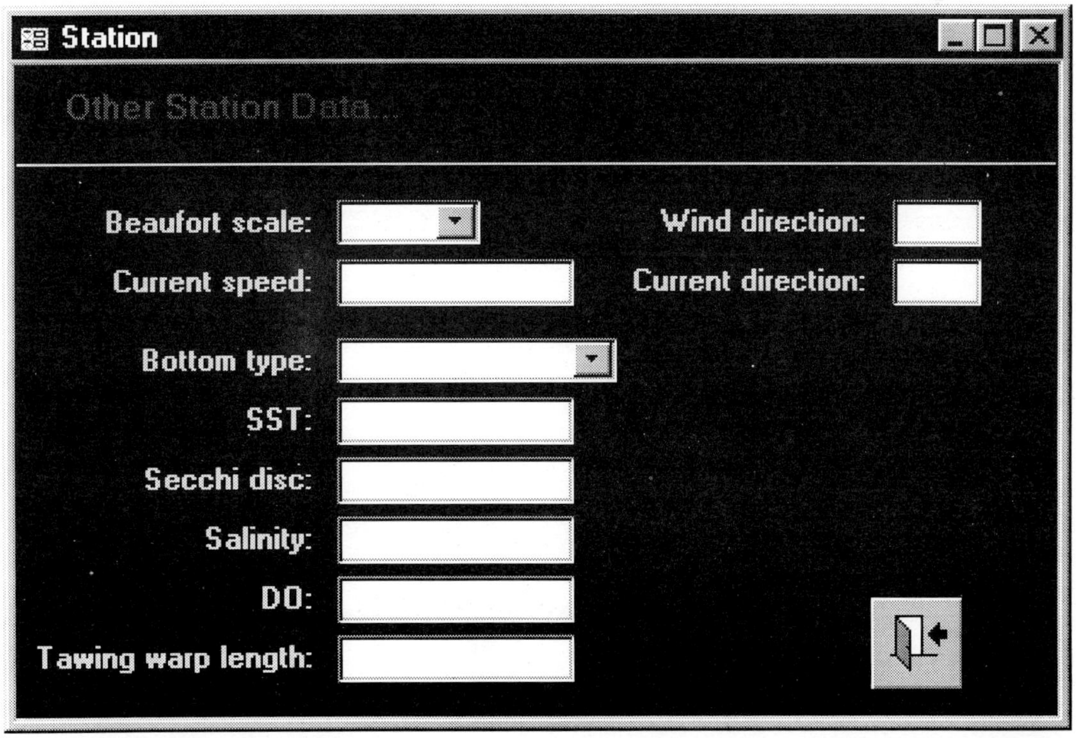

Fig.7b. Subform for station-related data.

130

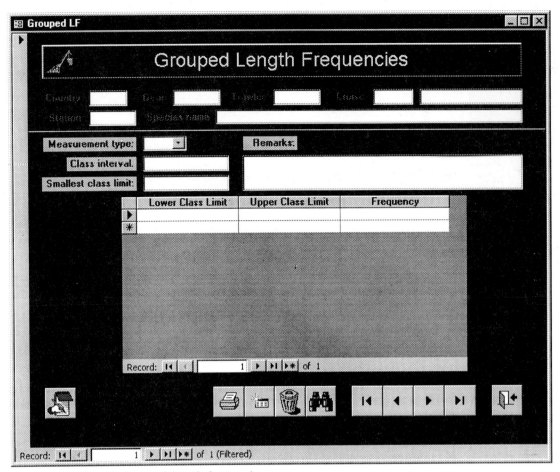

Fig. 8. Entry form for grouped length frequencies.

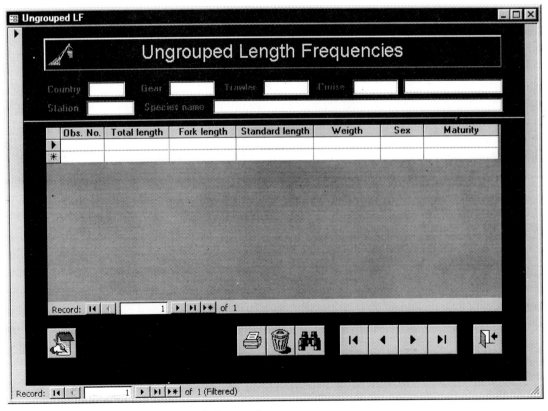

Fig. 9. Entry form for ungrouped length frequencies.

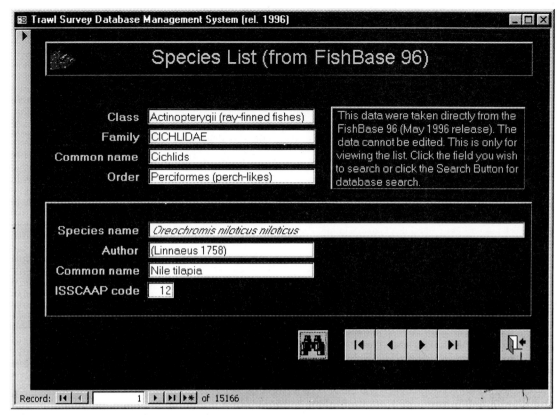

Fig.10. The View form containing species information imported from FishBase 97.

View Country List

This command button displays key country information, to be updated with each new release of FishBase. Clicking any of the fields opens a dialog box that allows users to search the database.

Import/Export Data

The *Import/Export Data* routine allows users to import *NAN-SIS* data files, or to export any part of a table (or a combination of tables) in *TrawlBase* to a spreadsheet file. The difference between this routine and the routine available in the *Trawl Survey* forms is that users are not limited to exporting data for community analyses.

The following are the available options when exporting a data file:

1. Microsoft Excel format (ver. 2.0 to 4.0, or 5.0);
2. LOTUS (wk1);
3. LOTUS (wks);
4. LOTUS (wk3); and
5. FiSAT (ver. 1.10)

Acknowledgments

The authors thank Mr. E. Garnace for assistance in developing the NAN-SIS import routine and Mr. F. Torres, Jr. for initial testing of the prototype.

References

Aitchinson, J. 1955. On the distribution of a positive random variable having a discrete probability mass origin. J. Am. Stat. Assoc. 50: 901-908.

Beverton, R.J.H. and S.J. Holt. 1957. On the dynamics of exploited fish populations. Fish. Invest. London Ser. 2 (19): 533 p.

Bianchi, G. 1991. Demersal assemblages of the continental shelf and slope edge between the gulf of Tehuantepec (Mexico) and the Gulf of Papayago (Costa Rica). Mar. Ecol. (Prog. Ser.) 81: 101-120.

Bianchi, G. 1996. Demersal fish assemblages of trawlable grounds off northwest Sumatra, p.123-130. *In* D. Pauly and P. Martosubroto (eds.) Baseline studies of biodiversity: the fish resources of Western Indonesia. ICLARM Stud. Rev. 23, 312 p.

Branson, P. 1987. Fishermen's handbook. Fishing News Books Ltd., Farnham, Surrey, England.

132

Clarke, K.R. and R.M. Warwick. 1994. Change in marine communities: an approach to statistical analysis and interpretation. Plymouth Marine Laboratory, UK.

Cochran, W.G. 1977. Sampling techniques (3rd edition). John Wiley and Sons, Inc., New York.

Coronado, G.U. and R. Froese. 1994. MAPPER: a low-level geographic information system. ICLARM Software 9. 23 p.

Dates, C.J. 1981. An introduction to database systems (3rd edition). Addison-Wesley Pub., Reading, Massachusetts, USA.

English, S., C. Wilkinson and V. Baker (eds.) 1994. Survey manual for tropical marine resources. Australian Institute of Marine Science. Townsville, Qld., Australia.

Froese, R. and D. Pauly (eds.) 1997. FishBase 97: concepts, design and data sources. ICLARM, Manila.

Gayanilo, F.C., Jr., P. Sparre and D. Pauly. 1996. The FAO-ICLARM stock assessment tools (FiSAT) user's guide. FAO Comput. Info. Ser. (Fish.) No. 8. FAO, Rome. 126 p.

Gulland, J.A. 1979. Report of the FAO/IOP workshop on the fishery resources of the Western Indian Ocean South of the Equator. IOFC/DEV/79/45. FAO, Rome.

Hill, M.O. 1979. TWINSPAN – a FORTRAN program for arranging multivariate data in an ordered two-way table by classification of individuals and attributes. Cornell University, Ithaca, New York.

McManus, J.W. 1985. Descriptive community dynamics: background and an application to tropical fishes management. University of Rhode Island. Narragansett, R.I. 217 p. Ph.D. dissertation.

McManus, J.W. 1986. Depth zonation in a demersal fishery in Samar Sea, Philippines, p. 483-486. In J.L. Maclean, L.B. Dizon and L.V. Hosillos (eds.) The First Asian Fisheries Forum. Asian Fish. Soc., Manila.

McManus, J.W. and D. Pauly. 1990. Measuring ecological stress: variations on a theme by R.M. Warwick. Mar. Biol. 106: 305-308.

McManus, J.W. and M.C.A. Ablan (eds.) 1996. ReefBase: a global database on coral reefs and their resources. ICLARM, Manila. 150 p.

Munro, J.L. and R. Thompson. 1983. Areas investigated, objectives and methodology, p. 15-25. In J.L. Munro (ed.) Carribean coral reef fishery resources. ICLARM Stud. Rev. 7, 276 p.

Pauly, D. 1980. A new methodology for rapidly acquiring basic information on tropical fish stocks: growth, mortality and stock-recruitment relationships, p. 154-172. In S. Saila and P. Roedel (eds.) Stock assessment for tropical small-scale fisheries. Proc. of a workshop held 19-21 September 1979 at the University of Rhode Island. Int. Cent. Mar. Res. Dev., University of Rhode Island, Kingston, R.I.

Pauly, D. 1984a. Methods for assessing the marine stocks of Burma, with emphasis on the demersal species. BUR/77/003. FAO Field Doc. 6. FAO, Rome. 22 p.

Pauly, D. 1984b. Fish population dynamics in tropical waters: a manual for use with programmable calculators. ICLARM Stud. Rev. 8, 325 p.

Pauly, D. 1996. Biodiversity and the retrospective analysis of demersal trawl surveys: a programmatic approach, p. 1-6. In D. Pauly and P. Martosubroto (eds.) Baseline studies of biodiversity: the fish resources of Western Indonesia. ICLARM Stud. Rev. 23, 312 p.

Pennington, M. 1983. Efficient estimator of abundance for fish and plankton surveys. Biometrics 39: 281-286.

Pielou, E.C. 1977. Mathematical ecology. John Wiley and Sons, New York.

SCSP. 1978. Report of the workshop on the demersal resources of the Sunda Shelf, 31 October-6 November 1977, Penang, Malaysia. Part II. SCS/GEN/77/13. South China Sea Fisheries Development and Coordinating Programme, Manila.

Shindo, S. 1973. General review of the trawl fishery and demersal fish stocks of the South China Sea. FAO Fish. Tech. Pap. 120. 49 p.

Smith, S.J. 1981. A comparison of estimators of location for skewed populations with application to ground fish trawl surveys, p. 154-163. In W.G. Doubleday and D. Rivard (eds.) Bottom trawl surveys. Can. Pub. Fish. Aquat. Sci. 58.

Som, R.K. 1973. A manual of sampling techniques. Heinemann Educational Books, Ltd., London.

Strømme, T. 1992. NAN-SIS: Software for fishery survey data logging and analysis. User's manual. FAO Comput. Info. Ser. (Fish.) No. 4.

ter Braak, C.J.F. 1991. CANOCO – a FORTRAN program for canonical community ordination by (partial) (detrended) (canonical) correspondence analysis, principal components analysis and redundancy analysis (version 3.10) ITI-TNO. Wageningen, the Netherlands. (manual and update notes.)

Thompson, W.F. and F.H. Bell. 1934. Biological statistics of the Pacific halibut fishery. Rep. Int. Fish. (Pacific Halibut) Comm. 8.

Vakily, J.M. 1992. Assessing and managing the marine fish resources of Sierra Leone, West Africa. Naga, ICLARM Q. 15(1): 31-35.

Warwick, R.M. 1986. A method for detecting pollution effects on marine macrobenthic communities. Mar. Biol. 92: 557-562.

Ecological Community Structure Analysis: Applications in Fisheries Management[1]

J. McMANUS

International Center for Living Aquatic Resources Management
MCPO Box 2631, Makati City 0718, Philippines

McMANUS, J.W. 1997. Ecological community structure analysis: applications in fisheries management, p. 133-142. *In* G. Silvestre and D. Pauly (eds.) Status and management of tropical coastal fisheries in Asia. ICLARM Conf. Proc. 53, 208 p.

Abstract

This paper describes some of the analytical options for the application of ecological community analysis to fisheries management. The methods of community analysis are useful in determining the boundaries of assemblages of fish, which may be used as the basis for assigning particular parts of the fishery to specific groups of fishers, gear types and harvest pressures. They can also be useful in identifying the effects of environmental change on a fish community. A very valuable use of these techniques is in determining fundamental patterns of species abundances within harvested ecosystems which can be a basis for identifying the effects of fishing and for reconstructing the nature of the original ecosystems. Through extensive analyses of the dynamic variability of assemblages and their responses to fishing pressure, it may be possible to manage fisheries to optimize the ecosystems for resilience, sustainability and productive yield. An example of the application of community analyses to a trawl fishery in the Philippines is presented.

Introduction

Modern fisheries science evolved principally from early 20th century analyses of the population dynamics of individual species of commercially important fish and invertebrates. The emphasis has generally been on applied population ecology, with extensions to the multispecies case (Pauly 1979; Pauly and Murphy 1982). More recently, emphasis has been placed on system ecological approaches, involving energy, nutrient and trophodynamic analyses (Sherman et. al 1990; Christensen and Pauly 1993). However, a third, intermediary approach has seen only occasional applications in fisheries management, and holds considerable promise as a tool for facilitating the management of multispecies fisheries. Community ecology arose primarily from vegetation ecology, and focuses on identifying patterns in the abundances of species in geographical space, their variations over time, and the causes thereof. The methods of community analysis are particularly useful in determining the boundaries of assemblages of fish that can be considered to be "Assemblage Production Units

(APUs)" (Tyler et al. 1982), and used as the basis for assigning particular parts of the fishery to specific groups of fishers, gear types and harvest pressures. They can also be useful in identifying the effects of environmental change on a fish community, as with pollution or siltation (Shepard et al. 1992). A third, and ultimately very valuable, use of these techniques is in determining fundamental patterns of species abundances within harvested ecosystems as a basis for identifying the effects of fishing and for reconstructing the nature of the original ecosystems. Finally, it may be possible, through extensive analyses of the dynamic variability of assemblages and their responses to fishing pressure, to manage fisheries to optimize the ecosystems for resilience, sustainability and productive yield.

There is a small but growing literature on applications of ecological community analysis to fisheries situations (e.g., Fager and Longhurst 1968; Martosubroto 1982; Ralston and Polovina 1982; Tyler et al. 1982; McManus 1985,1986, 1989 and 1996; Nañola

---

[1] ICLARM Contribution No. 1391.

et al. 1990; Bianchi 1992, 1996; Federizon 1992; Bianchi et al. 1996; McManus et al. 1996). This paper briefly describes some of the analytical options, and ends with an application example to a trawl fishery in the Philippines (to illustrate potential analyses which can result from increased availability of trawl survey data).

Community Structure Analysis

The term 'community structure' refers to patterns in the abundance of individuals, biomasses or importance values among species. Generally, the patterns are studied with respect to two-or occasionally three-dimensional space, based on sampling at a particular time. This type of analysis can clarify the patterns of diversity related to habitat heterogeneity and so may be considered to be a form of beta diversity analysis. An extension to this analysis of a 'static' community structure is the analysis of patterns in the way the community varies over time, or 'community structure dynamics' (McManus 1985).

One can distinguish between inferential and exploratory statistical analysis. Inferential analysis usually involves the choice of one hypothesis among two or more preconceived hypotheses and an assessment of the probability of having made the wrong choice. Exploratory analysis is often thought of as a process of generating hypotheses through various manipulations and simplifications of the data set. Proponents of exploratory analysis often suggest that the analysis should be carried out with no preconceived hypotheses, to avoid limiting the hypotheses which may be generated.

Increasingly, analyses are carried out with the express purpose of identifying patterns in tables of ecological data, wherein the rows represent species, the columns represent sample sites, and each cell contains the abundance of a species in a site. Ideally, the data are generated from a sampling design based on specific hypotheses developed from prior experience as derived from studies of similar ecological communities. The analyses, therefore, are also based on a set of preconceived analyses, although the possibility exists of generating new explanatory hypotheses and finding convincing evidence in favor of the acceptance of a newly generated hypothesis. There are even statistical analyses which can be conducted to determine the probability that an identified pattern or value for an index could have been the result of random data, often involving computer-intensive randomization. This hybrid of inferential and exploratory analyses has been described as 'pattern analysis', and involves the use of an open-ended set of hypotheses (Williams 1976; McManus et al. 1996).

The fundamental analysis of species abundance data with respect to the factors believed to cause it to vary is known as 'direct gradient analysis' (Whittaker 1970). For example, one may graph the abundance of a species of fish against depth or sediment grain size. Statistical analyses may be correlational, perhaps with respect to a distribution model such as a Gaussian frequency curve for cases wherein a species maximum along a gradient is suspected to exist (Gauch 1982).

Direct gradient analysis may not reveal patterns which are best identified through multivariate analysis. Here, we will briefly discuss ordination, classification and skewer analysis.

Ordination

Suppose that a trawl catch involves 10 individuals of species A and 20 of species B. One can plot the position of that haul as a point on a graph wherein the x-axis (for example) is the amount of species A (=10), and the y-axis is the amount of species B (=20). Thus, it is possible to plot sample units in *species space*, i.e., the Cartesian space defined by axes representing abundance by species. A set of such sample units from the same ecological community plotted similarly will generally form a cloud of points, or *data cloud*. For a community of three species, it is possible to plot the cloud in a three-dimensional graph, perhaps with 'pins' indicating the positions of the points in one of three planes. However, it is difficult to plot sample units in a species space defined by four or more axes. It is often not possible to represent the cloud of points defined by three or more species into a two-dimensional graph without distortion. The problem is similar to that of cartographers, whose two-dimensional maps of the three-dimensional Earth invariably involve distortion of distances. However, several methods are available to provide for minimal distortion in the graphing of points from spaces of high dimension into graphs of two or three dimensions. In the context of ecological analysis, these techniques are known as *ordination analysis*. Common techniques particularly well-suited to ecological abundance data include Correspondence Analysis and Multidimensional Scaling. These techniques are described in a variety of recent texts, including Pielou (1984a), Jongman et al. (1995) and others. Canonical Correspondence Analysis is a method in which ecological data are analyzed simultaneously with environmental factors such as salinity and temperature, and permits these factors to be plotted on the ordination of species and sample units (ter Braak 1987).

The most common ordination method, principal components analysis (PCA), works by rotating an axis into such a direction into the multidimensional cloud of points (e.g., sites), until the projection of those sites onto the axis explains the greatest possible variance ('spreads out' the points best). This becomes the principal axis. The second axis is limited to being at right angles to the first, but rotated through the dimensions until it explains most of the residual variance. A third, fourth and fifth axis can be constructed accordingly, but this is difficult to see in one's mind. The data cloud projected onto those axes and plotted on a piece of paper then represents the most variance possible in the data cloud in a two-dimensional picture.

Correspondence analysis (CA) and the nearly identical reciprocal averaging (RA) works similarly to PCA. However, the final axes are not rotated to account for each relative to the other. The final graph can include both species and sites on the same plane, something which is not possible in simple PCA.

Multidimensional scaling (MDS) differs from these methods in that the objective is to more clearly display the data in a given set of dimensions with as little distortion as possible. Thus, there is no rotation of axes. Rather, the data are shifted into place in an algorithm that minimizes the distortion in the distances between the points as it proceeds. Mathematically, this involves the creation of a distance matrix. Frequently, the data are transformed to ln(x+1) to minimize effects of spurious abundances, and then converted to a Euclidean distance matrix for analysis. As with PCA, a major portion of the variance tends to be associated with the first axis (although not in such an 'international manner'), and thus it is reasonable to plot the sites based on the first axis coordinates against environmental parameter values such as depth.

PCA is particularly problematic when dealing with nonlinearly related points, such as sites related by species abundances. Both PCA and MDS tend to form arches where one would expect straight lines reflecting known gradients. CA also arches, but less so, and avoids problems such as inversion of points at ends of a gradient. The arch can even be minimized through 'detrending' (see Bianchi 1996; Bianchi et al. 1996). Thus CA or detrended CA (DCA) is preferred by many ecologists. Other ecologists prefer MDS because of its philosophical orientation (minimizing distortion). Still others prefer to use both, to check that results are robust, i.e., not an artifact of a particular method. In the end, the choice may be simply one depending on the availability of working, reliable software.

Classification

An alternative approach to the analysis of data of high species dimensionality is classification or cluster analysis. These techniques involve the grouping together of similar sample units based on comparisons of the abundances of each species in one sample unit with those of all other sample units. In *hierarchical classification* methods, sample units are related through connected lines showing the relationships

among units, sets of units, sets of sets of units and so on up to the level of the whole sample. This diagram may be formed either by grouping from the unit to the whole as just described (*agglomerative clustering*), or by successively dividing the whole into smaller and smaller sets of units (*divisive classification*). An example of an agglomerative clustering method is the *unweighted pair groups method* (UPGM). A common method for divisive classification is *two-way indicator species analysis* (TWIA), as implemented in the TWINSPAN program (Hill 1979). These and many other classification methods are described in Gauch (1982), Pielou (1984a), Jongman et al. (1995) and others. Often, a particular grouping of data points revealed by classification analysis is indicated on an ordination of the same data through the use of differential graphing symbols such as 'x' for points in one group and 'o' for points in another group.

TWIA is a method for rearranging tables of species abundances by site and dividing the table into distinct sets of similar species (based on where they are found in what abundance) and similar sites (based on what species are found in them in what abundance). A successful classification yields data arranged in clear blocks within the table. In the early days of plant community ecology ('phytosociology'), such tables were created by tediously rearranging raw data tables by hand, with repetitive cutting and pasting of columns and rows. The TWIA approach, embodied in the TWINSPAN program of Hill (1979) revolutionized the field by automating the procedure.

TWIA begins by turning a table of abundances into a table of presence-absence values. This is accomplished by requiring an initial coding of all data into single-digit integer values, such as a 1-5 semiquantitative scale. A species x with a maximum value of 3 in a row, becomes species x1, x2 and x3, each with a maximum value of 1 in any column. The program then arranges the sites (columns) of a table based on the scores of each site on the first axis of a reciprocal averaging (basically correspondence analysis) ordination axis. This procedure groups similar sites together and lines them up along the axis representing the greatest variance among both sites and species taken simultaneously. The program then attempts to divide the table into left and right portions from the center of the columns. It analyzes the effect of this split in terms of how well the abundances of each species are clumped onto one side or the other of the division. It tries several such divisions, cutting left and right of the center, each time tallying up a score based on how well the species abundances are divided up. It then chooses the best division, and this becomes the highest level division of the classification. The program now repeats the ordination for only those sites (columns) on the left of the division, to find a good arrangement and division of this subset of columns. Then it does the same on the right of the first division. Now there are four classes of sites, as indicated by two levels of division. A third level of division would yield eight classes of sites, and so on. This is the basic procedure, although the program contains corrections and adjustments for rare species weightings, etc., which can be manipulated in the analysis.

Next, the program classifies the species (rows) in a manner similar to that of the sites (columns). Once the species classification has been completed, the program switches around the species (keeping the species group intact) in such a way that the table is in an optimal 'block' form, with blocks of high species abundances arranged along one or the other diagonal of the table. The table is then converted back to the integer values, and printed with indications of the hierarchy of site and species divisions. Additionally, the program indicates what species are particularly good indicators of any site division, and helps to identify 'border' species of uncertain position in a division of sites.

The method has the advantage over most agglomerative classifications that it is very easy in the end to explain *why* a particular set of sites or species were grouped together. This is generally obvious from studying the output table. A divisive 'top-down' approach, the method also avoids the problem of 'chaining', i.e., the tendency for a single class in agglomerative 'bottom-up' clustering to grow like a snowball, gathering new characteristics as it does, and thus obscuring the inherent structure of the data set. However, TWINSPAN does not always provide clear one-to-one matching between site and species groups due to failed divisions (based on various criteria for failure that the program uses). Even equal numbers of site and species groups may not match up well, such that the third group of sites may not correspond to the third group of species. This is because the method does not perform the two-way analyses simultaneously. Finally, there is no way to determine the probability that a given division could have been achieved by the mere fact that the matrix was sparse (containing lots of zero values) and patchy to begin with. Nonetheless, TWINSPAN is the most widely used program of its type currently available.

Skewer Analysis

Skewer analysis (Pielou 1984b) is an example of a technique which can be used in inferential, exploratory or pattern analysis modes. One begins with a data cloud in species space in its high-dimensional state—i.e., one dimension per species in the total sample. One might wish to know if there is a significantly nonrandom ordering of points in a particular sequence of the points. For example, one might be investigating if a series of trawl sample units taken at different depths exhibit a gradual change in species composition as one moves from shallow to deep. One would then, conceptually, label the points in the data cloud according to depth. Skewer analysis then involves running tens or hundreds of axes at random directions through the species space. Each time, the points in the data cloud are *mapped* onto the axis, i.e., for each point, a line is drawn in the shortest possible distance to the axis, and the new point on the axis is a mapping of the original point in species space. This mapped set of points is compared with the natural ordering 1,2,3 ... etc. using a rank correlation coefficient. If there is a significant ordering of the data as hypothesized, then some of the random axes will show mappings of the points which correlate very highly with the natural ordering. The distribution of the absolute values of correlation coefficients generated in the analysis will be distributed primarily around some non-zero correlation value. If the ordering of points you hypothesized existed was in reality a random ordering of points, then the values will be clustered near zero (the right half of a normal distribution if one uses absolute values of coefficients). In that case, one can determine that the hypothesized gradient does not exist. In the case of the sample units taken over depth, it would mean that there is no reason to believe that depth (for the depths studied) has any effect in determining continuous differences in community composition. Skewer analysis requires special software that is not widely available. A program for such analyses, written by the present author, is available on a diskette in the recent book by Gallucci et al. (1996).

Dynamic Approaches

The basic unit of data in an ordination, classification or skewer analysis is a two-dimensional table. A time series of community data may consist of a block of tables of sites (sample units) by species, each representing the community at a specific time—a 'time slice'. The third axis of the block is time, and extends across the tables. Analyses can be performed on each time slice independently. This results in ordinations and classifications which can be viewed in sequences as if they were frames in a movie.

Alternatively, the data can be summarized into three tables of average values along each of the site, species and time axes, resulting in tables of average times by species, sites by times and sites by species, respectively (Williams et al. 1982). Of these, the most useful set of analyses is often those based on the times by species table. An ordination of 'times' in species space is a representation of the mean states of the community as it varies in composition over time. By connecting the states in sequence with arrows, one can begin to understand the dynamics of community change. These and similar analyses can be used to identify dynamic behaviors such as cyclic change (possibly indicating seasonality), attraction to a limited, bounded region of species space (such as may be caused by an attractor or climax state), motion within bounds, departures and returns to a bounded state (indications of system resilience), and shifts among bounded states (community phase shifts).

Unfortunately, the dimensional reduction involved in ordination analysis may obscure dynamic patterns. An alternative approach which does not involve dimensional reduction is Ordered Similarity Analysis. This consists of analyses of pairwise distances (or similarities) among community states. The original method by Pielou (1979) has been expanded by McManus (1985) and utilized subsequently by Licuanan (1991) to analyze coral reef data. The method facilitates the identification of dynamic community

behaviors such as full cycles, reversing cycles, constant and sporadic progression, acceleration and deceleration.

Example of Application: Samar Sea Trawl Survey

The University of the Philippines College of Fisheries conducted trawl surveys in the Samar Sea of the Philippines eleven times from March 1979 to May 1980 (McManus 1985). The analysis using TWINSPAN indicated that there were distinct deep (to 100 m) and shallow fish assemblages, divided at approximately 40 m depth, but that several species of fish were also relatively ubiquitous with respect to depth (Table 1). The results were confirmed through detrended correspondence analysis (Fig. 1), and differed relatively little when the analyses were conducted on individual time slices or on a table averaged over time (times by species). The classification results are

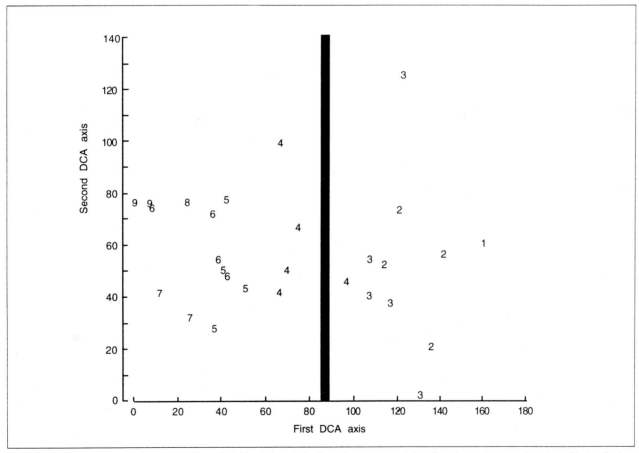

Fig.1. Detrended correspondence analysis (DCA) of sites based on the mean weight of each species caught at each site during 11 samplings in a 16-month period in the Samar Sea, Philippines. Weights were log-transformed before analysis. Each number on the graph represents the lower depth class limit (x 10 m) for a single study site. The position of the number indicates the relationship of that site to the other sites based on scores from the first two axes produced by DCA. These two axes provide an optimal representation of the affinities among the sites based on both site and species spaces. The sites generally range from deep to shallow along the first DCA axis. This indicates that depth was the primary gradient in the community, and that the first axis is an approximate representation of that gradient. The second axis represents one or more sources of residual variations not readily explained in terms of available environmental data. It provides a graphic representation of variations not related to the primary gradient, and gives a basis for judging the overall influence of this gradient. The thick solid line shows the division into deep and shallow primary subcommunities indicated by TWINSPAN analysis. The optimal division fell within the 30 and 40 m depth classes. (Source: McManus 1985).

138

Table 1. Dominant species and diversity by subcommunity in the Samar Sea, Philippines. Percents and cummulative % based on total catches (kg.) in each site group. Species ranges are ubiquitous (U), shallow (S) or deep (D) as per TWINSPAN classification of mean site. Trends are increasing (+) or neutral (0) in abundances over the 16 month study period as per classification of mean times by species table. Shannon diversities and evenness indices were computed as per Pielou (1977), and are approximations intended only for comparisons between subcommunities. Lower evenness of the deeper community is largely attributable to dominance by *Leiognathus bindus*. *L. splendens* and *L. equulus* characterize the shallow subcommunity. Deep stations characteristically include *Saurida undosquamis* and *Nemipterus nematophorus*. *L. bindus* dominates throughout. (Source: McManus 1985).

SHALLOW SUBCOMMUNITY (< 40 m)

Rank	Taxon	cum. %	%	trend
1	Leiognathus bindus	9	9	U 0
2	Leiognathus splendens	17	8	S 0
3	Loligo spp.	25	7	S 0
4	Leiognathus equulus	31	6	U 0
5	Rastrelliger brachysoma	34	4	U 0
6	Trichiurus lepturus	38	4	U 0
7	Saurida tumbil	41	3	U 0
8	Apogon spp.	44	3	U +
9	Upeneus sulphureus	46	3	U 0
10	Alepes djedaba	49	2	S +
11	Pentaprion longimanus	51	2	U 0
12	Selaroides leptolepis	54	2	S 0
13	Sepia spp.	56	2	U +
14	Leiognathus brevirostris	58	2	S 0
15	Stolephorus indicus	60	2	U 0
16	Rastrelliger kanagurta	62	2	U 0
17	Decapterus macrosoma	64	2	U 0
18	Stolephorus tri	65	2	S 0
19	Lagocephalus lunaris	67	2	U 0
20	Nemipterus japonicus	68	2	U 0

S	Total taxa	159
N	Total catch (kg)	18 400
	No. tows (1 hr. equiv.)	107
	Catch rate (kg/hr)	172
H	Shannon Index (bels)	3.9
B11H'/1n(S)	Evenness	0.77

DEEP SUBCOMMUNITY (40 + m)

Taxon	cum. %	%	trend
Leiognathus bindus	28	28	U 0
Pentaprion longimanus	35	7	U 0
Saurida undosquamis	41	6	D 0
Loligo spp.	45	5	U 0
Nemipterus nematophorus	49	4	D 0
Saurida undosquamis	53	4	U 0
Upeneus sulphureus	56	3	U 0
Decapterus macrosoma	59	3	U 0
Apogon spp.	62	3	U +
Fistularia spp.	64	2	D +
Priacanthus macracanthus	66	2	D +
Sepia spp.	69	2	U 0
Lagocephalus lunaris	71	2	U 0
Rastrelliger brachysoma	73	2	U 0
Rastrelliger kanagurta	75	2	U 0
Priacanthus tayenus	77	2	D +
Triglidae	78	2	D 0
Trichiurus lepturus	80	1	U 0
Upeneus mollucensis	81	1	D 0
Stolephorus indicus	82	1	U 0

Total taxa	166
Total catch	37 100
No. tows	201
Catch rate	185
Shannon Index	3.3
Evenness	0.66

ALL STATIONS

Taxon	cum. %	%	trend
Leiognathus bindus	22	22	U 0
Loligo spp.	27	6	U 0
Pentaprion longimanus	33	6	U 0
Saurida undosquamis	37	4	D 0
Saurida tumbil	41	3	U 0
Upeneus sulphureus	44	3	U 0
Nemipterus nematophorus	47	3	D 0
Leiognathus splendens	49	3	S 0
Rastrelliger brachysoma	52	3	U 0
Apogon spp.	54	3	U +
Decapterus macrosoma	57	2	D 0
Sepia spp.	59	2	U +
Trichiurus lepturus	61	2	U 0
Leiognathus equulus	63	2	S 0
Lagocephalus lunaris	65	2	U 0
Rastrelliger kanagurta	67	2	U 0
Fistularia spp.	69	2	D +
Priacanthus tayenus	71	2	U +
Priacanthus macracanthus	72	2	D +
Stolephorus indicus	74	1	U 0

Total taxa	172
Total catch	55 500
No. tows	308
Catch rate	180
Shannon Index	3.7
Evenness	0.71

Table 2. Sorted community table showing the mean abundances by site of the 20 dominant taxa in the shallow and deep subcommunities of the Samar Sea, Philippines. Within each category, sites and species are arranged by correspondence scores. TWINSPAN divisions are based on 172 taxa. Abundances are kg/hour ranked such that 1=0 to <2, 2=2 to <5, 3=5 to <10, 4=10 to <20, 5=20+. The community is primarily composed of ubiquitous species interspersed with important shallow and deep preferential species that define the two relatively homogenous subcommunities. (Source: McManus 1985).

	SHALLOW											DEEP																
Station numbers:	1	17	16	18	15	3	2	24	23	22	4	12	13	8	7	6	20	21	19	14	5	28	27	26	25	11	10	9
Depth midpoints (m):	16	27	35	25	38	31	23	32	25	40	40	81	75	74	61	53	55	46	65	56	45	65	57	49	44	92	97	82
Taxon																												
Stolephorus tri	3	3	2	1	2		2	2	1	2	1																	
Leiognathus splendens	5	2	4	2	3	5	2	2	2	1	5				1								1					
Leiognathus brevirostris	5	1	1	2		2	3	3	2		1				1	1			1	1								
Leiognathus equulus	4	3	2	2	2	3	3	4	3	3	5	1			1	1	2	1	1	1	1	1	1		1	3	2	
Alepes djedaba	2	2	1	1	2	3	1	1	2	1	5	1			1	1	2	1	1	1	1	1	1	1	1			
Selaroides leptolepis	2	2	1	2	2	2	2	3	3	2	2	1	1		1	1	2	1	1	2	2	2	2	2	3			
Upeneus molluccensis			1	1	1	1		1	1	1	1	1	2	3	2	1	1	2	3	1	1	1	3	1	1	3	2	5
Fistularia spp.		1	1	2	2	1	1	1	1	2	1	3	3	2	2	2	2	3	3	2	2	4	3	2	2	3	2	3
Saurida undosquamis	1	1	2	2	3	2	2	1	1	2	3	3	3	5	5	5	3	4	4	4	3	3	3	2	2	3	3	4
Priacanthus macracanthus			1	1	2		1	1	1	1	2	3	3	3	4	4	2	2	3	2	3	3	3	1	1	4	2	2
Nemipterus nematophorus		1	1	2		1		1	2	2		4	4	4	4	3	2	3	3	3	3	3	3	2	1	4	3	3
Triglidae					1		1	1	1	1		3	3	2	2	1	1	2	2	2	1	3	2	1	1	4	4	2
Trichiurus lepturus	3	2	2	2	4	3	2	2	2	2	4	2	2	2	2	2	4	1	1	1	2	1	4	1	1	1	1	1
Upeneus sulphureus	2	2	1	2	2	2	2	3	3	3	3	5	2	2	2	2	2	2	3	2	2	3	3	5	3	3	3	1
Stolephorus indicus	2	3	3	2	2	2	3	1	2	2	3	2	1	1	2	2	2	2	1	1	2	1	2	3	2	3	2	1
Lagocephalus lunaris	2	3	2	3	3	3	3	1	1	2	2	3	3	3	3	3	2	2	3	2	3	2	1	1	2	3	1	2
Sepia spp.	2	2	2	3	2	3	2	2	2	2	4	2	2	3	3	3	2	3	2	3	3	2	2	1	2	2	2	1
Saurida tumbil	2	3	3	3	3	3	3	2	3	3	3	3	4	3	4	4	4	3	3	3	3	3	3	3	3	2	2	2
Rastrelliger kanagurta	1	2	2	2	2	2	3	1	1	1	4	1	1	1	1	2	1	4	2	2	2	1	1	3	3	2	3	2
Rastrelliger brachysoma	2	2	2	3	2	3	1	3	4	4	3	4	3	3	1	4	3	3	2	2	2	3	2	3	3	1		2
Nemipterus japonicus	2	3	2	2	3	2	2	1	1	2	2	1	2	1	2	1	1	2	2	2	2	1	1	1	1	2	1	2
Loligo spp.	4	4	4	4	4	4	4	4	3	3	4	4	3	3	3	3	3	3	2	3	3	4	3	3	3	5	2	3
Apogon spp.	2	3	3	3	3	3	2	2	2	3	3	3	3	3	3	3	3	2	3	3	2	2	3	2	2	2	2	2
Priacanthus tayenus	1	1	1	2	1	2	2	1	2	2	3	2	2	2	2	2	3	3	2	2	2	3	3	2	1	1	3	1
Pentaprion longimanus	1	1	2	2	2	2	2	4	4	4	4	4	3	4	2	2	3	3	3	3	3	5	4	4	5	5	4	2
Leiognathus bindus	2	2	3	3	4	4	5	5	3	4		5	4	4	4	5	5	4	5	3	4	5	5	4	5	5	4	5
Decapterus macrosoma	1	5	2	1	1	1	1	1	1	1	1	1	2	2	1	1	1	1	1	1	1	3	3	2	2	3	5	3

139

140

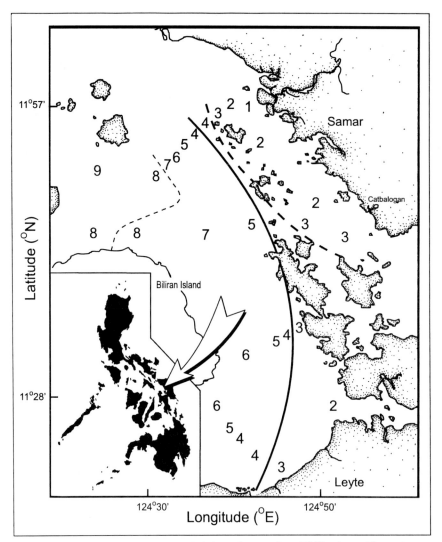

Fig. 2. Subcommunities of demersal fish in Samar Sea, Philippines. The primary division (solid line) separates the "shallow" from the "deep" fish subcommunity at around the 30-40 m depth contour. Numbers indicate depth of fishing station (x 10 m). (Source: McManus 1986).

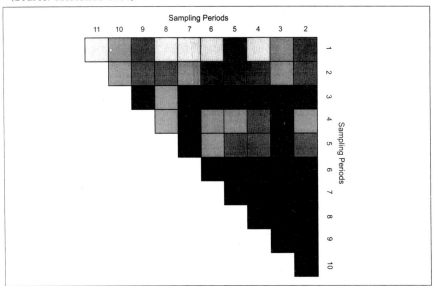

Fig. 3. OSMA; March 79 to 80. Community behavior from 60 to 110 m. Comparisons of average community states by percent similarity generalized to shadings where solid represents highest values. A strong pattern of deceleration over time is apparent (Q = 475). (Source: McManus 1985).

shown in geographic space in Fig. 2. The most abundant species were identified for each subgroup and overall (Table 2). The combination of these and other analyses indicated that a haul of fish caught at any given depth would likely be made up of roughly half the abundance in species specific to the depth zone, and half shared between depth zones. The immediate management conclusion from this was that while keeping trawlers below 40 m would optimally limit the effects of trawling on the shallow, artisanal fisheries, the separation would only concern about one-half of the catch in each case.

As a preliminary investigation of the dynamics of the community over time, the data were subjected to Ordered Similarity Matrix Analysis (OSMA) analysis overall and within three depth categories. The clearest pattern was that in samples from 60 to 100 m depth (Fig. 3). This showed that during the time of the study, the community varied less and less (became more stable, in the sense of constancy over time). However, both Pielou's Q test and skewer analysis indicated that the same assemblage exhibited a trend of net change over time. Further information on this analysis may be found in McManus (1985, 1986).

It is notable that subsequent analyses of trawl data covering a greater range of depth in the nearby Ragay Gulf (Federizon 1992) and off southwestern Indonesia (McManus 1989 and 1996; McManus et al. 1996) both indicated that a depth of 100 m was a far more optimal level for separating fisheries. However, the density of fish below 100 m tends to be very low, and the viability of a widespread demersal fishery restricted to depths greater than 100 m in this region would be in serious doubt (McManus 1996).

References

Bianchi, G. 1992. Demersal assemblages of tropical continental shelves. University of Bergen, Norway. 217 p. Ph.D. dissertation.

Bianchi, G. 1996. Demersal fish assemblages of trawlable grounds off northwest Sumatra, p.123-130. In D. Pauly and P. Martosubroto (eds.) Baseline studies of biodiversity: the fish resources of Western Indonesia. ICLARM Stud. Rev. 23, 312 p.

Bianchi, G., M. Badrudin and S. Budihardjo. 1996. Demersal assemblages of the Java Sea: a study based on the trawl surveys of the R.V. Mutiara 4, p. 55-61. In D. Pauly and P. Martosubroto (eds.) Baseline studies of biodiversity: the fish resources of Western Indonesia. ICLARM Stud. Rev. 23, 312 p.

Christensen, V. and D. Pauly, Editors. 1993. Trophic models of aquatic ecosystems. ICLARM Conf. Proc. 26, 390 p.

Fager, E.W. and A.R. Longhurst. 1968. Recurrent group analysis of species assemblages of demersal fish in the Gulf of Guinea. J. Fish. Res. Bd. Canada 25(70): 1405-1421.

Federizon, R.R. 1992. Description of the subareas of Ragay Gulf, Philippines, and their fish assemblages by exploratory data analysis. Aust. J. Mar. Freshwat. Res. 43: 379-391.

Galucci V.F., S.B. Saila, D.J. Gustafson and B.J. Rothschild (eds.) 1996. Stock assessment: quantitative methods and applications for small-scale fisheries. CRC Lewis, N.Y.

Gauch, H.G. 1982. Multivariate analysis in community ecology. Cambridge University Press, N.Y.

Hill, M.O. 1979. TWINSPAN–A FORTRAN program for arranging multivariate data in an ordered two-way table by classification of the individuals and attributes. Cornell University, Ithaca, N.Y.

Jongman, R.H.G., C.J.F. ter Braak and O.F.R. van Tongeren (eds.) 1995. Data analysis in community and landscape ecology. Cambridge University Press, N.Y.

Licuanan, W.Y. 1991. Temporal changes in the cover of life-forms in Puerto Galera, Mindoro Island, western Philippines: preliminary results. Proc. Regional Symp. on Living Resources in Coastal Areas, Marine Science Institute, University of the Philippines, Quezon City, 87-96 p.

Martosubroto, P. 1982. Fishery dynamics of the demersal resources of the Java Sea. Dalhousie University. 238 p. Ph.D. dissertation.

McManus, J.W. 1985. Descriptive community dynamics: background and an application to tropical fisheries management. University of Rhode Island, Narragansett, R.I. 217 p. Ph.D. dissertation.

McManus, J.W. 1986. Depth zonation in a demersal fishery in the Samar Sea, Philippines, p. 483-486. In J.L. Maclean, L.B. Dizon and L.V. Hosillos (eds.) The First Asian Fisheries Forum. Asian Fisheries Society, Manila.

McManus, J.W. 1989. Zonation among demersal fishes of Southeast Asia: the southwest shelf of Indonesia, p. 1011-1022. In Proceedings of the Sixth Symposium on Coastal and Ocean Management/ASCE, 11-14 July 1989. Charleston, South Carolina.

McManus, J.W. 1996. Marine bottomfish communities from the Indian Ocean coast of Bali to Mid-Sumatra, p. 91-101. In D. Pauly and P. Martosubroto (eds.) Baseline studies of biodiversity: the fish resources of Western Indonesia. ICLARM Stud. Rev. 23, 312 p.

McManus, J.W., C.L. Nañola, A.G.C. del Norte, R. B. Reyes, J.N.P. Pasamonte, N.P. Armada, E.D. Gomez and P.M. Aliño. 1996. Coral reef fishery sampling methods, p. 226-270. In V.F. Galucci, S.B. Saila, D.J. Gustafson and B.J. Rothschild (eds.) Stock assessment: quantitative methods, and applications for small-scale fisheries. CRC Lewis, N.Y.

Nañola, C.L., J.W. McManus, W.L. Campos, A.G.C. del Norte, R.B. Reyes, J.P.B. Cabansag and J.N.D. Pasamonte. 1990. Spatio-temporal variations in community structure in a heavily fished forereef slope in Bolinao, Philippines, p. 377-380. In R. Hirano and I. Hanyu (eds.) The Second Asian Fisheries Forum. Asian Fisheries Society, Manila.

Pauly, D. 1979. Theory and management of tropical multispecies stocks: a review, with emphasis on Southeast Asian demersal fisheries. ICLARM Stud. Rev. 1, 35 p.

Pauly, D. and G.I. Murphy, Editors. 1982. Theory and management of tropical fisheries. ICLARM Conf. Proc. 9, 360 p.

Pielou, E.C. 1977. Mathematical ecology. John Wiley and Sons, New York.

Pielou, E.C. 1979. Interpretation of paleoecological similarity matrices. Paleobiology 5(4): 435-443.

Pielou, E.C. 1984a. The interpretation of ecological data: a primer on classification and ordination. Wiley, N.Y.

Pielou, E.C. 1984b. Probing multivariate data with random skewers: a preliminary to direct gradient analysis. Oikos 42: 161-165.

142

Ralston, S. and J.J. Polovina. 1982. A multispecies analysis of the commercial handline fishery in Hawaii. Fish. Bull. 80(3): 435-448.

Sherman, K., L.M. Alexander and B. D. Gold (eds.) 1990. Large marine ecosystems: patterns, process, and yields. Amer. Assoc. Adv. Sci., Washington, D.C. 376 p.

Shepard, A.R.D., R.M. Warwick, K.R. Clarke and B.E. Brown. 1992. An analysis of fish community responses to coral mining in the Maldives. Environ. Biol. Fish. 33(4): 367-380.

ter Braak, C.J.F. 1987. The analysis of vegetation-environment relationships by canonical correspondence analysis. Vegetatio 69: 69-77.

Tyler, A.V., W.L Gabriel and W.J. Overholtz. 1982. Adaptive management based on structure of fish assemblages of northern continental shelves, p.149-156. In M.C. Mercer (ed.) Multispecies approaches to fisheries management advice. Can. Spec. Publ. Fish. Aquat. Sci. 59.

Whittaker, R.H. 1970. Communities and ecosystems. MacMillan, N.Y.

Williams, W.T., Editor. 1976. Pattern analysis in agricultural science. Elsevier Scientific Publishing Co., N.Y.

Williams, W.T., H.J. Clay and J.S. Bunt. 1982. The analysis, in marine ecology, of three-dimensional data matrices with one dimension of variable length. J. Exp. Mar. Biol. Ecol. 60: 189-196.

Appendix I

ADB Workshop on Sustainable Exploitation of Tropical Coastal Fish Stocks in Asia
2- 5 July 1996
The Hyatt Regency, Manila, Philippines

Program of Activities

Tuesday, 2 July 1996

8:00 a.m. Registration

9:00 a.m. Opening Ceremonies
Master of Ceremonies:
Len R. Garces

Welcome Address
Modadugu V. Gupta
Director, International Relations

9:20 a.m. Message
Dr. Muhammad A. Mannan
Manager, Forestry and Natural
Resources Division

9:40 a.m. Opening Remarks
Daniel Pauly
Principal Science Adviser
ICLARM

10:00 a.m. Photo Session
Coffee Break

10:30 a.m. Workshop Overview
Geronimo T. Silvestre
Research Scientist
ICLARM

Session 1: Status of Tropical Coastal
Fisheries in Asia
Chairperson: *Geronimo T. Silvestre*
Rapporteur: *Len R. Garces*

10:50 a.m. Regional Overview Paper
Geronimo T. Silvestre
ICLARM

11:30 a.m. Country Report: Philippines
Noel C. Barut
Bureau of Fisheries and
Aquatic Resources
Philippines

12:00 noon Lunch Break

1:15 p.m. Country Report: Bangladesh
Giasuddin Khan
Department of Fisheries
Bangladesh

1:45 p.m. Country Report: Indonesia
Bambang Edi Priyono
Directorate General of Fisheries
Indonesia

2:15 p.m. Country Report: Malaysia
Abu Talib b. Ahmad
Department of Fisheries
Malaysia

2:45 p.m. Country Report: Sri Lanka
Rekha Maldeniya
National Aquatic Resources
Agency, Sri Lanka

3:15 p.m. Coffee Break

3:30 p.m. Country Report: Thailand
Monton Eiamsa-ard
Department of Fisheries
Thailand

4:00 p.m. Country Report: Vietnam
Pham Thuoc
Research Institute for Marine
Products, Vietnam

4:30   p.m.   Open Forum

5:00   p.m.   End of Session

Wednesday, 3 July 1996

Session 2: Coastal Fisheries Management:
Issues, Strategies and Actions
Chairperson: *Daniel Pauly*
Rapporteur: *Geronimo T. Silvestre*

9:00   a.m.   Program Perspective: FAO
Purwito Martosubroto
Fisheries and Aquaculture
Division, FAO, Rome

9:20   a.m.   Program Perspective: FAO
Bay of Bengal Programme
Purwito Martosubroto
Fisheries and Aquaculture
Division, FAO, Rome

9:40   a.m.   Program Perspective:
SEAFDEC-MFRDMD
Dr. Mansor bin Mat Isa
Marine Fishery Resources
Development and Management
Department
SEAFDEC, Malaysia

10:00 a.m.   Open Forum

10:30 a.m.   Coffee Break

10:45 a.m.   Working Group Discussion:
Issues Impacting Sustainable
Exploitation of Tropical Coastal
Fish Stocks in Asia

12:00 noon   Lunch Break

1:15   p.m.   Working Group Discussion:
Management Strategies and
Actions for Sustainable
Exploitation of Tropical Coastal
Fish Stocks in Asia

3:00   p.m.   Coffee Break

3:15   p.m.   Working Group Discussion:
Scope for Regional Collaborative
Efforts in Research and
Management

5:00   p.m.   End of Session

Thursday, 4 July 1996

Session 3: Retrospective Analysis of Trawl
Surveys: Potential for Regional
Collaborative Efforts
Convenor: *Daniel Pauly*
Rapporteur: *Felimon Gayanilo, Jr.*

9:00   a.m.   Retrospective Analysis of Trawl
Surveys and Database Guidelines
Daniel Pauly
ICLARM

9:30   a.m.   Trawl Survey Database Prototype:
An Overview
Felimon Gayanilo, Jr.
ICLARM

10:00 a.m.   Community Analyses of Fish and
Fishery Data
John W. McManus
ICLARM

10:30 a.m.   Coffee Break

10:45 a.m.   Working Group Session:
Database Prototype

12:00 p.m.   Lunch Break

1:15   p.m.   Working Group Session:
Database Prototype Improvement

3:00   p.m.   Coffee Break

3:15   p.m.   Working Group Session:
Indicative Planning for Regional
Collaboration

5:30   p.m.   End of Session

Friday, 5 July 1996

Session 4: Summary and Recommedations
Chairperson: *Geronimo T. Silvestre*
Rapporteur: *Len R. Garces*

9:00   a.m.   Synopsis of Main Workshop
Results and Recommendations
Daniel Pauly
ICLARM

9:30   a.m.   Open Forum

10:00 a.m. Closing Program
Master of Ceremonies:
Len R. Garces

Closing Remarks
Mr. Weidong Zhou
Project Specialist
ADB

10:15 a.m. Remarks
Selected Country Representatives

10:45 a.m. Closing Remarks
Dr. Daniel Pauly
ICLARM

11:00 a.m. ICLARM Visit

12:00 noon Lunch at Hyatt

1:30   p.m. Tour of ICLARM Library

3:00   p.m. Fishbase Overview

4:00   p.m. Reefbase Overview

5:00   p.m. End of Visit

Appendix II

ADB Workshop on Sustainable Exploitation of Tropical Coastal Fish Stocks in Asia
2 - 5 July 1996
The Hyatt Regency, Manila, Philippines

Participants and Resource Persons

Bangladesh

Mr. Md. Giasuddin Khan
Principal Scientific Officer
Marine Fisheries Survey
Management and Development Project
Department of Fisheries
Chittagong 4100, Bangladesh
Tel : (880-31) 504206; (880-31) 504438
Fax: (880-31) 504206; (880-31) 503873

Mr. Md. Alamgir
Senior Scientific Officer
Fisheries Research Institute
Marine Fisheries and Technology Station
Motel Road, Laboni
Cox's Bazar, Bangladesh
Tel : (880-341) 3855; (880-341) 3976
Fax: (880-341) 3202; (880-341) 3783

Indonesia

Dr. Bambang Edi Priyono
Chief, Sub Directorate of Fisheries Potential
Resources DGF, Agriculture Department
    Building B, 6th Floor, Jl. Harsono
Rm. No. 3; Ragunan, Pasar Minggu
Jakarta, Indonesia
Tel : (62-21) 7811672
Fax: (62-21) 7803196 (DGF)
    (62-21) 7890624 (BADP)

Malaysia

Mr. Abu Talib bin Ahmad
Research Officer
Department of Fisheries

8th and 9th Floor, Wisma Tani,
Jln. Sultan Salahuddin,
50628 Kuala Lumpur
Malaysia
Tel : (60-3) 2982011 ext 3645
Fax: (60-3) 2910305
E-mail: abuahm01@dof.moa.my
    dof09@smtp.moa.my

Mr. Alias Man
Research Officer
Fisheries Research Institute
11960 Batu Maung
Penang, Malaysia
Tel : (60-4) 6263925/6
Fax: (60-4) 6262210
E-mail: aliman01@dof.moa.my
    fri@ipp.po.my

Philippines

Mr. Noel C. Barut
Supervising Aquaculturist
Bureau of Fisheries and Aquatic Resources
Department of Agriculture, 860 Arcadia Bldg.
Quezon Avenue, Quezon City 1100
Philippines
Tel : (63-2) 9287075; 9265428
Fax: (63-2) 9291249

Mr. Mudjekeewis D.Santos
Aquaculturist
Bureau of Fisheries and Aquatic Resources
Department of Agriculture
860 Arcadia Bldg.,Quezon Avenue,
Quezon City 1100
Philippines
Tel : (63-2) 9287075/ 9265428
Fax: (63-2) 9291249

Sri Lanka

Mrs. Rekha Maldeniya
Fisheries Officer
National Aquatic Resources Agency
Croco Island, Mattakkuliya
Colombo 15, Sri Lanka
Tel : (94-1)522000; 522006
Fax: (94-1) 522932

Thailand

Mr. Monton Eiamsa-ard
Senior Marine
Fishery Biologist
Southern Marine
 Fisheries Development Centre
Vichianchom Rd., Amphoe Muang
Songkhla, Thailand 90000
Tel : (66-74) 312595
Fax: (66-74) 312495

Ms. Supatra Amornchairojkul
Fisheries Engineering Division
Department of Fisheries
Bangkok 10900, Thailand
Tel/Fax: (66-2) 5798567

Vietnam

Dr. Pham Thuoc
Deputy Director
Research Institute of Marine Products
170 LeLai Street
HaiPhong City, Vietnam
Tel : (84-31) 846-656
Fax: (84-31) 836-812
E-mail: Chung.rimp@Vietap.tool.nl.

Dr. Nguyen Long
Head
Fishing Technology Division
Research Institute of Marine Products
170 LeLai Street
HaiPhong City, Vietnam
Tel : (84-31) 846664
Fax: (84-31) 836-812
E-mail: Chung.rimp@Vietap.tool.nl

FAO

Dr. Veravat Hongskul
Senior Fishery Officer and APFIC Secretary
FAO-Regional Office for Asia and the Pacific

Maliwan Mansion,
39 Phra Atit Road
Bangkok 10200, Thailand
Tel : (66-2) 2817844
Fax: (66-2) 2800445
E-mail: VeravaHongskul@field.fao.org

Dr. Purwito Martosubroto
Marine Resources Service
Fishery Resources Division
FAO of the United Nations
Viale delle Terme di Caracalla
00100 Rome, Italy
Tel : (39-6) 5225-6469
Fax: (39-6) 5225-3020
E-mail: Purwito.Martosubroto@fao.org

SEAFDEC-MFRDMD

Dr. Mansor bin Mat Isa
Research Officer
Marine Fishery Resources Development
 and Management Department
Southeast Asian Fisheries Development Center
Chendering, 21080 Kuala Terengganu
Malaysia
Tel : (60-9) 6175135
Fax: (60-9) 6175136
E-mail: seafdec@po.jaring.my

PCAMRD

Ms. Ester C. Zaragoza
Supervising Science Research Specialist
Marine Research Division
Philippine Council for Aquatic and Marine
 Research and Development
Los Baños, Laguna, Philippines
Tel : (6-49) 536-0015 to 19
Fax: (6-49) 536-4077
CompuServe: 104540, 3506

PBSP

Dr. Rudolf Hermes
Marine Biologist
Coastal Resources Management
Philippine Business for Social Progress
Intramuros, Manila
Philippines
Tel : (63-2) 527-7741 to 51
Fax: (63-2) 527-3743

PRIMEX

Dr. Elvira Ablaza
Executive Director
Pacific Rim Innovation and Management
 Exponents, Inc. (PRIMEX)
502 Manila Luxury Condominium
Pearl Drive, Pasig
Metro Manila, Philippines
Tel : (63-2) 633-9052/3717,(63-2) 634 5135/7339
Fax: (63-2) 634-7340
Telex: 66485 PINTR PN

ADB

Mr. Weidong Zhou
Fisheries Specialist
Forestry and Natural Resources (East)
Asian Development Bank
6 ADB Avenue, Mandaluyong City, Philippines
Tel :   (63-2) 632-4414, (63-2) 631-7961
Fax: (63-2) 636-2444,(63-2) 636-2400
E-mail: wzhou@mail.asiandevbank.org

Dr. Thomas Gloerfelt-Tarp
Fisheries Specialist
Forestry and Natural Resources Division
Agriculture and Social Sectors Department (West)
Asian Development Bank
6 ADB Avenue, Mandaluyong City, Philippines
Tel :   (63-2) 632-6722
Fax: (63-2) 636-2300
E-mail: tgloerfelt-tarp@mail.asiandevbank.org

ICLARM

Dr. Daniel Pauly
Principal Science Adviser
International Center for Living Aquatic
Resources Management
3F Bloomingdale Bldg.
205 Salcedo Street, Legaspi Village
1229 Makati City, Philippines
Tel :   (63-2) 818-0466/ 9283, (63-2) 817-5255/ 5163
Fax: (63-2) 816-3183
E-mail: Pauly@fisheries.com

Dr. John W. McManus
Program Leader, Aquatic Environments Program
Project Leader, ReefBase Project
E-mail: J.McManus@cgnet.com

Mr. Geronimo T. Silvestre
Research Scientist
E-mail: G.Silvestre@cgnet.com

Mr. Felimon C. Gayanilo, Jr.
Fisheries Science Programming Specialist
E-mail: F.Gayanilo@cgnet.com

Mr. Len R. Garces
Research Associate
E-mail: L.Garces@cgnet.com

Ms. Rowena Andrea V. Santos
Research Assistant
E-mail: B.Santos@cgnet.com

Mr. Dan Bonga
Research Assistant
E-mail: D.Bonga@cgnet.com

Appendix III

Preliminary Compilation of Trawl Surveys Conducted in Tropical Asia[1]

Fisheries scientists and managers often complain about the limited availability of data and information inputs for the management decision-making process. Yet, there are numerous surveys of fisheries resources, undertaken at significant costs, which remain largely underutilized or untapped. A number of trawl surveys have been conducted in tropical Asia some for research/academic interests, but most for assessing fisheries potential and development. However, the data from most of these surveys remain underutilized, usually having been analyzed only in terms of total stock density, then stored in the files in various fisheries agencies/institutes. Pauly (1996) highlighted the utility of retrospective analyses of these surveys for studies of biodiversity and for improving fisheries management, using newer techniques of analysis. He emphasized the significance of standardization and computerization of these surveys in overcoming the "shifting baseline syndrome" which frequently afflicts fisheries assessment and management (Pauly 1995). Indeed, in the case of the Philippines, the work of Silvestre et al. (1986) illustrates how the use of these surveys can provide better insight of fisheries status and related fisheries policies.

In view of the potential of these trawl surveys, a compilation of the trawl surveys conducted in South and Southeast Asia was undertaken before the workshop. Contributions to the compilation were made by: M.G. Khan, M. Alamgir and M.N. Sada (Bangladesh), B.E. Priyono and B. Sumiono (Indonesia), A. Abu Talib and M. Alias (Malaysia), N.C. Barut ,M.D. Santos and L.R. Garces (Philippines), R. Maldeniya (Sri Lanka), M. Eiamsa-ard and S. Amornchairojkul (Thailand), P. Thuoc and N. Long (Vietnam), and M.I. Mansor (SEAFDEC). These contributions were supplemented by searches conducted by Dan Bonga and Len Garces (ICLARM) using ICLARM Library bibliographic databases (i.e., ASFA, NAGA and LIBRI). ASFA (Aquatic Sciences and Fisheries Abstracts) is an international bibliographic database covering information supplied from the

Aquatic Sciences and Fisheries Information System (ASFIS). ASFIS is maintained by the Food and Agriculture Organization of the United Nations (FAO), the Intergovernmental Oceanographic Commission (IOC) of UNESCO, the UN Office for Ocean Affairs and the Law of the Sea (OALOS), the UN Environment Program and a network of research centers throughout the world (Compact Cambridge 1991). Entries from ASFA covered the period 1978 to March 1996. LIBRI and NAGA are reference databases maintained by ICLARM, the former covering reference materials and collections of the ICLARM library and the latter covering the extensive reference citations of the information section of Naga (the ICLARM quarterly publication). Also, the extensive computerized reprint/report collection of Daniel Pauly (with over 30 000 entries on tropical fishery and related topics) was searched. Bibliographic searches were made using the following descriptors: 'trawl' and related terms (i.e., trawls, trawling and trawler), 'demersal' and 'fishery resources'. Each descriptor was tied to a particular country or region, e.g., India, Indian Ocean, Sunda Shelf. Positive responses were further screened to delete irrelevant entries, i.e., those not dealing with trawl and other demersal surveys in our reference area (e.g., areas outside of tropical Asia, studies on socio-economic issues, policies and management intervention resulting from trawl and other fishery surveys, taxonomic description of rare/first record species caught by trawl, icthyoplankton and zooplankton studies based on trawling, etc.).

Table III.1 gives the resulting compilation of surveys conducted in tropical Asia. A total of 301 independent surveys have been identified thus far and included in the table. The geographical distribution of these surveys is as follows: Bangladesh - 30; Brunei - 7; Indonesia - 25; Malaysia - 58; Philippines - 41; Sri Lanka - 10; Thailand - 40; Vietnam - 29; Burma - 4; India - 35; Pakistan - 12; South China Sea area - 5; and southern Indonesia/northern Australia - 3. It was acknowledged by Workshop participants that the

[1] Compiled by D. Bonga, L. Garces and G. Silvestre.

surveys enumerated in the table do not include all the surveys actually carried out in South and Southeast Asia, and that many more remain unpublished in the files of various agencies/institutes. For example, entries in the table without reference/sources are unpublished surveys that the various contributors are personally aware of. The Workshop participants were in agreement that more systematic and exhaustive searches would substantially improve the list given in Table III.1 and resolved to make such efforts part and parcel of an expanded regional collaboration to follow-up the Workshop results.

Despite its preliminary nature, the Workshop participants deemed the compilation given in Table III.1 as a good start. Sample data from several surveys in the table were brought by the individual Workshop participants. These sample data were used to test the prototype version of TrawlBase (Gayanilo, et al., this vol.) and explore how retrospective analyses could provide insights on ways to improve fisheries management in South and Southeast Asia.

Table III-1. Summary of information of trawl surveys conducted in tropical Asia, by country/area.

BANGLADESH

Area/survey no.	Period (date)	Vessel	Trawl gear[a]	Trawl operations	Results/notes	References
1	Nov - Dec 1958	Chosui Maru				Chowdhury (1983)
2	Jan - Apr 1960	Kagawa Maru		• 99 hauls	• extent of trawled area: 550 sq. miles • mean catch rate:127 kg·hour^{-1}	
3	Dec 1961 - Dec 1962	Kinki Maru			• sites: estuaries of Sundarbans, off the coast of greater Khulna, Barisal and Chittagong • other gears: shrimp net and floating gill net	
4	1962 - 1966	Jalwa	overhang bottom trawl net		• depth range: 5-58 m • mean catch rate: 135 kg·hour^{-1}	Cushing (1971); Chowdhury (1983), see also Bain (1965)
5	Apr 1963	Anton Bruun			• exploratory surveys in Bay of Bengal and Arabian Sea, see notes/results in India (whole coastline)	Hida and Pereyra (1966)
6	1968 - 1971	Sagar Sandhani/ Meen Sandhani	Mexico type shrimp trawl		• extent of trawled area: whole shelf, 26 000 km^2 • depth range: 10-75 m • abundance estimate: 264 000 - 373 000 t	West (1973)
7	Nov 1969 - Jan 1970	R.T.M. Lesnoy				Anon. (1970)
8	Jan 1972 - Nov 1972	Tamango SRTM 499				
9	Oct 1976 - Oct 1977	M.V. Santamonica/ M.V. Orion 8	shrimp trawl	• 1 902 hauls	• extent of trawled area: 4 200 sq. miles • catch rates; 40 kg·hour^{-1} (shrimp); 49 kg·hour^{-1} (quality fish)	Rachid (1983)
10	Mar 1979	Fisheries Research Vessel No. 2	bottom trawl	• 20 operations	• depth range: 10-82 m • other gears: drift gill net and 50 x 15 m net • catch rate: 649 kg·hour^{-1}	
11	Nov - Dec 1979	R.V. Dr. Fridtjof Nansen	shrimp trawl HL: 96 ft GR: 63 ft	• 49 hauls	• extent of trawled area: whole shelf: 42 000 km^2 • depth range: >10-200 m • other gears used: pelagic trawl, gillnet and long-line • other information collected: general water quality parameters, plankton and bathymetry • mean catch rates: 19-672 kg·hour^{-1} (all depths) • stock density: 164-6 451 kg·km^2 (all depths) • abundance estimate: 160 000 t	Chowdhury et al. (1979); Saetre (1981)
12	May 1980	R.V. Dr. Fridtjof Nansen	shrimp trawl HL: 134 ft	• 61 hauls	• extent of trawled area: whole shelf: 42 000 km^2 • depth range: >10-200 m • other gears used: pelagic trawl, gillnet and long-line • other information collected: general water quality parameters, plankton and bathymetry • mean catch rates: 62-995 kg·hour^{-1} (all depths) • stock density 453-3 267 kg·km^2 (all depths) • abundance estimate: 92 000 t	Chowdhury et al. (1980); Saetre (1981)

[a] HL: headline; HR: headrope; FR: footrope; GR: groundrope; MS: codend mesh size; TS: towing speed; TD: trawling duration; WS: wing span; GT: gross tons; HP: horsepower; LOA: length overall

152

Area/survey no.	Period (date)	Vessel	Trawl gear	Trawl operations	Results/notes	References
13	1981 - 1982	BFDC/ other commercial trawler	single-rig stern trawl/ twin-rigged shrimp trawl		• landing survey • standing stock of demersal fish: 54 900 to 92 000 t	Penn (1983)
14	Jul 1983	R.V. Anusandhani				Khan (1983)
15	Dec 1983	R.V. Anusandhani	midwater trawl rigged as bottom trawl		• extent of trawled area: northern Bay of Bengal • abundance estimate: 150 000 t	Humayon et al. (1983); Khan et al. (1983)
16	15 - 25 Sep 1984	R.V. Anusandhani	Engel type MS: 32 mm	• 42 hauls • TD: 0.5 hour	• depth range: 15-100 m • mean catch rate: 312 kg·hour^{-1}; highest at 10-20 m depths decreasing with depth • dominant species included Sciaenidae, catfish (Ariidae) and rays which made up >50% of total weight of catch • demersal biomass: 200 000-300 000 t; around 2 000 to 3 500 t were prawns	White and Khan (1985a)
17	3 - 13 Oct 1984	R.V. Anusandhani	Engel type MS: 32 mm	• 45 hauls • TD: 0.5 hour	• depth range: 10-105 m • mean catch rate: 226 kg·hour^{-1}; highest at 10-20 m depths, decreasing with depth • most dominant species include Indian mackerel (*Rastrelliger kanagurta*), croaker, catfish and rays which made up >50% of total weight of catch • biomass: 100 000-180 000 t; around 1 700 to 5 500 t were prawns	White and Khan (1985b)
18	20 - 31 Oct 1984	R.V. Anusandhani	Engel type MS: 32 mm	• 43 hauls • TD: 0.5 hour	• depth range: 10-105 m • mean catch rate: 174 kg·hour^{-1}; highest at 20-30 m • dominant species include croakers, catfish, goatfish and rays which made up approx. 40% of total weight of catch • biomass: 115 000-180 000 t; around 6 500 t were prawns	White and Khan (1985c)
19	9 - 20 Nov 1984	R.V. Anusandhani	Engel type MS: 32 mm	• 44 hauls • TD: 0.5 hour	• depth range: 15-102 m • mean catch rate: 234 kg·hour^{-1}; highest at 10-30 m depths • dominant species include Indian mackerel, threadfin bream, catfish, goatfish and croaker • biomass: 170 000-255 000 t; around 2 300 t were prawns	White (1985a)
20	27 Nov - 6 Dec 1984	R.V. Anusandhani	Engel type MS: 32 mm	• 40 hauls • TD: 0.5 hour	• depth range: 16-105 m • mean catch rate: 296 kg·hour^{-1}; highest at depths >100 m and lowest at 21-30 m depth • dominant species include Japanese threadfin bream, catfish, goatfish, hairtail and white pomfret • penaeid prawns were abundant at 10-20 m • biomass: 160 000-340 000 t; around 4 300 t were prawns	White (1985b)

153

Area/survey no.	Period (date)	Vessel	Trawl gear	Trawl operations	Results/notes	References
21	13 - 21 Dec 1984	R.V. Anusandhani	Engel type MS: 32 mm	• 41 hauls • TD: 0.5 hour	• depth range 15-105 m • mean catch rate: 274 kg·hour⁻¹; highest at 10-20 m depths • dominant species were round and hard scads (*Decapterus maruadsi* and *Megalaspis cordyla*) croakers, hairtail, threadfin bream, catfish and goatfish, • biomass: 230 000 t, prawns <1%	White (1985c)
22	6 - 16 Jan 1985	R.V. Anusandhani	Engel type MS: 32 mm	• 45 hauls • TD: 0.5 hour	• depth range: 15-107 m • mean catch rate: 198 kg·hour⁻¹ • dominant species include catfish, croaker, hairtail, scads, threadfin bream and lizard fish • biomass: 40 000-294 000 t; about 100 t were prawns	White (1985d)
23	31 Jan - 11 Feb 1985	R.V. Anusandhani	Engel type MS: 32 mm	• 49 hauls • TD: 0.5 hour	• depth range: 15-113 m • mean catch rate: 274 kg·hour⁻¹; rates at 10-20 m depths four times higher than deepest zone • dominant species include scads, croaker, hairtail, Japanese threadfin bream, catfish and goatfish • biomass: 220 000 t; about 3 700 t were prawns	White (1985e)
24	17 Feb - 24 Feb 1985	R.V. Anusandhani	Engel type MS: 32 mm			White (1985f)
25	19 - 24 May 1985	R.V. Anusandhani	Engel type MS: 32 mm			White (1985g)
26	Sep 1984 - Jun 1986	R.V. Anusandhani	Engel type MS: 32 mm	• 544 hauls • TD: 0.5 hour	• extent of trawled area: whole shelf; 31 340 km² • depth range: 10->100 m • mean stock density: 2 914-7 926 kg·km² (all depths and all species of fish) • biomass: 161 000 t	Lamboeuf (1987)
27	Nov 1985 - Jan 1987	R.V. Anusandhani			penaeid shrimp trawl survey	Mustafa et al. (1987)
28	Oct 1988 - Apr 1989	R.V. Anusandhani	shrimp trawl net HR: 27.8 m GR: 33 m MS: 45 mm	• 83 hauls • TD: 1.0 hour	• extent of trawled area: whole shelf • depth range: 20-100 m • other information collected: length-frequency, oceanographic parameters • biomass estimate: 1 548 t (penaeid shrimp) • max. stock density: 141 kg·km² at 51-80 m	Khan et al. (1989a)
29	Jun 1988 - Jan 1989	R.V. Machhranga	high opening Engel type HR: 29 m GR: 37 m MS: 32 mm	• 31 hauls • TD: 1.0 hour	• extent of trawled area: shallow portion of the shelf; 6 870 km² • depth range: <20 m • other information collected: length-frequency, oceanographic parameters • biomass estimate: 17 000 t (fish), 135 t (shrimp)	Khan et al. (1989b)

Area/survey no.	Period (date)	Vessel	Trawl gear	Trawl operations	Results/notes	References
30	1989 - 1991		shrimp trawlnet HR: 15.2 m GR: 18.6 m MS: 45 mm Engel high opening trawl; HR: 75.5 m GR: 18.6 m MS: 32 mm		• depth range: 10-80 m • catch rates: 6.6 kg·hr⁻¹ (shrimp in shrimp trawl); 74 kg·hr⁻¹ (finfish in shrimp trawl); 53 kg·hour⁻¹ (fish in finfish trawl)	Mustafa and Khan (1993)

BRUNEI DARUSSALAM

Area/survey no.	Period (date)	Vessel	Trawl gear	Trawl operations	Results/notes	References
1	Jun - Dec 1949 Jan - Sep 1950	M.V. Saripah M.F.V. Tenggiri both 23.1 m LOA	30 ft (9.2 m)		• other gears used: Danish seine, fish pots and longlines	Beales et al. (1982)
2	1 - 8 Dec 1955	F.R.V. Manihine 36.3 LOA, 2x220 hp	'Peter Carey' otter trawl; HR: 22 m MS: 89 mm	• 8 hauls • TD: variable • TS: 3 knots	• depth range: 36-198 m • longline, drift net, troll and lift net	Beales et al. (1982)
3	14 - 16 Mar 1968 1 - 13 Jul 1968 14 - 27 Sep 1968 9 - 23 Nov 1968	M.F.V. Arapan Tei M.F.V. Berjaya M.F.V. Berjaya M.F.V. Berjaya both 12.3 m LOA 60 hp	'prawn trawl' HR: 17.6 m MS: 38 mm	• 115 hauls • TS: 2.0-2.5 knots • TD: 1.0-1.5 hour	• depth range: 11-46 m	DOF (1968)
4	20 - 21 Apr 1972	F.R.V. Penyelidek I F.R.V. Penyelidek II both 23 m LOA former 325 hp and the latter 365 hp	'Engel type' trawl, HR('effective'): 20 m MS: 40 mm	• 24 hauls • TS: 2.8 knots • TD: 1.0 hour	• depth range: 29-88 m	Mohammed Shaari et al. (1976b)
5	Feb 1979 - Nov 1981	K.P. Lumba-lumba 15.2 m LOA, 287 hp	'Boris Goshawk' trawl, HR: 17.4 m HR('effective'): 9.1m MS: 35 mm	• 281 hauls • TS: 3.0 knots • TD: 0.5 hour	• depth range: 10-180 m	Beales et al. (1982)
6	1980 - 1986				• independent trawl monitoring	Halidi (1987)

155

Area/ survey no.	Period (date)	Vessel	Trawl gear	Trawl operations	Results/notes	References
7	Jul 1989 - Jun 1990	K.P. *Lumba-lumba* 15.2 m LOA 287 hp	'Boris Goshawk' trawl, HR: 17.4 m HR ('effective'): 9.1 m MS: 35 mm	• 143 hauls • TS: 3.0 knots • TD: 0.5 hour	• extent of trawled area: 7 396 km² • depth range: 10-100 m • stock density: 8.3 t·km⁻² • biomass: 61 300 t	Silvestre and Matdanan (1992)

INDONESIA (JAVA SEA and SOUTHERN SOUTH CHINA SEA)

Area/ survey no.	Period (date)	Vessel	Trawl gear	Trawl operations	Results/notes	References
1	12 Nov - 1 Dec 1972	R.V. *Oh Dae San*	HR: 50 m GR: 63 m	• 51 hauls • TS: 3.0-3.5 knots • TD: variable	• South coast of Java • depth range: 20-290 m	-
2	11 - 22 Dec 1973	M.S. *Dah Choan*	HR: 152 ft GR: 155 ft	• 53 hauls • TD: 1.5-2.0 hours	• South China Sea (Karimata Strait) • depth range: 20-80 m	-
3		M.S. *Jeou Ta*	HR: 152 ft	• 69 hauls • TD: 1.5-2.0 hours	• South China Sea (Karimata Strait) • depth range: 20-80 m	-
4	Nov 1974 - Jul 1976	R.V. *Mutiara IV*	Thailand trawl	• 441 hauls • TD: 1.0 hour	• extent of trawled area: Java Sea and Southern South China Sea; 717 562 km² • depth range: 10-70+ m • other information collected: length-frequency, mesh selection experiment, zooplankton and benthos sampling; back-and-forth fishing experiment • biomass: 259-389 t • stock density: 0.8-5.2 t·km⁻²	Pauly et al. (1996)
5	Jun - Dec 1976	R.V. *Mutiara IV*	Thailand trawl	• 189 hauls • TS: 3.0 knot • TD: 1.0 hour	• extent of trawled area: Java Sea, south of latitude 03°S between the southeastern coast of Sumatra, the north coast of Java, the southern coast of Kalimantan and eastwards to the edge of the continental shelf in southern Strait of Makassar; 467 000 km² • depth range: 10-70+ m • total mean catch rate for the monsoon survey was 262 kg·hour⁻¹ and from 'aimed' fishing, 163 kg·hour⁻¹ • the standing stock was estimated at 1.2 million t, of which 0.7 million t was estimated for offshore zone with potential annual yield of around 0.3 million t	Losse and Dwiponggo (1977)
6	Jan - Dec 1977	R.V. *Mutiara IV*		• 212 hauls	• Java Sea (nine subareas); 297 500 km² • depth range: 10-50 m • mean catch rate: 212 kg·hour⁻¹ • stock density: 2.03 t·km⁻² - 6.6 t·km⁻² • other information collected: catch composition by sub-area, catch rate by depth	Dwiponggo and Badrudin (1978a, 1978b)

Area/survey no.	Period (date)	Vessel	Trawl gear	Trawl operations	Results/notes	References
7	Jan - Dec 1978	R.V. Mutiara IV	Thailand trawl MS: 20, 40 and 60 mm	• 167 hauls • TS: 3.0 knots • TD: 0.5-1.0 hour	• Java Sea sub-areas (southeast coast of Sumatra, north coast of west and central Java and south coast of Kalimantan and offshore area around Bawean and Masalembu Island) • depth range:10-90 m • mean total catch rate: 143 kg·hour⁻¹ • high mean catch rate recorded at north coast of Java • good fishing grounds were found at 10-30 m depth	Dwiponggo and Badrudin (1979a, 1979b)
8	Jan - Dec 1979	R.V. Mutiara IV	Thailand trawl HL: 34.6 m FR: 42.2 m MS: 40 mm	• 351 hauls • TS: 3.0 knots • TD: 0.5-1.0 hour	• Java Sea, six inshore sub-areas • depth range: 10-50 m • catch rate of food fish: 159 kg·hour⁻¹ • good fishing grounds were located at depths <40 m • other information: catch composition by sub-area, by depth and by season	Dwiponggo and Badrudin (1980a, 1980b)
9	Jun - Jul 1978	R.V. Mutiara IV	Thailand trawl HL: 34.6 m FR: 42.2 m MS: 40 mm	• 64 hauls • TS: 3.0 knots • TD: 1.0 hour	• southern South China Sea between longitude 104° E and West Kalimantan (Borneo) and between latitudes 03'S and 02'N; 287 033 km² • depth range: 10-70 m • mean catch rate: 119 kg·hour⁻¹ • mean stock density: 1.8 t·km⁻² • potential annual yield: 200-300 thousand t • large areas of offshore fishing grounds (>30 m sounding depth) provided low catch rates; highest catch rates were obtained in near-shore waters off Kalimantan and around islands off the east coast of Sumatra • prevailing oceanographic conditions were discussed	Sudradjat and Beck (1978)
10	11 Jan - 1 Dec 1979	R.V. Mutiara IV	HR: 34.6 m GR: 42.2 m	• 344 hauls • TS: 3.0 knots • TD: 1.0 hour	• Java Sea • depth range: 10-70 m	—
11	Oct 1982	R.V. Mutiara IV	HR: 34.6 m GR: 23.4 m	• 6 hauls • TS: 2.5 knots • TD: 1.0 hour	• South coast of Java • depth range: 20-50 m	—
INDONESIA (MALACCA STRAIT)						
1	Jul - Sep 1973	M.V. Mutiara I	double rig shrimp trawl HR: 23.44 m GR: 27.84 m MS: 30 mm	• 61 hauls • TS: 2.5-3.5 knots • TD: 2.0 hour	• Malacca Strait • depth range: 10-40 m • mean catch rates: 37-78 kg·hour⁻¹ (all depths) • stock density: 0.5-1.1 t·km⁻²	Martosubroto et al. (1996)
2	Jul - Sep 1973	M.V. Mutiara II	double rig shrimp trawl HR: 19.0 m GR: 22.06 m MS: 30 mm	• 87 hauls • TS: 1.5-2.5 knots • TD: 2.0 hour	• Malacca Strait • depth range: 10-40 m • mean catch rates: 41-65 kg·hour⁻¹ • stock density: 0.9-1.4 t·km⁻²	

Area/survey no.	Period (date)	Vessel	Trawl gear	Trawl operations	Results/notes	References
3		M.V. Larasati	HR: 18.4 m GR: 22.4 m	• 5 hauls • TS: 2.5-3.5 knots • TD: 2.0 hour	• Malacca Strait • depth range: 10-40 m	-
4	Jan - Mar 1975	M.V. Mutiara IV	Thailand trawl HR: 36 m MS: 40 mm	• 40 hauls • TS: variable • TD: 2.0 hour	• Malacca Strait • depth range: 10-70+ m • mean catch rates: 45-158 kg·hour^{-1} • stock density: 0.7-2.5 t·km^{-2}	Martosubroto et al. (1996)
5	Apr 1983	R.V. Bawal Putih I	HR: 34.6 m GR: 42.2 m	• 10 hauls • TS: 3.0-3.5 knots • TD: 1.0 hour	• Malacca Strait • depth range: 20-60 m	-
			HR: 24.1 m GR: 29.0 m	• 34 hauls • TS: 3.0-3.5 knots • TD: 1.0 hour		

INDONESIA (SOUTHEASTERN PART OF INDIAN OCEAN)

Area/survey no.	Period (date)	Vessel	Trawl gear	Trawl operations	Results/notes	References
1	6 - 30 Aug 1980	R.V. Dr. Fridtjof Nansen	shrimp trawl 1 500 hp HR: 134 ft	• 48 hauls	• north and west Sumatra • depth range: 10-300 m • average catch rates: 27-112 kg·hour^{-1}, all depths • estimated standing demersal stock was 65 000 t and about 80% of the demersal fish was found within 50 m depths (acoustic estimate of standing demersal stock was 124 000 t) • Leiognathidae was the most dominant catch Trichiurus haumela and various Carangidae were caught mostly in shallow areas; the families Nemipteridae, Mullidae and Synodontidae were frequently occurring while Lutjanidae and Serranidae were restricted along the edge of shelf • other gears used: pelagic trawl (22 hauls) and echo recordings • other information collected: bottom conditions, hydrography and nutrients	Aglen et al. (1981a), Bianchi (1996)
2	Aug 1979 - Jul 1981	F.R.V. Jurong	high opening bottom otter trawl MS: 40 mm	• 515 hauls • TS: 3.0 knots • TD: 1.0 hour	• extent of trawled area: central coast of Sumatra to Java and Bali; 70 000 km^{2} • depth range: 50-200+ m • other gear/information used: pelagic trawl (154 hauls), echosounding • mean catch rates: 18-1 495 kg·hour^{-1} (n= 301, all depths and seasons) • biomass: 43 000-140 000 t	Lohmeyer (1996)

158

Area/ survey no.	Period (date)	Vessel	Trawl gear	Trawl operations	Results/notes	References
3	Nov 1980 - Oct 1981	R.V. Bawal Putih 2	Engel type, high opening bottom trawl MS: 40 mm	• 121 hauls	• southeastern part of Indian ocean (in the vicinity of Lombok, Sumbawa, Sumba, Flores and Timor); 3 270 km^2 • depth range: <100 m • other information: temperature profiles by bathythermographs • catch rates: 211-646 kg·hour^{-1} • stock density: 1.9-4.5 t·km^2 • total potential fish yield: 163 950 t·yr^{-1}	Martosubroto et al. (1996)
4	Jun 1982	R.V. Bawal Putih I	HR: 32.6 m GR: 40.2 m HR: 40.6 m GR: 43.7 m	• 9 hauls • TS: 3.0 knots • TD: 1.0 hour • 9 hauls • TS: 3.0 knots • TD: 1.0 hour	• Arafura Sea (sub-area Dolak) • depth range: 15-40 m	-
5	Feb - Mar 1984	M.V. Aman-10	HR: 25.3 m GR: 29.0 m	• 52 hauls • TS: 2.4-3.0 knots • TD: 2.5-3.3 hour	• Arafura Sea (sub-area Bintuni Bay) • depth range: 10-30 m	-
6	27 Mar - 5 May 1985	R.V. Bawal Putih II	HR: 33.8 m GR: 37.7 m	• 38 hauls • TS: 2.5-3.0 knots • TD: 1.0-1.5 hour	• Arafura Sea (sub-area Dolak, Timika) • depth range: 10-40 m	-
7	May - Oct 1991	R.V. Bawal Putih II		• 476 hauls • TS: 3.0 knots • TD: 2.0-3.0 hour	• Arafura Sea (sub-area Aru) • depth range: 10-40	-
8	5 Nov - 21 Dec 1992	M.V. Baruna Jaya IV	HR: 16.4 m GR: 18.1 m	• TS: 3.0 knots • TD: 1.0 hour	• Arafura Sea (sub-area Dolak) • depth range: 30-250 m	-
9	Sep - Oct 1995	R.V. Bawal Putih I	HR: 34.6 m GR: 42.2 m	• 40 hauls • TS: 2.5 knots • TD: 0.5-1.0 hour	• depth range: 15-90 m • West Sumatra (sub-area Bengkulu)	-

MALAYSIA (West Coast)

1	1965	M.V. Selayang wooden trawler 16.75 m LOA 95 hp	otterboard trawl	• 31 hauls	• off Penang • depth range: 10-50 m • range of mean catch rate: 50-234 kg·hour^{-1} highest at 30 m depth and lowest at 50 m • other information: catch rates and percent composition of species by depth, bathymetric distributions of fish fauna, increase of mean lengths of fish with depth and relationship of catch with hydrographic factors	Pathansali et al. (1967)

Area/survey no.	Period (date)	Vessel	Trawl gear	Trawl operations	Results/notes	References
2	1965,1966, 1967,1970	Dhanarat later renamed Fishery Research No. 2	Japanese type trawl net HR: 74 ft GR: 105 ft MS: 2.9 in	• 321 hauls • TS: 2.5 knots • TD: 1.0 hour	• from Satul down to about latitude 5°N (south of Penang) 16 425 sq. miles • depth range: > 100 m • catch rate: 454 kg·hour⁻¹ (1965) • standing stock: 186 000 - 271 000 t (1965)	Isarankura (1971)
3	12 Dec 1970 - 22 Jan 1971	Penyelidek I (JENAHAK) Wooden stern trawler 23 m LOA, 55 GT, 325 hp diesel engine	Standard German otter trawl MS: 40 mm	• 81 hauls • TS: 2.8 knots • TD: 1.0 hour	• Pulau Langkawi to Pulau Penang (sub-area I) to Pulau Pangkor (sub-area II) • depth range: 10-60 m • mean catch rate:131 kg·hour⁻¹; sub-area I more productive (152 kg·hour⁻¹) than sub-area II (108 kg·hour⁻¹) • catch rates of 300->400 kg·hour⁻¹ were reported at 30-40 m depths at sub-area I • mean stock density: 1.3 t·km⁻² (1.5 t·km⁻² at sub-area I; 1.0 t·km⁻² at sub-area II) • other information collected: oceanography, bottom soil samples and length-frequency data	Mohammed Shaari et al. (1974)
4	30 Nov 1971 - 11 Jan 1972	Penyelidek II (MERAH) Wooden stern trawler 23 m LOA, 365 hp diesel engine	Standard German otter trawl MS: 40 mm	• 97 hauls • TS: 2.8 knots • TD: 1.0 hour	• extent of trawled area: Pulau Langkawi to Pulau Pangkor; 5 880 km² • depth range: 10-60 m • mean catch rate:142 kg·hour⁻¹; sub-area I more productive (165 kg·hour⁻¹) than sub-area II (122 kg·hour⁻¹) • stock density : 1.4 t·km² • other information: catch per unit effort by species, length frequencies	Lam et al. (1975a)
5	13 - 17 Jun 1972	Jenahak	German net HR: 13.30 m GR :15.85 m MS: 30 mm	• 95 hauls • TS: 2.5 knots • TD: 1.0 hour	• Prawn resource survey • depth range: 5-50 m • average catch rate: 43 kg·hour⁻¹: (prawns, 3.54 kg·hour⁻¹; food fish, 13.6 kg·hour⁻¹; 'trash fish', 26 kg·hour⁻¹) • mean catch rate highest at Cape Rachado - Johore area (59 kg·hour⁻¹) but the best area for prawns was obtained at Penang - Langkawi area (4.1 kg·hour⁻¹) • other information: mean catch rates by depth, prawn species composition, lengths and body weights	Ong and Weber (1977)
6	19 - 23 Jun 1972	Jenahak	German net HR: 16.3 m GR: 19.90 m MS: 35 mm	• 58 hauls • TS: 2.5 knots • TD: 1.0 hour	• Prawn resource survey • depth range: 5-50 m • other information: lengths and body weights	A. Abu Talib and M. Alias (pers. comm.)
7	4 - 30 Apr 1973	Jenahak Wooden stern trawler 23 m LOA, 85 GT, 325 hp diesel engine	Standard German otter trawl MS: 40 mm	• 96 hauls • TS: 2.8 knots • TD: 1.0 hour	• off Pulau Pangkor to southwestern tip of Johore • depth range: 21-60 m • mean catch rates:125 kg·hour⁻¹, highest at west coast of Johore (172 kg·hour⁻¹) and lowest at areas between Kuala Sepang and Kuala Kesang (65 kg·hour⁻¹) • other information: species composition of commercial fish, productivity of the grounds and length-frequency data	Mohammed Shaari et al. (1976a)

Area/survey no.	Period (date)	Vessel	Trawl gear	Trawl operations	Results/notes	References
8	16 Nov - 11 Dec 1974	*Jenahak*	Standard German otter trawl	• 90 hauls • TS: 3.0 knots • TD: 1.0 hour	• Pulau Langkawi to Pulau Pangkor • depth range: 10-60 m • other information: length-frequency data • mean catch rate: 92 kg·hour^{-1}; rates indicated 30% decrease from previously reported values in sub-areas I and II • higher percentage of squids and cuttle fish caught than the previous surveys; Nov. was cephalopod season	Mohammed Shaari and Chai (1976)
9	17 Oct - 18 Nov 1978	*Jenahak*	Standard German otter trawl	• 114 hauls • TS: 2.8 knots • TD: 1.0 hour	• Pulau Langkawi to Pulau Pangkor • depth range: 10-60 m • mean catch rates: 69 kg·hour^{-1}, further decline from previous surveys, sub-area I could only yield 50-100 kg·hour^{-1} • high catch rate recorded at 21-30 m depth (105 kg·hour^{-1}) • other information: catch composition, catch rates by fish group/families and length-frequency data	Lui (1981)
10	5 - 15 Jul 1980	*R.V. Dr. Fridjoft Nansen*	shrimp trawl HL: 134 ft	• 35 hauls	• between latitude 6°N near the Thai-Malaysia border to 3°N off Port Kelang and extending seawards to the median line between Malaysia and Sumatra • depth range: 10-100 m • average catch rates of 50-60 kg·hour^{-1}, no strong variation with depth • total standing demersal stock: 34 500 t (acoustic estimate was 300 000 t of pelagic fish and 30 000 t demersal fish) • other information collected: fish distribution and catch composition, bottom conditions and hydrography • vessel was equipped for acoustic surveying, bottom and midwater trawling, hydrography and plankton observations	Aglen et al. (1981b)
11	20 Nov - 23 Dec 1980	*Jenahak*	Standard German otter trawl	• 92 hauls • TS: 2.8 knots • TD: 1.0 hour	• depth range: 10-60 m • other information: length frequency	A. Abu Talib and M. Alias (pers. comm.)
12	22 Apr - 14 Jul 1981	*Pelaling* Wooden hull 15.85 m LOA 42 GT, 160 hp	Local design net TL: 35.14 m HR: 23.5 m GR: 24.7 m MS: 40 mm	• 117 hauls • TS: 2.2 knots • TD: 1.0 hour	• Prawn resource survey • depth range: 1-50 m • prawn catch rate: 2.18 kg·hour^{-1} in 1981 0.59 kg·hour^{-1} in 1984	Nasir and Johari (1990)
13	12 Apr - 29 Aug 1984			• 137 hauls • TS: 2.2 knots • TD: 1.0 hour	• reduction in number of fishing grounds with high catch rates • other information: individual total lengths and body weights	Ahmad Adnan (1987)

Area/survey no.	Period (date)	Vessel	Trawl gear	Trawl operations	Results/notes	References
14	13 Oct - 15 Nov 1981	*Jenahak*	Standard German otter trawl	• 76 hauls • TS: 2.8 knots • TD: 1.0 hour	• Pulau Langkawi to Pulau Pangkor • depth range: 60-120 m • mean catch rate: 55 kg·hour⁻¹ • other information: catch composition, catch rates by fish groups/families and length-frequency data	Ahmad Adnan (1987)
15	1986	*Mersuji*		• 44 hauls • TD: 1.0 hour	• other information: length-frequency data	
16	6 Jul - 19 Sep 1987	*Pelaling*	Local design net	• 148 hauls • TS: 1.2 knots • TD: 1.0 hour	• Prawn resource survey • depth range: 5-50 m • other information: lengths and body weights	A. Abu Talib and M. Alias (pers. comm.)
17	23 Oct - 19 Dec 1988	*Kerapu IV* Wooden hull 110 hp, 40 GT	Local design net	• 104 hauls • TS: 1.2 knots • TD: 1.0 hour	• Prawn resource survey • depth range: 5-50 m • other information: total lengths and body weights	A. Abu Talib and **M. Alias** (pers. comm.)
18	1988	*Mersuji*		• 41 hauls • TD: 1.0 hour	• other information: length-frequency data	-
19	1989, 1990	*Mersuji*		• 28 hauls • TD: 1.0 hour	• other information: length-frequency data	-
20	31 Jul - 19 Dec 1990	*Pelaling*	Local design net	• 68 hauls • TS: 2.0 knots • TD: 1.0 hour	• Prawn resource survey • depth range: 5-50 m • other information: lengths and body weights	A. Abu Talib and M. Alias (pers. comm.)
21	1990, 1991	*Mersuji*		• 58 hauls • TD: 1.0 hour	• other information: length-frequency data	-
MALAYSIA (East Coast)						
1	1926	*S.T. Tonggol*			• depth range: 10-60 m	Birtwistle and Green (1927, 1928); Menavesta (1970)
2	1955	*F.R.V. Manihine*				Ommaney (1961 *In* Menavesta 1970)
3	1 Mar - 11 Apr 1967	*Pramong II* Wooden stern trawler 23 m LOA 320 hp	German type otterboard trawl net MS: 40 mm	• 153 hauls • TD: 1.0 hour	• Thai border to Johore • depth range: 10-60 m • mean catch rate: 428 kg·hour⁻¹ (range: 23-3 000 kg·hour⁻¹) • Sub-area III (off Kuantan and north of P. Tioman) had highest mean catch rate of 745 kg·hour⁻¹ while subarea lowest at sub-area IV (off Johore),183 kg·hour⁻¹ • 21-30 m depth yielded highest mean catch (646 kg·hour⁻¹); other depths compare as follows: 31-40>10-20>41-50>51-60 • other information collected: catch composition, comparison of catch rates between fish for human consumption and 'trash fish', estimate of catch value, bottom characteristics, fish tagging and fish biology studies	Marine Fisheries Laboratory, Department of Fisheries, Ministry of Agriculture, Bangkok, Thailand and Fisheries Research Institute, Fisheries Division, Ministry of Agriculture and Cooperatives, Malaysia (1967)

Area/survey no.	Period (date)	Vessel	Trawl gear	Trawl operations	Results/notes	References
4	10 Aug - 4 Oct 1970	*Penyelidek I* Wooden stern trawler 85 GT, 23 m LOA 325 hp	Engel type trawl HR: 20 m MS: 40 mm	• 151 hauls • TS: 2.8 knots • TD: 1.0 hour	• Thai border to Johore • depth range: 10-65 m • mean catch rate: 516 kg·hour[-1]; highest catches were recorded at Pahang with a mean of 835 kg·hour[-1]; lowest at Johore with a mean of 210 kg·hour[-1] • 21-30 m depths yielded highest mean catch (816 kg·hour[-1]); other depths compare as: 10-20>31-40>51-60 • other information: data on catch rates according to commercial fish and 'trash fish' as well as catch rates according to species, length frequencies, tagging of fish, soil/bottom sample, oceanography (temperature, salinity, dissolved oxygen and pH)	Pathansali et al. (1974)
5	23 Mar - 26 May 1971	*R.V. Jenahak (Penyelidek I)*	Engel type trawl HR: 20 m MS: 40 mm	• 150 hauls • TS: 2.8 knots • TD: 1.0 hour	• Thai border to Johore • depth range: 10-60 m • other information: length frequencies, tagging of fish, soil/bottom sample, oceanography (temperature, salinity, dissolved oxygen and pH) • mean catch rate: 167 kg·hour[-1]	Jothy et al. (1975)
6	14 Aug - 20 Sep 1972	*R.V. Jenahak*	Engel type trawl HR: 20 m MS: 40 mm	• 145 hauls • TS: 2.8 knots • TD: 1.0 hour	• Thai border to Johore • depth range: 10-80 m • mean catch rate: 255 kg·hour[-1], highest at sub-area II (southern two-thirds of Terengganu), 377 kg·hour[-1] and lowest at sub-area III (Pahang), 217 kg·hour[-1] • other information: catch rates by depth zones, catch rate of food and 'trash fish' and catch rate by species, length-frequency data	Lam et al. (1975b)
7	13 Jul - 12 Aug 1974	*R.V. Jenahak*	Standard German otter trawl net	• 97 hauls • TS: 2.8 knots • TD: 1.0 hour	• depth range: 10-60 m • mean catch rate : 238 kg·hour[-1]; highest at sub-area II (southern two-thirds of Terengganu), 358 kg·hour[-1] and lowest at sub-area I (coast of Kelantan), 145 kg·hour[-1] • commercial fish catch highest at 31-60 m depths (155.1-233 kg·hour[-1]) • other information: occurrence of the frequently occurring families and species of commercial value were discussed	Lamp and Mohammed Shaari (1976)
8	Jun - Jul 1981	*R.V. Jenahak*	Standard German otter trawl net	• 78 hauls • TS: 2.8 knots • TD: 1.0 hour	• Sub-areas I to IV surveyed • depth range: 10-60 m • mean catch rate: 160 kg/hr[-1]; commercial fish, 57.8 kg·hour[-1] and 'trash fish', 102 kg·hour[-1] • other information: catch rates and percent dominance of fish groups by depth zones and subareas, length-frequency data	Ahmad Adnan (1990)

Area/survey no.	Period (date)	Vessel	Trawl gear	Trawl operations	Results/notes	References
9	Nov 1969 - Oct 1973	M.V. Changi 387 GT research vessel of the Marine Fisheries Research Department, SEAFDEC		• 321 hauls	• examined population parameters of ten selected demersal fish and squid species taken off the southern east coast at sub-area VII (Pahang, 216 hauls) and sub-area VIII (Johore, 105 hauls) • catch and effort data (by area) and the distribution of *Lutjanus* spp., Mullidae, *Nemipterus* spp., Pomadasyidae, *Priacanthus* spp., *Pristipomoides* spp., *Scomberomorus* spp. and squid and cuttlefish were provided • stock density: 4.5 t·km⁻² (area VII) and 6.3 t·km⁻² (area VIII) • standing stock: 19 000 t (area VII) and 15 000 t (area VIII)	A. Abu Talib and Hayase (1984b)
10	10 - 25 Jun 1980	R.V. Dr. Fridtjof Nansen	shrimp trawl HL: 134 ft	• 60 hauls	• from Singapore to off Tumpat coast • areas divided into coastal areas (10-25 m depth zones) and offshore areas (26-100 m depth zones) • average catch rates: coastal areas: 160 kg·hour⁻¹ offshore areas: 51 kg·hour⁻¹ • demersal stock: 80 000 t (32 000 t at coastal areas and 48 000 t at offshore areas) • other information provided were pelagic trawl to echo-recording results	Aglen et al. (1981b)
11	1983	R.V. Mersuji	Standard German otter trawl net	• 115 hauls • TS: 2.8 knots • TD: 1.0 hr	• other information: length-frequency data	
12	1984	R.V. Mersuji	Standard German otter trawl net	• 120 hauls • TS: 2.8 knots • TD: 1.0 hr	• other information: length-frequency data	
13	Jul 1984 and Aug 1985	Unipertama I	WS(effective): 6 m MS: 38.1 mm	• 15 hauls • TS: 2.3-2.4 knots • TD: 1.0 hr	• extent of trawled area: off Terengganu; 1 050 km² • depth range: <20 m • other information: length-frequencies, biomass estimates using ECOPATH • stock density: 1 216 kg·km⁻²	Chan and Liew (1986)
14	7 - 17 May 1985	R.V. Kagoshima Maru 68.93 m LOA 1 293 GT	otter trawl HL: 53.07 m and 42.64 m MS: 38.1 mm and 31.8 mm	• 13 hauls • TD: 1.0 hr	• results of *Expedisi Matahari '85* in the offshore waters of Malaysian EEZ off the coast of Terengganu, 8 147 km², 90-220 km from the coast, between latitude 05°10'N - 06°23'N and longitude 103°30'E - 140°36'E • depth range: 60-80 m • dominant catches were Lutjanidae, Nemipteridae, Synodontidae, Priacanthidae, Balistidae, Carangidae, Mullidae, Loliginidae and Sepiidae • biomass: 2 762 t • total catch landed: 937 kg	Mohamed et al. (1986)

164

Area/ survey no.	Period (date)	Vessel	Trawl gear	Trawl operations	Results/notes	References
15	1988	R.V. Mersuji	Standard German otter trawl net	• 120 hauls • TS: 2.8 knots • TD: 1.0 hour	• other information: length-frequency data	-
16	1990	R.V. Mersuji	Standard German otter trawl net	• 60 hauls • TS: 2.8 knots • TD: 1.0 hour	• other information: length-frequency data	-
17	1991	R.V. Mersuji	Standard German otter trawl net	• 62 hauls • TS: 2.8 knots • TD: 1.0 hour	• other information: length-frequency data	-
18	1995	R.V. Mersuji	Standard German otter trawl net	• 100 hauls • TS: 2.8 knots • TD: 1.0 hour	• other information: length-frequency data	-

MALAYSIA (BOTH EAST AND WEST COASTS)

Area/ survey no.	Period (date)	Vessel	Trawl gear	Trawl operations	Results/notes	References
1	1971 - 1980				• assessment of demersal resources taken from total landings by trawls and demersal landing from all types of fishing gears in the west and east coasts of Malaysia; divided into eight sub-areas • data analysis of catch by trawl vs catch by all types of fishing gear were done for demersal fish and prawns • results showed that the trawl was the sole or major gear used for demersal fishes while prawn were also strongly exploited by other gears at the same rate target species of trawls have changed from demersal fish to prawns after 1972 • MSY of demersal fish: 168 810 $t \cdot year^{-1}$ (west coast) and 56 870 t (east coast) • MSY of prawns: 66 340 $t \cdot year^{-1}$ (west coast) and 4 060 $t \cdot year^{-1}$ (east coast)	A. Abu Talib and Hayase (1984a)

SARAWAK, SABAH AND BRUNEI

Area/ survey no.	Period (date)	Vessel	Trawl gear	Trawl operations	Results/notes	References
1	29 Mar - 1 May 1972	R.V. Jenahak Wooden vessel 23 m LOA, 85 GT 325 hp; and R.V. Merah Wooden vessel 23 m LOA, 80 GT 365 hp	Standard German otter trawl MS: 40 mm Polyethylene	• 292 hauls • TS: 2.8 knots • TD: 1.0 hour	• off the coasts of Sarawak, Brunei and west coast of Sabah • depth range: 10-60 m • mean catch rates: 318 $kg \cdot hour^{-1}$ • other information: catch composition, average catch rates by sub-areas and depths and bottom samples	Mohammed Shaari et al. (1976b)

Area/survey no.	Period (date)	Vessel	Trawl gear	Trawl operations	Results/notes	References
2	17 Aug - 18 Sep 1973	R.V. Jenahak	Standard German otter trawl MS: 40 mm Polyethylene	• 112 hauls • TS: 2.8 knots • TD: 1.0 hour	• off the coast of Sarawak, Tanjung Datu to Miri • depth range: 10-60 m • mean catch rates: 210 kg·hour⁻¹ (→ $210\ \text{kg}\cdot\text{hour}^{-1}$) • highest mean catch rates 21-30 m depths ($372\ \text{kg}\cdot\text{hour}^{-1}$) • 'trash fish' contributed 27% to total catch by weight • other information: catch composition of commercial fish, locations of productive grounds and bottom characteristics	Mohammed Shaari et al. (1976c)
3	16 Jul - 8 Aug 1975	R.V. Jenahak and R.V. Merah	Standard German otter trawl MS: 40 mm Polyethylene	• 174 hauls • TS: 2.8 knots • TD: 1.0 hour	• off the coast of Sarawak, Tanjung Datu to Miri • depth range: 10-60 m • mean catch rates: $200\ \text{kg}\cdot\text{hour}^{-1}$ • 'trash fish' contributed 39% to total catch by weight • other information: average catch rates by sub-areas and depth, catch composition of commercial fish, distribution of lutjanids and bottom samples	Lui et al. (1976)
4	29 Jun - 12 Aug 1977	R.V. Merah	Standard German otter trawl MS: 40 mm Polyethylene	• 105 hauls • TS: 2.8 knots • TD: 1.0 hour	• depth range: 10-60 m • other information: bottom sample and length frequencies	A. Abu Talib and M. Alias (pers. comm.); Bejie undated; (unpublished)
5	10 Jul - 10 Aug 1979	R.V. Merah	MS: 38.1 mm	• 120 hauls • TS: 2.8 knots • TD: 1.0 hour	• depth range: 10-50 m • other information: surface water temperature, bottom characteristics and 'trash fish' composition	A. Abu Talib and M. Alias (pers. comm.); Bejie (1980; unpublished)
6	12 May - 25 Jul 1980	R.V. Malong Wooden vessel 83 GT 190 hp, Diesel Double rigger with boom trawler	Double-rigged HR: 15.7 m MS: 31.5 mm Polyethylene	• 64 hauls • TS: 2.0 knots • TD: 1.0 hour	• prawn resource survey • coast of Sarawak (Pulau Burong to Kuala Bintulu) • depth range: 4-20 m • mean catch rate: 95.05 kg/haul (food fish, 40.51 kg/haul; prawns, 2.08 kg/haul; trash fish, 52.52 kg/haul) • most abundant prawn species were *Penaeus indicus* and *Parapeneopsis hardwickii* • other information: prawn species composition, general biology of prawns, length by sex and grounds suitable for trawling	Bejie (1985a)
7	26 Jun - 8 Aug 1980	R.V. Merah	MS: 38.1 m	• 133 hauls • TS: 2.8 knots • TD: 1.0 hour	• depth range: 10-60 m • other information: bottom sample, surface water temperature and salinity	A. Abu Talib and M. Alias (pers. comm.); Gambang (1981; unpublished)
8	8 - 28 Apr 1981	R.V. Malong	Double-rigged HR: 15.7 m MS: 31.5 mm Polyethylene	• 75 hauls • TS: 2.8 knots • TD: 1.0 hour	• Prawn resource survey • coast of Sarawak, Semantan to Miri • depth range: 4-28 m • mean catch rate: $126.71\ \text{kg}\cdot\text{hour}^{-1}$ (fish + prawn); $4.25\ \text{kg}\cdot\text{hour}^{-1}$ (prawn only) • other information: catch rate by species and depth, by haul and sub-area, prawn species composition, length by sex, salinity and sea surface temperature	Bejie (1985b)

Area/survey no.	Period (date)	Vessel	Trawl gear	Trawl operations	Results/notes	References
9	16 May - 6 Jul 1981	R.V. Merah	MS: 38.1 mm	• 120 hauls • TS: 2.8 knots • TD: 1.0 hour	• extent of trawled area: coast of Sarawak; 25 000 km² • depth range: 10-60 m • mean catch rate: 162 kg·hour⁻¹ (53% were commercial fish; 47%, 'trash fish') • Sub-area II (Bintulu to Miri) had highest catch rates, 217 kg·hour⁻¹ and lowest at sub-area I (Tg. Datu - Tg. Sirik), 125 kg·hour⁻¹ • highest catch rate was at 41-50 m depth, but at 40 m towards shallow water, rates increased • most abundant groups were Rajiformes, Mullidae and Leiognathidae • other information: mean catch rates by substrate, catch composition of commercial fish, trawlable grounds, hauls with distinct catch, bottom sample, surface water temperature and salinity	Gambang (1986a)
10	11 Mar - 13 Jun 1982	R.V. Malong	Double-rigged HR: 15.7 m MS: 31.5 mm Polyethylene	• 82 hauls • TS: 2.8 knots • TD: 1.0 hour	• Prawn resource survey • coast of Sarawak, Semantan to Miri • depth range: 6-33 m • catch rate: 3.9 kg·hour⁻¹ (prawn only) • other information: catch rates by groups, prawn species composition, length by sex; salinity and surface sea temperature	Beije (1986)
11	20 Apr - 10 Jun 1982	R.V. Merah	MS: 38.1 mm	• 103 hauls • TS: 2.8 knots • TD: 1.0 hour	• coast of Sarawak; 17 420 km² • depth range: 10-60 m • catch rate: 171 kg·hour⁻¹ (commercial: 80 kg·hour⁻¹; trash: 91 kg·hour⁻¹) • stock density: 3.98 t·km⁻² (commercial); 4.88 t·km⁻² ('trash') • total biomass: 48 957 t • Rajiformes and Tachysuridae dominated the catch • other information: mean catch rates by depth, mean catch rates by substrate, catch composition of commercial fish, trawlable grounds, bottom characteristics, surface water temperature and salinity	Gambang (1986b)
12	16 Aug - 18 Oct 1983	R.V. Merah	MS: 38.1 mm	• 93 hauls • TS: 2.8 knots • TD: 1.0 hour	• depth range: 10-60 m • other information: bottom sample, surface water temperature and salinity	A. Abu Talib and M. Alias (pers. comm.)
13	12 May - 17 Jun 1987	R.V. Merah	MS: 38.1 mm	• 89 hauls • TS: 2.8 knots • TD: 1.0 hour	• coast of Sarawak • depth range: 10-60 m • mean catch rate: 79 kg·hour⁻¹; 46% were commercial fish and 54% were 'trash fish' • change in most abundant groups, Clupeidae, Lamniformes and Tachysuridae dominated • other information: mean catch rates by sub-area, depth zone and substrate, catch composition of commercial fish, bottom sample, surface water temperature and salinity	Gambang (1987)

Area/survey no.	Period (date)	Vessel	Trawl gear	Trawl operations	Results/notes	References
14	1990	R.V. Malong HR: 15.7 m MS: 31.5 mm Polyethylene	Double-rigged	• 26 hauls • TS: 2.8 knots • TD: 1.0 hour	• Prawn resource survey • other information: length by sex; salinity and surface sea temperature	A. Abu Talib and M. Alias (pers. comm.)
15	1991	R.V. Malong	Double-rigged HO: 15.7 m MS: 31.5 mm Polyethylene	• 24 hauls • TS: 2.8 knots • TD: 1.0 hour	• Prawn resource survey • other information: length by sex; salinity and surface sea temperature	A. Abu Talib and M. Alias (pers. comm.)
16	1992	R.V. Malong	Double-rigged HO: 15.7 m MS: 31.5 mm Polyethylene	• 24 hauls • TS: 2.8 knots • TD: 1.0 hour	• Prawn resource survey • other information: length by sex; salinity and surface sea temperature	A. Abu Talib and M. Alias (pers. comm.)
17	5 Apr - 22 May 1991	R.V. Manchong 27.5 m Fiberglass 150 GT 900 hp Diesel	High-opening otter trawl HO: 20.5 m MS: 28 mm GR with rubber-discs	• 46 hauls • TS: 3.6 knots • TD: 1.0 hour	• depth range: 20-60 m • other information: bottom samples, surface water temperature, salinity and length-frequency data	A. Abu Talib and M. Alias (pers. comm.)
18	1996	M.V. SEAFDEC and K.K. Manchong			• surveyed Sarawak, Brunei and Sabah	M.I. Mansor (pers. comm.)
PHILIPPINES (exploratory surveys in different areas)						
1	1845 - 1847	Galathea Danish vessel			• surveyed Manila Bay, Dinagat and Surigao	Warfel and Manacop (1950); Sebastian (1951); Ronquillo (1959)
2	1872 - 1877	British H.M.S. Challenger Deep Sea Expedition			• surveyed Philippine Deep Seas	Warfel and Manacop (1950); Sebastian (1951); Ronquillo (1959)
3	1907 - 1909	Albatross Philippine Expedition sponsored by the US Bureau of Fisheries			• Extensive investigation of marine life in the islands	Warfel and Manacop (1950); Sebastian (1951); Ronquillo (1959)
4	1909	English Steam trawler	Otter trawl		• Explored trawlable grounds from Manila to Visayas	Warfel and Manacop (1950); Sebastian (1951); Ronquillo (1959)
5	1940	Experiment of the Philippine Bureau of Fisheries	Otter trawl			Warfel and Manacop (1950)

168

Area/survey no.	Period (date)	Vessel	Trawl gear	Trawl operations	Results/notes	References
6	1947 - 1949	*Theodore N. Hill* 400 hp and *David Star Jordan*	'Eastern type net' MS: 4.5 in and 'Western type' MS: 2.5 in	• 157 hauls	• exploratory trawl fishing • 24 areas Lingayen Gulf, West of Bataan, Manila Bay approach, Manila Bay, Tayabas Bay, Mangarin Bay, Ragay Gulf, Burias Pass, Alabat Sound, Tabaco Bay, Samar Sea, Carigara Bay, San Pedro Bay, Leyte Gulf, West Visayan Sea, Guimaras Strait, Panay Gulf, Panguil Bay, Sibuguey Bay and Off Taganak Island • depth range: 4-180 m • catch rates: 2.3-115 kg·hour^{-1} (mean for all areas: 68.6 kg·hour^{-1}) • greatest catch rates taken at 11-36 m depth • 98 species (30 families) caught, slipmouths predominated	Warfel and Manacop (1950)
7	1951 - 1952	Danish Deep Sea *Galathea* Expedition			• Trawled deepest part of Philippine trench	Megia et al. (1953)
PHILIPPINES (MANILA BAY)						
1	1956	Fishing craft, 2.5-9 hp	'Dug-out' trawl	• 8-10 hauls per night • TS: 1-1.5 knots • TD: 20-30 min	• Manila Bay • depth range: 0.5-3 m	Manacop and Laron (1956)
2	Nov 1956 - Oct 1958		Otter trawl, MS: 3.2-2.8 cm		• Manila Bay between Cavite and Bataan • depth range: 4-16 m	Tiews and Caces-Borja (1959)
3	Apr 1957 - Oct 1958		'Dug-out' or Baby trawl MS: 11 mm, 3/4 in	• 703 hauls • TD: 30 min	• Manila Bay between Cavite and Bataan • depth range: 1-4 m	Tiews and Caces-Borja (1959)
4	1957 - 1959	*M.V. Ildefonso I* 18.77 GT, 80 hp; *M.V. Ildefonso IV* 48.37 GT, 80 hp; *M.V. Leonor V* 60.43 GT, 160 hp; *M.V. Dona Lina D* 83.85 GT, 325 hp	Otter trawl (mestizo type) MS: 1-1.5 in		• Manila Bay • catch rates: 16 kg·hour^{-1} ('57); 13 kg·hour^{-1} ('58); 12 kg·hour^{-1} ('59) • annual mean landings: 7 165 t ('57); 6 724 t ('58); 6 724 t ('59)	Ronquillo et al. (1960)
5	1960 - 1962	commercial trawls			• Manila Bay	Caces-Borja et al. (1972)
6	Dec 1978	Commercial medium-sized trawlers, 40-44 GT, 225 hp	German Trawl A (244x160 mm), German Trawl B (418x160 mm), *Norwegian Star Trawler*, Norwegian net (locally made)	• 21 hauls • TD: 2-4 hours	• Manila Bay	Caces-Borja (1972)

Area/ survey no.	Period (date)	Vessel	Trawl gear	Trawl operations	Results/notes	References
7	Nov 1970 - Feb 1971	*F.B. Carlos Renato II* 44.9 GT, 2x225 hp and *F.B. Maria Cynthia II* 40.01 GT, 2x250 hp *Norwegian Star* Trawl net Horsenet (locally made) German Trawl (418-160 mm)	German (Herman Engel type) trawl (234x160 mm)	• TS: 2.5-3.5 knots • TD: 0.5-3.0 hours	• Central Manila Bay	De Jesus (1976)
8	15 - 29 Feb 1980	Fishing boat 8.18 GT	Otter trawl	• 3 hauls	• Manila Bay (Bulacan, Batangas and Cavite coasts) • depth range: 6-14 m • stock density: 0.98 t·km^{-2} • biomass: 1 760 t	Bautista and Rubio (1981)
9	Nov 1992 - Oct 1993	Baby trawler FUSO 2DR5 LOA: 13.5 m Fishing boat with outriggers	Otter trawl MS: 220 mm (stretched)	• TD: 1.0 hour	• Manila Bay: 1 782 km^2 • depth range: 10-30 m • 16 fishing stations • CPUE: 13.8 kg·hour^{-1} • biomass: 1 730 t • stock density: 0.96 t·km^{-2}	MADECOR and National Museum (1994)
10	Sep 1995 - May 1996	commercial trawler FUSO 4DR5 LOA: 13.5 m	4 seam otter trawl HR: 61.5 m MS: 4.2 cm	• 31 hauls • TS: 3.8 knots • TD: 1.0 hour	• Manila Bay approaches • CPUE: 64 kg·hr^{-1}	Pura et al. (1996b)
PHILIPPINES (SAN MIGUEL BAY)						
1	1957/58	*Arca I* and *Arca II*		• >100 hauls	• San Miguel Bay • stock density: 5.20 t·km^{-2} • trawlable biomass: 4 370 t	daily reports of a private operator to BFAR Research Division *In* Pauly (1982)
2	Jul 1967	*R.V. Maya Maya*		• 2 hauls	• San Miguel Bay • stock density: 3.91 t·km^{-2} • trawlable biomass: 3 280 t	logbook of *R.V. Maya Maya* (BFAR Research Division) *In* Pauly (1982)
3	20 Apr - 17 May 1975 9 - 10 Nov 1975	Commercial Boat (Baby trawl) 35 ton, 9 hp			• San Miguel Bay	Legasto et al. (1975a)
4	Sep 1977	'a baby trawl'		• 6 hauls	• San Miguel Bay • stock density: 3.49 t·km^{-2} • trawlable biomass: 2 930 t	Manuscript, BFAR Research
5	Jul 1979	*F.B. Gemma*		• 3 hauls	• San Miguel Bay • stock density: 1.84 t·km^{-2} • trawlable biomass: 1 560 t	Manuscript, BFAR Research Division *In* Pauly (1982)
6	Feb 1980	*F.B. Sandeman*		• 25 hauls	• San Miguel Bay • stock density: 1.89 t·km^{-2} • trawlable biomass: 1 590 t	Manuscript, BFAR Research Division *In* Pauly (1982)

Area/ survey no.	Period (date)	Vessel	Trawl gear	Trawl operations	Results/notes	References
7	1980 - 1981	average small trawler			• San Miguel Bay • stock density: 2.13 t·km⁻² • trawlable biomass: 1 790 t	Vakily (1982)
8	Sep 1992 - Jun 1993	*F.B. Ryan IV* 1.93 GT, 65 hp.	4-seam bottom trawl MS: 0.9 cm	• 56 hauls • TS: 3.0 knots • TD: 1.0 hour	• San Miguel Bay; 1 115 km² • depth range: 7.4-18 m • 9 fishing stations • CPUE: 32.3 kg·hour⁻¹ • biomass: 1 636 t • stock density: 1.96 t·km⁻²	Cinco et al. (1995)

PHILIPPINES (OTHER AREAS SOUTHEAST OF LUZON)

Area/ survey no.	Period (date)	Vessel	Trawl gear	Trawl operations	Results/notes	References
1	Dec 1967	Japanese type two-boat trawler, 100 GT	4-seam type, MS: 51.2 mm		• Tayabas Bay and Samar Sea • depth ranges: 35-38 m; 36-39 m	Encina (1972)
2	18 - 28 Apr 1972	Commercial boat Baby trawler 30 hp	MS: 4 cm (stretched)	• 4 hauls • TS: 1.0-2.0 knots • TD: 0.5 hour	• Sorsogon Bay • depth range: 10-20 m	Ordoñez et al. (1975)
3	13 - 26 Mar 1974	*R.V. Researcher* 508.59 GT	Norwegian type otter trawl	• 2 hauls • TD: 2.25 hour	• extent of trawled area: Tayabas Bay (area north of Marinduque Is. and Mompog Pass); 4 000 km² • depth range: 25-45 m	Ordoñez et al. (1977)
4	Oct 1975	Commercial Boat 10 hp	Baby trawl MS: 1.0 cm	• 6 hauls • TS: 2.0-3.0 knots • TD: 0.5-1.0 hour	• Maqueda Bay, Villareal Bay and Zumarraga Channel	Legasto et al. (1975b)
5	Nov 1981 - Jan 1983	*R.V. Sardinella*	HR: 35.3 m FR: 50.6 m MS: 180 mm (stretched)		• extent of trawled area: Ragay Gulf, Burias Pass, Ticao Pass and waters north of Samar Sea; 10 830 km² • catch rates: 156-350 kg·hour⁻¹ (mean: 257 kg·hr⁻¹) • stock density: 2.25-2.85 t·km⁻² • standing stock: 1 704-6 073 t	Mines (1984)
6	Mar 1994 - Jan 1995	Mini trawler 0.5 GT, 16 hp	2-seam trawl MS: 1.38 cm & 1.17 cm	• 84 hauls • TS: 2.0 knots • TD: 15-30 min	• Sorsogon Bay; 256 km² • CPUE: 6.64 kg·hour⁻¹ • stock density: 1.04 t·km⁻² • biomass: 143 495 t	Cinco and Perez (1996)

PHILIPPINES (VISAYAN SEA)

Area/ survey no.	Period (date)	Vessel	Trawl gear	Trawl operations	Results/notes	References
1	Apr 1976 - Mar 1977	*M.V. Albacore* 190 GT, 600 hp	4-seam trawl, MS: 1.5 cm (stretched)	• 144 hauls • TD: 2.0 hour	• extent of trawled area: Visayan Sea; 5 184 km² • depth range: 14-140 m	Aprieto and Villoso (1979)
2	Jul 1976 - Mar 1977	*M.V. Albacore* 190 GT, 600 hp	2-seam high opening otter trawl MS: 15 mm (stretched)	• 107 hauls	• depth range: 20-40 m • Visayan Sea: 5 184 km²	Ordoñez (1985)
3	16 - 20 May 1979	Commercial vessel 60.5 GT	Otter trawl	• 17 hauls • TD: 1.0 hour	• Visayan Sea	Gonzales et al. (1981)

PHILIPPINES (SAMAR SEA AND CARIGARA BAY)

Area/ survey no.	Period (date)	Vessel	Trawl gear	Trawl operations	Results/notes	References
1	Mar 1979 - Mar 1980	M.V. Albacore 190 GT, 600 hp	2-seam bottom trawl MS: 40 mm (stretched)	• 301 hauls • TS: 3.0 knots • TD: 1.0 hour	• Samar Sea and Carigara Bay; 3 050 km² • depth range: 10-100 m • mean CPUE: 174 kg·hour^{-1} • standing stock: 3.1 t·km^{-2}	Armada and Silvestre (1981)
2	11 - 12 May 1982	R.V. Sardinella 411 GT	2-seam high opening bottom trawl MS: 40 mm (stretched)	• TS: 4.0 knots • TD: 15-40 min	• Samar Sea • depth range: 18.5-85 m	Silvestre et al. (1986)
3	Dec 1983	R.V. Albacore 600 hp	2-seam high opening bottom trawl MS: 40 mm	• 28 hauls • TS: 3.0 knots • TD: 1.0 hour	• Samar Sea • depth range: 9.9-90 m • mean CPUE: 91.8 kg·hour^{-1} (period I) and 191.8 kg·hour^{-1} (period II) • density: 1.0 t·km^{-2} (period I) and 2.1 t·km^{-2} (period II)	Armada et al. (1983)
4	Oct 1991 - Aug 1992		Baby trawl	• 69 hauls	• Carigara Bay; 512 km²	Calumpong et al. (1993)
5	Jul 1995 - Apr 1996	Baby trawler 85 hp, 36 GT LOA: 10 m	2 seam bottom trawl HR: 10 m MS: 2.0 cm	• 40 hauls • TS: 2.5 knots • TD: 1.0-2.0 hour	• Carigara Bay; 512 km² • depth range: 10-50 m • eight fishing stations • CPUE: 12 kg·hour^{-1} • biomass: 533 t • stock density: 1.04 t·km^{-2}	Pura et al. (1996a)

PHILIPPINES (OTHER AREAS)

Area/ survey no.	Period (date)	Vessel	Trawl gear	Trawl operations	Results/notes	References
1	13 - 16 Dec 1969	F.B. Maharlika de Visayas 83.8 GT, 380 hp	Star trawl net 560x170 mm Horsenet 600x4 knots (4 inch mesh)	• 16 hauls • TS: 3.0-3.7 knots • TD: 3.0 hour	• Central Asid Gulf (near Circe Bank) • depth range: 12-54 m	De Jesus and Maniulit (1973)
2	Oct 1975	R.V. Researcher 508.59 GT	Norwegian type otter trawl	• TD: 0.5 hour	• Maiampaya Sound	Barut (pers. comm)
3	1977 - 1988	R.V. Researcher 508.59 GT	Norwegian type otter trawl		• Malampaya Sound, Bacuit Bay, Imman Bay, Manila Bay and Ulugan Bay	Unprocessed, unpublished documents
4	Feb 1978 - Jan 1979	M.V. Albacore 190 GT, 600 hp	Otter trawl	• TD: 2.0 hour	• Lingayen Gulf, 1 900 km² • depth range: 30-90 m	Aprieto and Villoso (1982)
5	Apr 1983 - Apr 1984		Medium trawl Large trawl	• 97 hauls each • TD: 2.0-4.0 hour	• extent of trawled area in Lingayen Gulf; 2 109 km²	Mines (1986)
6	Mar - Dec 1991		Small otter trawl MS: 15 cm	• TD: 1.0 hour	• Panguil Bay • depth range: 10-15 m	MSU (1993)

Area/survey no.	Period (date)	Vessel	Trawl gear	Trawl operations	Results/notes	References
SRI LANKA						
1	1920 - 1923	*R.V. Lilla* 126' LOA, 500 hp	Bottom trawl HR: 70' (21.8 m) FR: 100' (31.2 m) mesh 5" (2 cm) tapering to 1" (0.4 cm) codend	• 395 hauls • TS: 7 mi·hour[-1] • TD: 1.0 hour	• around Sri Lanka including Wadge and Pedro Banks • depth range: mainly 4-35 m • other information collected: echosounding, topography and bottom profiling hydrology, length frequencies, sex and stage of maturity, stomach contents • catch rates: 0.9-203.5 kg·hour[-1]	Pearson and Malpas (1926) *In* Sivasubramaniam and Maldeniya (1985)
2	1954 - 1957	*R.V. Northstar* 45 LOA, 80 hp			• Inshore waters around Sri Lanka • catch rates: bottom longlining: 1.5-17.3 kg/100 hooks hand lining: 0.2-9.9 kg-line-hour[-1] shark longline: 84.8-637 kg/100 hooks	Jean (1957) *In* Sivasubramaniam and Maldeniya (1985)
3	Jun 1955 - Oct 1956	*R.V. Canadian* 45' LOA, 80 hp	Bottom trawl HR: 40' (12.5 m) FR: 50' (19.7 m) Yankee 35 MS: 3 3/4" (1.5 cm)	• TS: 1.0-2.0 knots	• Inshore waters around Sri Lanka • catch rates: 2.2-27.9 kg·hour[-1]	Jean 1957 *In* Sivasubramaniam and Maldeniya (1985)
4	Nov 1963 - Apr 1967	*R.V. Canadian* 45' LOA, 80 hp	Shrimp trawl HR: 30' MS: 35mm		• North and East coasts • catch rates: 49-472.8 kg·hour[-1]	Sivasubramaniam and Maldeniya (1985)
5	May - Jun 1967	*R.V. Myliddy* 30 m LOA, 240 hp	Granton trawl MS: 80 mm Shrimp trawl MS: 40 mm	• TS: 1.5-3.0 (prawn trawl), 3.0-4.0 (fish trawl) • TD: 1.0 hour	• North and East coasts • depth range: 2-80 fathom (12-48 m) • other information: acoustic recordings, water temperature and salinity • catch rates: 60-1200 kg·hour[-1] (8 stations, fish and shrimp trawls)	Berg (1971) *In* Sivasubramaniam and Maldeniya (1985)
6	Mar - Dec 1972	*S.R.T.M. Optimist*	Bottom trawling in deep waters	• 400 trawls	• northwest and northeast of Sri Lanka and the Wadge Bank • depth range: 200-350 m • other information: hydrology, composition of seawater, length frequencies in trawl and gill net fisheries • total catch including fish: 22 344 kg; 4 781 kg (prawn catch); 5 892 kg (lobster catch) • catch rates: 175.24 kg·hour[-1] (prawns, lobsters, fish); 46.21 kg·hour[-1] (lobsters); 37.40 kg·hour[-1] (prawns)	Demidenko (1972) *In* Saetersdal and de Bruin (1979)
7	Nov 1975 - Jun 1978	*R.V. Hurulla* 11 m length, 96 hp	Bottom trawl MS: 32-36 mm	• TS: 6.0 knots • TD: 1.0 hour	• depth range: 3-28 m • Palk Bay and Gulf of Mannar • other information collected: economics of operations • catch rates: 64-123 kg·hour[-1] (4 stations)	Henriksson (1980) *In* Sivasubramaniam and Maldeniya (1985)

Area/survey no.	Period (date)	Vessel	Trawl gear	Trawl operations	Results/notes	References
8	Aug - Sep 1978	R.V. Dr. Fridtjof Nansen	shrimp trawl HL: 96 ft (32 m)		• five areas: Negombo-Galle, Hambantota, east coast, Trincomalee to Mullaitivu and Pedro Bank; 5 000 nm^2 • main survey effort provided distribution of observed echo intensity of fish; average density of fish biomass in the whole area was 100 t·km^{-2}; • biomass: 500 000 t • annual potential yields: 70 000 t of demersal fish and 100 000 t of pelagic fish • most important demersals were snappers, groupers, breams and trevallys • other information collected: bottom conditions and hydrography	Saetersdal and de Bruin (1979)
9	Apr - Jun 1979	R.V. Dr. Fridtjof Nansen	shrimp trawl HL: 96 ft (32 m)		• six areas: northwest coast, southwest coast from Negombo to Galle, Hambantota Banks, east coast and Batticaloa Banks, Trincomalee-Mullaitivu and Pedro Banks • main survey effort provided distribution of observed echo intensity of fish • other gears: mesh pelagic trawl and bottom long lines • other information collected: bottom conditions and hydrography • demersal resources at Pedro Bank based on trawl survey: 8 000 t; based on acoustic survey: 11 000 t • highest echo intensities at the northwest coast north of Negombo • biomass of demersals: 330 000 t	Blindheim et al. (1979)
10	Jan - Feb 1980	R.V. Dr. Fridtjof Nansen	shrimp trawl HL: 96 ft (32 m)		• six areas: northwest coast, southwest coast from Negombo to Galle, Hambantota Banks, east coast and Batticaloa Banks, Trincomalee-Mullaitivu and Pedro Banks • main survey effort provided distribution of observed echo intensity of fish • other gears used: 1 600 mesh pelagic trawl bottom longline • other information collected: hydrography and bottom sediment • biomass of demersals: 250 000 t • at Pedro Bank, 8 000 t demersals were estimated using trawl survey and 15 000 t using acoustic records	Blindheim and Foyn (1980)

THAILAND (GULF OF THAILAND)

Area/survey no.	Period (date)	Vessel	Trawl gear	Trawl operations	Results/notes	References
1	1961	R.V.2	Otterboard	• 133 hauls • TS: 2.5 knots • TD: 1.0 hour	• stations: I-IX (Gulf of Thailand) • depth range: 10-50 m • catch rate: 298 kg·hour^{-1}	-
2	1963	R.V.2	Otterboard	• 200 hauls • TS: 2.5 knots • TD: 1.0 hour	• stations I-IX • depth range: 10-50 m • catch rate: 256 kg·hour^{-1}	-
3	1964	R.V.2	Otterboard	• 122 hauls • TS: 2.5 knots • TD: 1.0 hour	• stations I-IX • depth range: 10-50 m • catch rate: 226 kg·hour^{-1}	-

Area/survey no.	Period (date)	Vessel	Trawl gear	Trawl operations	Results/notes	References
4	1965	R.V.2	Otterboard	• 192 hauls • TS: 2.5 knots • TD: 1.0 hour	• stations I-IX • depth range: 10-50 m • catch rate: 179 kg·hour^{-1}	-
5	1966	R.V.2	Otterboard	• 712 hauls • TS: 2.5 knots • TD: 1.0 hour	• stations I-IX • depth range: 10-50 m • average catch rate: 131 kg·hour^{-1} • analysis of catch composition by species of economical value in relation to the depth, area and comparison of catches in the morning and afternoon, and by seasons (dry, wet) were made	Ritragsa et al. 1968
6	1967	R.V.2	Otterboard	• 713 hauls • TS: 2.5 knots • TD: 1.0 hour	• stations: I-IX and by seasons (dry, rainy) • depth range: 10-50 m • catch rate: 115 kg·hour^{-1}	-
7	1968	R.V.2	Otterboard	• 719 hauls • TS: 2.5 knots • TD: 1.0 hour	• stations I-IX • depth range: 10-50 m • average catch rate: 106 kg·hour^{-1} • highest catch rate of 132 kg·hour^{-1} at 44 m depth; lowest catch of 89 kg·hour^{-1} at 10-19 m depth • detailed analysis of catch rate by depth in each sub-area, comparison of catch rate between each sub-area, comparison of fish of economic value and other fish, catch composition by weight, catch rate by grid square and by region, catch rates in relation to time of the day (and night) were presented	Ritragsa et al. (1969); Ritragsa and Pramokchutima (1970)
8	1969	R.V.2	Otterboard	• 720 hauls • TS: 2.5 knots • TD: 1.0 hour	• stations I-IX • depth range: 10-50 m • average catch rate: 103 kg·hour^{-1}	-
9	1970	R.V.2	Otterboard	• 718 hauls • TS: 2.5 knots • TD: 1.0 hour	• stations I-IX • depth range: 10-50 m • other information: length-frequency data • catch rate: 97 kg·hour^{-1}	-
10	1971	R.V.2	Otterboard	• 720 hauls • TS: 2.5 knots • TD: 1.0 hour	• stations I-IX • depth range: 10-50 m • other information: length-frequency data • catch rate: 66 kg·hour^{-1}	-
11	1972	R.V.2	Otterboard	• 720 hauls • TS: .22.5 knots • TD: 1.0 hour	• stations I-IX • depth range: 10-50 m • other information: length-frequency data • catch rate: 63 kg·hour^{-1}	-
12	1973	R.V.2	Otterboard	• 718 hauls • TS: 2.5 knots • TD: 1.0 hour	• stations I-IX • depth range: 10-50 m • other information: length-frequency data • catch rate: 52 kg·hour^{-1}	-

Area/ survey no.	Period (date)	Vessel	Trawl gear	Trawl operations	Results/notes	References
13	1974	R.V.2	Otterboard	• 540 hauls • TS: 2.5 knots • TD: 1.0 hour	• stations I-IX • depth range: 10-50 m • other information: length-frequency data • catch rate: 58 kg·hour[-1]	-
14	1975	R.V.2	Otterboard	• 480 hauls • TS: 2.5 knots • TD: 1.0 hour	• stations I-IX • depth range: 10-50 m • other information: length-frequency data • catch rate: 47 kg·hour[-1]	-
15	1976	R.V.2	Otterboard	• 261 hauls • TS: 2.5 knots • TD: 1.0 hour	• stations I-IX • depth range: 10-50 m • other information: length-frequency data • catch rate: 57 kg·hour[-1]	-
16	1977	R.V.2	Otterboard	• 579 hauls • TS: 2.5 knots • TD: 1.0 hour	• stations I-IX • depth range: 10-50 m • other information: length-frequency data • catch rate: 47 kg·hour[-1]	Eiamsa-ard et al. (1979); Eiamsa-ard and Dhamniyom (1979)
17	1978	R.V.2, R.V.9	Otterboard	• 442 hauls • TS: 2.5 knots • TD: 1.0 hour	• stations I-IX • depth range: 10-50 m • other information: length-frequency data • catch rate: 52 kg·hour[-1]	Eiamsa-ard and Dhamniyom (1980)
18	1979	R.V.2, R.V.9	Otterboard	• 235 hauls • TS: 2.5 knots • TD: 1.0 hour	• stations I-IX • depth range: 10-50 m • other information: length-frequency data • mean catch rate: 512 kg·hour[-1] • catch rate highest at station V (75 kg·hour[-1]) and lowest at station III (24 kg·hour[-1])	Eiamsa-ard and Dhamniyom (1981); Eiamsa-ard (1981)
19	1980	R.V.2, R.V.9	Otterboard	• 245 hauls • TS: 2.5 knots • TD: 1.0 hour	• stations: I-IX • depth range: 10-50 m • other information: length-frequency data • mean catch rate: 48 kg·hour[-1] • economic fish (23 kg·hour[-1]), juveniles of economic fish (9 kg·hour[-1]) and 'trash fish' (16 kg·hour[-1]) • dominant species were cephalopods, Nemipteridae, Priacanthidae, Synodontidae and Carangidae • catch rate highest at station V (51 kg·hour[-1]) and lowest at station III (13 kg·hour[-1])	Charnprasertporn (1982)
20	1981	R.V.2, R.V.9	Otterboard	• 159 hauls • TS: 2.5 knots • TD: 1.0 hour	• stations I-IX • depth range: 10-50 m • other information: length-frequency data	-
21	1982	R.V.2, R.V.9	Otterboard	• 211 hauls • TS: 2.5 knots • TD: 1.0 hour	• stations I-IX • depth range: 10-50 m • other information: length-frequency data	-

Area/ survey no.	Period (date)	Vessel	Trawl gear	Trawl operations	Results/notes	References
22	1983	*R.V.2, R.V.9*	Otterboard	• 328 hauls • TS: 2.5 knots • TD: 1.0 hour	• stations I-IX • depth range: 10-50 m • other information: length-frequency data • catch rate: 30 kg·hour[1]	-
23	1984	*R.V.2, R.V.9*	Otterboard	• 172 hauls • TS: 2.5 knots • TD: 1.0 hour	• stations I-IX • depth range: 10-50 m • other information: length-frequency data	-
24	1985	*R.V.2, R.V.9*	Otterboard	• 228 hauls • TS: 2.5 knots • TD: 1.0 hour	• stations I-IX • depth range: 10-50 m • other information: length-frequency data	-
25	1986	*R.V.2, R.V.9*	Otterboard	• 260 hauls • TS: 2.5 knots • TD: 1.0 hour	• stations I-IX • depth range: 10-50 m • other information: length-frequency data	-
26	1988	*R.V.2, R.V.9*	Otterboard	• 125 hauls • TS: 2.5 knots • TD: 1.0 hour	• stations I-IX • depth range: 10-50 m • length-frequency data	-
27	1989	*R.V.2, R.V.9*	Otterboard	• 179 hauls • TS: 2.5 knots • TD: 1.0 hour	• stations I-IX • depth range: 10-50 m • other information: length-frequency data	-
28	1990	*R.V.9*	Otterboard	• 21 hauls • TS: 2.5 knots • TD: 1.0 hour	• stations VII-IX • depth range: 10-50 m • other information: length-frequency data	-
29	1991	*R.V.9*	Otterboard	• 21 hauls • TS: 2.5 knots • TD: 1.0 hour	• stations VII-IX • depth range: 10-50 m • other information: length-frequency data	-
30	1992	*R.V.9*	Otterboard	• 21 hauls • TS: 2.5 knots • TD: 1.0 hour	• stations VII-IX • depth range: 10-50 m • other information: length-frequency data	-
31	1993	*R.V.9*	Otterboard	• 22 hauls • TS: 2.5 knots • TD: 1.0 hour	• stations VII-IX • depth range: 10-50 m • other information: length-frequency data	-
32	1994	*R.V.9*	Otterboard	• 24 hauls • TS: 2.5 knots • TD: 1.0 hour	• stations VII-IX • depth range: 10-50 m • other information: length-frequency data	-
33	1995	*R.V.9*	Otterboard	• 23 hauls • TS: 2.5 knots • TD: 1.0 hour	• stations VII-IX • depth range: 10-50 m • other information: length-frequency data	-

Area/ survey no.	Period (date)	Vessel	Trawl gear	Trawl operations	Results/notes	References

34	May - Jun 1984 and Sep 1984	M.V. Paknam	Otterboard	• 20 hauls
				• TS: 3.0-3.2 knots
				• TD: 1.5-2.5 hour
					• central part of Gulf of Thailand
					• depth range: 40-70 m
					• other gears used: tuna longline and drift net
					• other information collected: hydrography and plankton samples
					• May-Jun survey:
					  stock density: 110-1 900 kg·km^{-2}
					  standing stock (all demersal fishes): 12 892 t standing stock (all useful demersal): 3 416 t
					• Sep survey:
					  stock density: 280-1 600 kg·km^{-2}
					  standing stock (all demersal fishes): 6 252 t standing stock (all useful demersals): 3 380 t
						SEAFDEC (1985)

35	Sep 1995 and May 1996	M.V. SEAFDEC, Pramong IV and K.K. Mersuji			• Gulf of Thailand	M.I. Mansor (pers. comm.)

THAILAND (WEST COAST, ANDAMAN SEA)

Area/ survey no.	Period (date)	Vessel	Trawl gear	Trawl operations	Results/notes	References
1	1965,1966, 1967,1970	Dhanarat later renamed Fishery Research No. 2	Japanese type trawl net HR: 74 ft GR: 105 ft MS: 2.9 in	• 353 hauls • TS: 2.5 knots • TD: 1.0 hour	• west coast of Thailand; 13 725 sq. miles • depth range: > 100 m • catch rate: 449 kg·hour^{-1} (1965) • standing stock: 154 000 - 224 000 t (1965)	Isarankura (1971)
2	Jul 1980	R.V. Dr Fridtjof Nansen			• west coast of Thailand	Aglen et al. (1981c)
3	1 - 15 Nov 1981	M.V. Nagasaki-Maru Thai-Japanese-SEAFDEC joint survey	bottom trawl	• 15 hauls • TD: 1.0-3.0 hour	• Indian ocean coast of Thailand; 11 576 km^2 • depth range: 35-92 m • standing stock (area surveyed): 17 750 t estimated standing stock (whole region, 44 000 km^2): 50 800 t • potential yield of whole region: 52 700 t·year^{-1}	Hayase (1983)
4		F.O.R.V. Sagar Sampada	bottom and pelagic trawls	• 30 trawl stations	• Andaman-Nicobar Seas • exploratory survey on cephalopods' occurrence at the area	Sreenivasan and Sarvesan (1989)
5	Sep 1987		deep-sea trawl		• Andaman sea • list of 68 fish, 10 decapods and six cephalopods species given	Suppachai-Ananpongsuk (1989)

VIETNAM

Area/ survey no.	Period (date)	Vessel	Trawl gear	Trawl operations	Results/notes	References
1	1960				• Southeastern Vietnam; 135 800 km^2 • depth range: 0-50 m • potential yield: 212 000 t	Aoyama (1973 In Yeh 1981)
2	1970				• Southeastern Vietnam; 135 800 km^2 • depth range: 0-50 m • potential yield: 451 000 t	SCS (1978 In Yeh 1981)

Area/ survey no.	Period (date)	Vessel	Trawl gear	Trawl operations	Results/notes	References
3	1970 - 1977	Taiwanese commercial trawlers	pair trawl		• Southeastern Vietnam; 135 800 km² • depth range: 0-50 m • potential yield: 250 000 t·year⁻¹	Yeh (1981)
4	Jun 1977 - Jun 1978	*Bien Dong* 1 500 hp		• 111 hauls • TD: 2.0-3.0 hour	• Tonkin Gulf (Vinh Bac Bo); 7 872 km² • depth range: 20-100 m • other information collected: pelagic trawl, hydrography and echo survey • mean catch rate: 148 kg·hour⁻¹ • biomass: 580 000 - 740 000 t	Godo (1979)
5	Jul 1978 - Dec 1979	*Bien Dong* 1 500 hp		• 100 hauls	• Southeastern Vietnam; 16 793 km² • depth range: 0-150 m • other information collected: hydrography and echo survey • mean catch rate: 284 kg·hour⁻¹ • stock density: 8.3-22.2 kg·ha⁻¹ • biomass: 222 700 t	Godo (1980)
6	Sep - Dec 1978	*Bien Dong* 1 500 hp		• 33 hauls • TD: 1.0-2.0 hour	• Tonkin Gulf • depth range: 20-60 m • Pelagic trawl	-
7	Jan, May - Jul, Sep - Nov 1979	*Bien Dong* 1 500 hp		• 80 hauls • TD: 1.0-2.0 hour	• Southeastern Vietnam • depth range: 30-80 m • Pelagic trawl	-
8	Jan - Feb, May, Jul, Oct, Dec 1979	*Aelita* 800 hp		• 1 482 hauls • TD: 1.0-2.0 hour	• Central and South Vietnam • depth range: 20-80 m	-
9	Jan - Apr 1979	*Kalper* 3 800 hp		• 507 hauls • TD: 1.0-2.0 hour	• depth range: 40-300 m • Pelagic trawl	-
10	Jan - Jul 1979	*Elsk* 1 000 hp		• 504 hauls • TD: 1.0-2.0 hour	• depth range: 20-100	-
11	Apr - Jun 1979	*Yalta* 1 350 hp		• 578 hauls • TD: 1.0-2.0 hour	• depth range: 20-150 m	-
12	Jun - July, Nov - Dec 1979 Jan - Mar, Aug 1980	*Nauka* 1 350 hp		• 808 haul • TD: 1.0-2.0 hour	• depth range: 20-150 m	-
13	Dec 1979; Jan 1980; Jan - Mar, Jun - Jul 1981	*Semen Volkov* 1 000 hp		• 803 hauls • TD: 1.0-2.0 hour	• depth range: 20-100 m	-
14	Jan - May, Jul 1980	*Bien Dong* 1 500 hp		• 63 hauls • TD: 1.0-2.0 hour	• Southeastern Vietnam • depth range: 30-80 m • Pelagic trawl	-

Area/survey no.	Period (date)	Vessel	Trawl gear	Trawl operations	Results/notes	References
15	Jul - Sep 1980 Jan - Mar 1981	*Marlin* 800 hp		• 362 hauls • TD: 1.0-2.0 hour	• depth range: 20-80 m	-
16	Nov - Dec 1980 Jan 1981	*Vozrozdenhie* 3 800 hp		• 109 hauls • TD: 1.0-2.0 hour	• depth range: 40-300 m	-
17	Nov - Dec 1981 Jan - Mar 1982	*Zavetinsk* 800 hp		• 644 haul • TD: 1.0-2.0 hour	• depth range: 20-80 m	-
18	Jul - Oct 1982	*Milogravodo* 2 300 hp		• 246 hauls • TD: 1.0-2.0 hour	• depth range: 30-150 m	-
19	Sep 1982	*Trud* 800 hp		• 154 hauls • TD: 1.0-2.0 hour	• depth range: 20-80 m	-
20	Jul - Sep 1983 May - Jun 1987	*Gerakl* 2 300 hp		• 347 hauls • TD: 1.0-2.0 hour	• depth range:: 30-150 m	-
21	Dec 1983 - Mar 1984	*Antiya* 1 000 hp		• 137 hauls • TD: 1.0-2.0 hour	• depth range: 20-100 m	-
22	Jan - Mar 1984	*Gidrobiolog* 1 000 hp		• 33 hauls • TD: 1.0-2.0 hour	• depth range: 20-100 m	-
23	Oct - Dec 1984	*Uglekamensk* 1 000 hp		• 55 hauls • TD: 1.0-2.0 hour	• depth range: 20-100 m	-
24	Dec 1984 Oct - Dec 1987	*Ochakov* 2 300 hp		• 808 hauls • TD: 1.0-2.0 hour	• depth range: 20-150 m	-
25	Nov - Dec 1985 Jan - Feb 1986	*Omega* 2 300 hp		• 162 hauls • TD: 1.0-2.0 hour	• depth range: 30-150 m	-
26	May - Jun 1986	*Shantar* 2 300 hp		• 61 hauls • TD: 1.0-2.0 hour	• depth range: 30-150 m	-
27	Aug - Dec 1987	*Mux Dalnhi* 2 300 hp		• 213 hauls • TD: 1.0-2.0 hour	• depth range: 30-150 m	-
28	Feb - Mar, May - Jun 1987	*Mux Tichi* 2 300 hp		• 194 hauls • TD: 1.0-2.0 hour	• South Vietnam • depth range: 30-150 m	-
29	Jan - May 1988	*Kizveter* 2 300 hp		• 205 hauls • TD: 1.0-2.0 hour	• depth range: 30-150 m	-

BAY OF BENGAL
BURMA (MYANMAR)

Area/survey no.	Period (date)	Vessel	Trawl gear	Trawl operations	Results/notes	References
1	1953 - 1955	*Taiyo Maru No. 11*	HR: 40 m MS: 2 inches		• off Burma in Gulf of Martaban • largest catches were made in depths around 54 m • most common fishes were croakers, grunts, snappers, leiognathids and lizardfishes • from one-fourth to one-third of the catch comprised small fishes of no commercial value	Ba Kyaw (1956)

Area/survey no.	Period (date)	Vessel	Trawl gear	Trawl operations	Results/notes	References
2	9 - 25 Sep 1966 10 - 28 Oct 1966 5 - 20 Jan 1968	*F.V. Linzin*	otterboard HR: 38.6 m GR: 53.7 m MS: 72 mm	• 170 hauls • total 711 trawling hours were expended	• southern Burma • mean catch rates: 226 kg·hour^{-1} (Sep 1966); 164 kg·hour^{-1} (Oct 1966); 282 kg·hour^{-1} (Jan 1963) • most important fish were four species of Sciaenidae; *Pomadasys hasta*, *Lutjanus* spp.., *Trichiurus savala*, *Ilisha filigera* and carangids • analysis of biological data on 23 species is given (ripeness of gonads, diet composition and condition coefficient)	Druzhinin and Hliang (1972)
3	Sep - Nov 1979 and Mar - Apr 1980	*R.V. Dr. Fridtjof Nansen*			• between border with Bangladesh in the north and with Thailand in the south (three divisions: Arakan coast, Delta area and the Tenasserim coast) • survey combined acoustic and exploratory fishing using bottom and mid-water trawls	Stromme et al. (1981)
4	1981 - 1983				• demersal fish resources of continental shelf and slope	Rijavec and Htun Htein (1984)

BAY OF BENGAL AND EAST COAST OF INDIA

Area/survey no.	Period (date)	Vessel	Trawl gear	Trawl operations	Results/notes	References
1	Nov 1959 - Oct 1960	*M.T. Ashok*			• exploratory trawling for demersal fishes, day and night, off the coasts of Andhra and Orissa between latitudes 17°40' and 20°10'N • depth range: 5-64 m • sharks, skates, rays, catfishes, miscellaneous large and small fishes and prawns showed diurnal variations • no definite trends were seen in areawise and depthwise analysis of fish groups	Rao and Krishnamoorthi (1982)
2	1959 - 1974	24 vessels of varying sizes and hp ranging from 42-578	shrimp and fish trawls of different specifications		• east coast of India (entire coastline of West Bengal, Orissa, Andhra Pradesh and the east coast of Tamil Nadu • standing stock: 3.88 t·km^{-2} • potential yield: 206 000 t·year^{-1} • other information: catch rates, catch composition by area and depth, relative abundance by area and depth, and seasonality of catch	Joseph et al. (1976a)
3	1959 - 1974	*FISH Tech. No. 7* wooden stern trawler; 12.16 m LOA and 7 GT	otter trawl HR: 18.36 m		• off Kakinada between latitude 16°50'N - 17°N and longitude 82°20'E - 82°40'E • depth ranges: 15-50 m (inshore) and 51-100 m (offshore) • average catch rate: 340 kg·hour^{-1}; catch rate found to be 5.7 times higher in the 51-100 m depth range • differences in catch composition between the two depth ranges were noted	Narayanappa et al. (1972)
4	1961 - 1970	*M.T. Ashok*			• northwestern Bay of Bengal, off Andhra-Orissa coast; • 41 478 km^2 • max. standing stock: 418 682 t • potential yield: 251 209 t·year^{-1}	Krishnamoorthi (1976)

Area/survey no.	Period (date)	Vessel	Trawl gear	Trawl operations	Results/notes	References
5	1961 - 1985		bottom trawl		• off northeast coast between 15°-21°N and 80°-88°E; 81 341 km² • exploratory fishery surveys • potential yield: 14 620 t·year⁻¹ (range: 1 396-23 429 t) • sharks and skates, rays, catfishes, mackerel, threadfin breams, jacks and goatfishes were underexploited • lizard fishes and croakers have reached optimum level of exploitation	Reuben et al. (1988)
6	1972 - 1978				• off Visakhapatnam between latitudes 15°40'N and 19°40'N • industrial fisheries based on exploratory surveys • depth range: 10-90 m	CMFRI (1980)
7	1973 - 1981	M.V. *Meenagaveshak* (17.5 m) and M.V. *Meenasitara* (17.5 m)	20 m trawl with a sweep of 0.102 km·hour⁻¹	TS: 2.75 knots	• off North Tamil Nadu - South Andhra coasts between latitude 10°40'N (off Velanganni) and 15°40'N (off Nizampatnam) • potential yield: 1.09 t·km⁻²·year⁻¹ • latitude-wise potential was estimated for certain major categories of demersal fishes and crustaceans. Silver bellies (leiognathids) ranked first (0.4 t·km⁻²) followed by 'perches' (0.1 t·km⁻²) for prawns, the most productive zone was located off Ramayapatnam in the north and off Porto Novo/Cuddalore in the south	Vivekanandan and Krishnamoorthi (1985)
8		M.V. *Matsya Vigyani*			• location of potentially rich fishing ground off West Bengal was reported • concentration of prawns at depths 50 -80 m • *Metapeneaus ensis* formed an important constituent, *M. monoceros* was also abundant • commercial feasibility of outrigger trawl fishing is indicated	Sudarsan and Joseph (1975)
9	1980 - 1981				• off Rameswaram • given catch trends and catch compositions	Pillai et al. (1983)
10	1981 - 1983	trawl survey conducted by Fishery Survey of India			• Wadge Bank; 12 505 km² • depth range: 0-200 m • stock assessment of cuttle fish, *Sepia pharaonis* total biomass: 2 060 t • 20-50 m depth contained 74% of biomass, lowest at 100-200 m depths	Philip and Ali (1989)
11					• off the coast of Visakhapatnam, Andhra Pradesh • squid and cuttlefish survey • *Sepia pharaonis* and *S. aculeata* contributed 50% to catch from Sep. to May	Mohan and Rayudu (1986)

Area/ survey no.	Period (date)	Vessel	Trawl gear	Trawl operations	Results/notes	References
12	Oct - Dec 1991	*Matsya Shikari*	shrimp trawl: 45 m fish trawl: 34 m	• total fishing effort: 73.50 hours	• upper east coast of India between latitude 18°N and 20°N • depth range: 0-100 m • catch rate (shrimp trawl): 614 kg·hour[-1] (0-50 m depth) • major fish catch (Dec): goat fishes and silver bellies • a single haul consisting of 1400 kg of *Ariomma indica* was recorded at 50-100 m depth	Anon. (1994a)
13	Oct - Dec 1991	*Matsya Darshini*			• along upper east coast between latitudes 16°N and 18°N • threadfin breams, goat fishes and ribbon fishes were major catches at 50-100 m depth zone	Anon. (1994b)
14	Oct - Dec 1991	*Matsya Jeevan*	fish trawl 27 m	• total fishing effort: 109 hours	• lower east coast of India (Tamil Nadu coast) between latitudes 10°N and 16°N • depthwise analysis of catch rates were presented; silver bellies were major fish catch in Dec	Anon. (1993)
ARABIAN SEA AND WEST COAST OF INDIA						
1	1948 - 1973	16 vessels; their hp ranged from 42 to 475	different types and sizes of fish trawls; starting 1959 vessels used two-seam Russian type trawl	• total fishing effort: 37 000 hours	• north west coast of India between latitude 15°N and 23°N and longitude 67°E and 74°E (entire coastline of Gujarat, Maharashtra and part of coastline of the Union territory of Goa, Daman and Diu; bulk of effort expended near Bombay and Veraval) • around 90% of the effort was expended in areas within 25 fathom depth, the depth zone 24-40 fathom were surveyed in part • latent demersal fishery potential at Marashtra was 120 000 t·year[-1] • standing stock: 4.38 t·km[-2] • abundant catches include elasmobranchs and 'dhoma' • other information: catch composition by area and depth, relative abundance by area and depth, anc seasonality of catch	Joseph (1974)
2	1957 - 1974	18 vessels of various designs and size with hp ranging from 42 to 240	fish and shrimp trawls with different specifications	• total fishing effort: 29 400 hours	• south west coast of India between latitudes 7°N and 15°N and longitude 73°E and 78°E (entire coast of Kerala and Karnataka and west coast of Tamil Nadu) • effort expended at 0-19 m and 40-59 m depth were 23% and 10% respectively, little effort at 60-79 m and 80-99 m • prawns in <20 m depths were abundant, elasmobranch in 40-59 m and catfish in 40-79 m • standing stock: 3.8 t·km[-2] • potential yield: 124 000 t·year[-1] • other information: catch composition by area and depth, relative abundance by area and depth, and seasonality of catch	Joseph et al. (1976b)

Area/survey no.	Period (date)	Vessel	Trawl gear	Trawl operations	Results/notes	References
3	1961 - 1967	11 vessels of the Government of India; their hp ranged from 42-300 and GT ranged from 9.95-123.24	otter trawls of different specifications	• total fishing effort: 1 896 hours	• northwestern part of India between latitudes 15°N to 23°10'N and longitudes 68°10'E and 73°50'E; 25,100 nautical sq. miles • vessels were grouped into : 201-300 hp; 101-200 hp and below 100 hp • earlier operations covered as many areas as possible but from Dec 1963, the operations involved systematic linear bottom trawling repeated at monthly intervals • annual average of catch was 391 568 kg (range: 309 133-499 519 kg) and of catch rates were 207 kg·hour^{-1} (range: 165-251 kg·hour^{-1}) • mixed sciaenids, 'dhoma', formed a third and rays about a quarter of the total catch • seasonal trends showed highest yields in the 4th quarter; species catch trend was also discussed • catch rates were high in 40-60 m depths	Virabhadra et al. (1972)
4	Nov 1963 - May 1966	Karwar-1, M1/M4, and INP 167 with 90, 48 and 24 hp engines respectively	shrimp trawl	• total fishing effort: 2 303 hours	• exploratory surveys near Karwar between latitudes 14°30'-15°00' and longitude 73°54'-74°20', roughly 10-15 nautical miles from shore; 900 sq. miles • overall catch rate: 192 kg·hour^{-1} • catches include *Leiognathus* spp. and *Opisthopterus tardoore*, accounting for more than 50% of the catch, others include *Lactarius lactarius*, sciaenids, sharks, rays and prawn and *M.F.V. Gulf Shrimp*; all boats are 14.93 m in length and powered 87 hp	Bapat et al. (1972)
5	1962 - 1963 and 1967 - 1968	M.F.V. Silver Pomfret, M.F.V. Indian Salmon	shrimp and cotton trawls		• South Gujarat, Veraval, Porbundar, Dwarka and Gulf of Kutch; 25 000 km² • depth range: 8-90 m • the sciaenids (*Otolithus ruber*), contributed 50%-75% of the catches, followed by rays and miscellaneous other fishes • new fishing grounds for prawns and lobster were located • the most productive area was Veraval, followed by Porbundar and Dwarka; the gulf of Kutch and South Gujarat did not contain potential trawling grounds	Survey and Research Division, Directorate of Fisheries, Government of Gujarat, India (1972)
6	1968 - 1970	M.T. Kalyani IV and M.T. Kalyani V		• total fishing effort: 100 hours	• Bombay-Saurashtra waters • areawise abundance and seasonal distribution of 10 categories of fishes in 8 major areas were examined	Nair (1976)
7	1975 - 1983	R.V. Dr. Fridtjof Nansen	pelagic trawl with 750 mm opening		• northern and western Arabian Sea • abundances measured by echosounders and echo-integrators • during daytime mesopelagic fish were found at 250-350 m depth, in night at the upper 100 m • total abundance: 100×10⁶ t • in Gulf of Oman, biomass range: 6-20×10⁶ t	Gjøesaeter (1984)

Area/ survey no.	Period (date)	Vessel	Trawl gear	Trawl operations	Results/notes	References
8	Mar 1976 - Mar 1977		bottom trawl		• off Mangalore between Suratkal and Someswar; 850 km² up to 50 m isobath • other information collected: hydrographical parameters (i.e., temperature, salinity, dissolved oxygen and density) • stock assessment studies indicate abundance of *Nemipterus japonicus, Saurida tumbil* and *Grammoplites scaber* • high correlations observed between temperature and dissolved oxygen, and fish catches during various months	Maliel (1979)
9	1980 - 1983	*Bari* oceanic fleet			• along the Somalian coasts of Indian Ocean • survey deals with identification of fishes caught • 97 fish species classified/identified	Vacarella et al. (1985)
10	Oct - Dec 1984 and Jan - Mar 1985		3 types of trawl		• off Veraval, Gujarat • depth range: 25-35 m and 35-45 m • abundance highest at 25-35 m depth from Oct to Dec • ribbon fish and *Lactarius* common from Oct to Dec • sciaenids and other quality fish and cephalopods common from Jan to Mar	Rao and Kunjipalu (1989)
11	1984 - 1987		bottom trawl		• northwest coast, between 15° and 22°N • depth range: 3 depth strata, <50; 50-100; 100-200 m • paper focused on *Priacanthus hamrur* • total biomass 88 560 t • max. sustainable yield 25 000 t·yr⁻¹	Biradar (1989)
12		trawl survey collected by 17.5 m and larger vessels			• Kerala and Karnataka ; between 8° and 15°N • depth ranges: small vessels, up to 50 m; larger vessel, down to continental slope • catch and effort standing stock and MSY parameters were estimated	Sudarsan et al. (1988)
13	Oct - Dec 1991		shrimp and fish trawl		• Karnataka, Goa and South Maharashtra	Naik et al. (1993)
14		*Matsya Nireekshani*			• along north west coast between 18°N and 23°N	Shamsudeen and Raj (1993)
INDIA (WHOLE COASTLINE)						
1	1963	R.V. Anton Bruun	13 m Gulf of Mexico shrimp trawl	• 104 hauls	• surveys in Bay of Bengal and in the Arabian Sea (Cruise 1 included survey at Andaman Islands, Thailand, Burma and East Pakistan while Cruise 4B covered India, West Pakistan, Gulf of Oman and Arabia) • in Bay of Bengal, highest catch rates were obtained in 15-37 m depths; in Burma and East Pakistan, stingrays, guitarfish and croakers were the most important groups • in Arabian Sea, highest catch rates were obtained in 38-73 m depths off the southeastern Arabian coast; stingrays, crabs, sciaenids, croakers, lizardfish, cardinal fish and threadfins were the most important groups	Hida and Pereyra (1966)

Area/survey no.	Period (date)	Vessel	Trawl gear	Trawl operations	Results/notes	References

2	Apr 1970 - Mar 1980	17.5 m commercial trawl vessel commisioned for the Exploratory Fisheries Project; 200 hp and 56.8 GT	24 m fish trawl and 28 m shrimp trawl		• along both the coasts of Indian sub-continent, from 24° to 7°N, spanning 273 000 km², west coast: 184 200 km² and east coast: 89 600 km²
					• depth range: up to about 80 m
					• potential yield:
					  whole region: 1.7×10^6 t·year⁻¹
					  west coast: 1.1×10^6 t·year⁻¹
					• relative abundance of important fish by region and depth were presented
					• along west coast, the most productive depth is at 40-79 m and at the east, at 0-39 m, except for Visakhapatnam, where the most productive zone was at 60-79 m
						Joseph (1980)

3	1968 - 1980				• exploratory surveys of cephalopod resources in Bombay-Gujarat: catch maximum at 7 609 kg·hour⁻¹ a year and catch rate of 7 kg·hour⁻¹	Silas et al. (1985)
4	1985 - 1986	F.O.R.V. Sagar Sampada	5 m Isaacs-Kiddward midwater trawl	• 364 stations	• deep scattering layers of EEZ of India study described distribution and abundance of fish fauna along various latitudes, depths and seasons	Menon (1990)
5	Feb 1985 - Aug 1988	F.O.R.V. Sagar Sampada	bottom trawl	• 378 hauls	• entire coastline of India • depth range: 30-450 m • data on catch and abundance and productive areas of edible crustacea	Suseelan et al. (1990)
6		F.O.R.V. Sagar Sampada	Isaacs-Kidd midwater trawl		• deep scattering layer of Indian EEZ • preliminary studies on cephalopods	Meiyappan and Nair 1990
7		F.O.R.V. Sagar Sampada			• Bay of Bengal: within 15°00' -19°00'N and 82°40'-87°25'E Arabian Sea; within 6°30' - 22°07'N and 64°16' - 78°00'E • survey for oceanic squids • occurrence of *Symplectoteuthis oualaniensis* at depths over 1 000 m • 95% of the squids taken at night	Nair et al. (1990)
PAKISTAN						
1	1921 - 1922	S.T. William Garrick			• off the coast of Sind • best catches made on mud from 10 to 30 fm	Cushing (1971)
2	1947 - 1950	M.V. Ala		• 314 hauls	• off the coast of Cape Monze (west of Karachi) • catch composition: elasmobranchs (25%), croakers (23.0%), perches (10%), threadfins (2.5%), catfish (14%), pomfrets (1.5%)	Qureshi (1955 *In* Cushing 1971)
3	Nov 1948 - Dec 1956	M.F.T. Ala		• 43 effective 19 trips	• mean catch rate: 57.3 kg·hour⁻¹	Hussain et al. (1972)

Area/survey no.	Period (date)	Vessel	Trawl gear	Trawl operations	Results/notes	References
4	Feb 1956 - 1959	M.F.V. Machhera		• 197 effective trips	• along the coast of former Sind, Karachi and Mekran • depth range: 18-55 m • mean catch rate: 140 kg·hour^{-1} • best catches taken at 18 to 27 m	Qureshi and Burney (1952 In Hussain et al. 1972)
5	1960 - 1967	M.F.V. Machhera 20.42 LOA 52.41 GT 120 hp	semi-balloon net HR: 21.34 m FR: 30.48 m MS: 1.5"	• 1 200 hauls	• surveyed Region I (Karachi and Sind coasts), Region II (Mekran coast), Region III (Swatch ground) and Region IV (Kori Great Bank) • variation of catch rates and catch composition by region were discussed	Hussain et al. (1972)
6	1975 - 1976	R.V. Dr. Fridtjof Nansen			• acoustic, oceanographic and trawl survey from Pakistan to Kenya under NORAD/UNDP/FAO	Kesteven et al. (1981); Venema (1983)
7	1977	R.V. Dr. Fridtjof Nansen			• acoustic, oceanographic and trawl survey in Pakistan under NORAD/PAKISTAN	Venema (1983)
8	1977	Thalassa			• demersal and oceanographic survey in Pakistan and Gulf of Oman	Venema (1983)
9	Jan - Jun 1977	R.V. Dr. Fridtjof Nansen			• acoustic, oceanographic and trawl survey in Pakistan	IMR (1978)
10	Sep 1983 - Jun 1984	R.V. Dr. Fridtjof Nansen			• in addition to fish biomass estimation by acoustic instruments, in each cruise, between 40 to 90 trawl fishing stations were worked for bottom fish assessments and to obtain fish samples • cruise tracks, hydrographic profiles and fishing stations were given	IMR (1986)
11	5 - 16 Sep 1983	R.V. Dr. Fridtjof Nansen			• acoustic survey of Pakistan waters, mapping the distribution and measuring abundance of pelagic, demersal and mesopelagic fish • temperature, salinity, density and oxygen data obtained given for four hydrographic sections • dominant species at each position are reported	Nakken (1983); Majid and Arshad (1985)
12	Jan - Mar 1985	R.V. Machhera and R.V. Tehkik			• data collected include distribution and abundance of commercially important demersal fish stocks, with respect to time and space, in Pakistan coastal waters; species composition (by length and sex)	Marine Fisheries Dept., Karachi (Pakistan) 1985
SOUTH CHINA SEA						
1	1970 - 1977	Taiwan commercial pair trawlers			• Sunda Shelf, 1.1x10^6 km^2 • max. standing stock: 4.4 million t·year^{-1} • MSY: 1.95 million t	Yeh (1981)
2	Mar - Apr 1967	Baek Du San 150 GT	otter trawl	• 45 hauls • TD: 1.0-1.5 hour	• central part of South China Sea, near Borneo and Malaysia • total catch was 48 171 kg; valuable fishing grounds occurred: near Kuching River, near Jesselton, near Balabac Island and near Sandakan • more than 100 species of fish were caught; most abundant were pomfret, sea bream and yellow croaker	Chang (1972)

Area/survey no.	Period (date)	Vessel	Trawl gear	Trawl operations	Results/notes	References
3	1970 - 1976	Taiwan commercial pair trawlers			• Sunda Shelf • data from catch statistics of commercial trawlers from Taiwan • authors made investigations on history of the fishery, species composition and distribution • abundant species: golden thread, lizard fish, goat fish, big eye, red snapper, cuttle fish and catfish • max. standing stock: 5 819 000 t·year[-1] • potential yield: 2 586 000 t·year[-1]	Liu et al. (1978)
4	1970 - 1980	Taiwanese pair trawlers			• Sunda Shelf • data from catch statistics of commercial trawlers from Taiwan and the national catch information compiled by FAO South China Sea Fisheries Development and Coordinating Programme • max. standing stock: 4 362 000 t·year[-1] • potential yield: 1 950 000 t • peak catches in the 1970s of (1 090 000 t) represented 56% of the potential	Yeh et al. (1981)
5	this report described past and "present" situations relative to the author's visit to five countries in the area in May-July 1973		the demersal species reported were those caught by demersal gears: trawls (otter, pair and beam); bottom gillnets and bottom longlines; and other traditional demersal gears		• whole South China Sea encompassing 3.7 million km² • divided into 12 sub-areas about half is in the South China basin in the middle of the South China Sea; two wide continental shelves are located in the area, the mainland shelf in the north and the Sunda Shelf in the south • about 15 000 trawlers operate in the area • catches of demersal species, numbers of vessels, etc. are given for Japan, China (Taiwan), Hongkong, China, North Vietnam, Vietnam Republic, Khmer, Thailand, Malaysia, Singapore, Indonesia and the Philippines • high densities shown in Malacca Strait, Tonkin Gulf, east coast of south Malay Peninsula and the coastal zones of Sarawak and Sabah • total stock size for the region: 301 million t • stock density: Zone A(<50 m):130 kg·hour[-1] Zone B(<500 m): 104 kg·hour[-1] Average (A+B): 114 kg·hour[-1] • estimated standing stock: 3.8 million t·year[-1] • overexploitation has taken place in Northern Mainland and in the Gulf of Thailand	Aoyama (1973)

188

Area/survey no.	Period (date)	Vessel	Trawl gear	Trawl operations	Results/notes	References
AUSTRALIAN WATERS, SOUTHERN INDONESIA						
1	1970 - 1976	Taiwan pair trawlers			• Australian waters • data from catch statistics of commercial trawlers from Taiwan • authors reported on history of the fishery, species composition and distribution • abundant species: golden thread, big eye, red snapper, lizard fish, porgies, grunt, squid, cuttle fish and pompanos • max. standing stock: 2 066 000 t • potential yield: 1 033 000 t·year[-1]	Liu et al. (1978)
2	1979 - 1981	exploratory fishing survey			• southern Indonesia and northwestern Australia • total of 1 100 species sampled with information on color photography, species name, morphological data and distribution; all records updated in FishBase 97 (Froese and Pauly 1997)	Gloerfelt-Tarp and Kailola (1984)
3	1980 - 1981	F.R.V. Soela 53 m and 1800 hp stern trawler	New Zealand Frank and Bryce net FR: 32 m MS: 10, 20 mm German Engel FR: 49 m MS: 10, 20 mm prawn net FR: 43 m		• Australian Fishing Zone Sector of the Timor-Arafura Seas and in Gulf of Carpentaria • the composition of the catches, with indications of the utility and abundance of the different species, total catch rates by depth and the catch rates by depth of 291 species of economic importance were presented • significance of cruise results to fisheries management and development in the region were briefly discussed	Okera and Gunn (1986)

References

Abu Talib A. and S. Hayase. 1984a. An assessment of demersal fisheries resources in Peninsular Malaysia, 1971-80. TrainingDepartment, SEAFDEC, Samutprakarn, Thailand. TD/CTP/27, 40 p.

Abu Talib A. and S. Hayase. 1984b. Population parameters of ten demersal species off the southern east coast of Peninsular Malaysia. Training Department, SEAFDEC, Samutprakarn, Thailand. TD/CTP/29, 30 p.

Aglen, A., L. Foyn, O.R. Godo, S. Myklevoll and O.J. Ostvedt. 1981a. A survey of the marine fish resources of the north and west coasts of Sumatra, August 1980. Reports on surveys with R.V. Dr. Fridtjof Nansen. Institute of Marine Research, Bergen, Norway. 55 p.

Aglen, A., L. Foyn, O.R. Godo, S. Myklevoll and O.J. Ostvedt. 1981b. Surveys of the marine fish resources of Peninsular Malaysia, June-July, 1980. Reports on surveys with the R.V. Dr. Fridtjof Nansen. Institute of Marine Research, Bergen, Norway. 69 p.

Aglen, A., L. Foyn, O.R. Godo, S. Myklevoll and O.J. Ostvedt. 1981c. A survey of the marine fish resources of the west coast of Thailand, July 1980. Reports on surveys with the R.V. Dr. Fridtjof Nansen. Institute of Marine Research, Bergen, Norway. 57 p.

Ahmad Adnan, N. 1987. Demersal fish resources in Malaysian waters (17). Sixth trawl survey of the coastal waters of west coast of Peninsular Malaysia, October-November 1981. Buletin Perikanan (Malays.) Fish. Bull. No. 47, 39 p. Department of Fisheries. Ministry of Agriculture, Malaysia.

Ahmad Adnan, N. 1990. Demersal fish resources in Malaysian waters (16). Fifth trawl survey of the coastal waters off the east coast of Peninsular Malaysia (June-July, 1981). Fish. Bull. Dep. Fish. Malays. Buletin Perikanan Jabatan Perikanan Malays. 60, 36 p.

Anon. 1970. The report of R.T.M. Lesnoy on the results of the scientific research expedition off the coast of East Pakistan (November 1969 - January 1970). Yugrybpromrazvedka, Ministry of Fishing Industry, Kerch, USSR. 58 p.

Anon. 1993. Results of demersal resources survey along the lower east coast between latitudes 10°N and 16°N. Seafood Export J. 25(10): 41-42.

Anon. 1994a. Results of demersal resources along upper east coast between latitudes 18°N and 21°N by Matsya Shikari. Seafood Export J. 25(11): 45-48.

Anon. 1994b. Results of demersal trawl survey along the upper east coast between latitudes 16°N and 18°N by Matsya Darshini. Seafood Export J. 25(12): 39-41.

Aoyama, T. 1973. The demersal fish stocks and fisheries of the south China Sea. SCS/DEV/73/3. 80 p. Food and Agriculture Organization of the United Nations Development Programme, Rome.

Aprieto, V.L. and E.P. Villoso. 1979. Catch composition and relative abundance of trawl-caught fishes in the Visayan Sea. Fish. Res. J. Philipp. 4(1): 9-18.

Aprieto, V.L. and E.P. Villoso. 1982. Demersal fish resources of Lingayen Gulf. Fish. Res. J. 7(2): 40-48.

Armada, N.B. and G.T. Silvestre. 1981. Demersal fish resource survey in Samar Sea and Carigara Bay. UP-NSDB Project Report. College of Fisheries, University of the Philippines in the Visayas. 56 p.

Armada, N.B., C. Hammer, J. Saeger and G.T. Silvestre. 1983. Results of the Samar Sea trawl survey, p. 1-46. In J. Saeger (ed.) Technical reports of the Department of Marine Fisheries No. 3, College of Fisheries, UP in the Visayas and GTZ, Federal Republic of Germany. 191 p.

Bain, K.H. 1965. Trawling operations in the Bay of Bengal. EPTA 2121, 38 p. FAO, Rome.

Ba Kyaw, U. 1956. Trawling results of Taiyo Maru No. 11 in Burmese waters. Proc. Indo-Pac. Fish. Counc. 6(2): 261-264.

Bapat, S.V., N. Radhakrishnan and K.N. Rasachandra Kartha. 1972. A survey of the trawl fish resources off Karwar, India. Proc. Indo-Pac. Fish. Counc. 13(3): 354-383.

Bautista, R.D. and A.H. Rubio. 1981. Preliminary report on the Manila Bay survey. Fish. Res. Div. Tech. Rep.

Beales, R., D. Currie and R. Lindley (eds.). 1982. Investigations into fisheries resources in Brunei. Monogr. of the Brunei Darussalam Mus.: (5): 204 p. Brunei Museum, Brunei.

Bejie, A.B. n.d. Demersal fish resources in Malaysian waters. Fourth trawl survey off the coast of Sarawak (29 June-12 August 1977). (Unpublished).

Bejie, A.B. 1980. Demersal fish resources in Malaysian waters. Fifth trawl survey off the coast of Sarawak (10 July-10 August 1979). (Unpublished).

Bejie, A.B. 1985a. First prawn resource survey along the coast of Sarawak (May-July 1980). Bul. Perikanan No. 30. 21 p. Dicetak Oleh Jabatan Percetakan Negara, Kuala Lumpur, Malaysia.

Bejie, A.B. 1985b. Second prawn resource survey along the coast of Sarawak (8-28 April 1981). Bul. Perikanan No. 31. 14 p. Dicetak Oleh Jabatan Percetakan Negara, Kuala Lumpur, Malaysia.

Bejie, A.B. 1986. Third prawn resource survey along the coast of Sarawak (18 March-13 June 1982). Bul. Perikanan No. 32. 23 p. Dicetak Oleh Jabatan Percetakan Negara, Kuala Lumpur, Malaysia.

Berg, S.E. 1971. Investigations on the bottom conditions and possibilities for marine prawn and fish trawling on the north and east coasts of Ceylon. Bull. Fish. Res. Stn. Ceylon 22(1/2): 53-88.

Bianchi, G. 1996. Demersal fish assemblages of trawlable grounds off northwest Sumatra, p. 123-130. In D. Pauly and P. Martosubroto (eds.) Baseline studies of biodiversity: the fish resources of Western Indonesia. ICLARM Stud. Rev. 23, 312 p.

Biradar, R.S. 1989. Estimates of stock density, biomass and maximum sustainable yield of Priacanthus hamrur (Forsskål) off the north west coasts of India, p. 55-65. In Studies on fish stock assessment in Indian waters. Spec. Publ. Fish. Surv. India, 2.

Birtwistle, W. and C.F. Green. 1927. Report on the workings of the S.T. Tongol for the period 28 May-31 December 1926. Pt. 1 by C.F. Green and Pt. 2 by W. Birtwistle. Government Printing Office, Singapore.

Birtwistle, W. and C.F. Green. 1928. Report on the workings of the S.T. Tongol for the year 1927. Pt. 1 by C.F. Green and Pt. 2 by W. Birtwistle. Government Printing Office, Singapore.

Blindheim, J., and L. Foyn. 1980. A survey of the coastal fish resources of Sri Lanka, No. 3, Jan-Feb 1980. Reports on the Survey with the R.V. Dr. Fridtjof Nansen. Fisheries Research Station, Colombo and Institute of Marine Research, Bergen, Norway.

Blindheim, J., G.H.P. de Bruin and G. Saetersdal. 1979. A survey of the coastal fish resources of Sri Lanka, No. 2, April-June 1979. Reports on the Survey with the R.V. Dr. Fridtjof Nansen. Fisheries Research Station, Colombo and Institute of Marine Research, Bergen, Norway.

Caces-Borja, P. 1972. On the ability of otter trawls to catch pelagic fish in Manila Bay. Philipp. J. Fish. 10(1/2): 39-56.

Caces-Borja, P., R. Bustillo and S. Ganaden. 1972. Further observations on the commercial trawl fishery of Manila

Bay (1960-1962), p. 631-637. *In* IPFC, Proceedings of the 13th Session, 14-25 October 1968, Brisbane, Queensland, Australia. Section III - Symposium on Demersal Fisheries. IPFC Proc. 13(3), 664 p.

Calumpong, H.P., L.J. Raymundo, E.P. Solis and R.O. de Leon (eds.) 1993. Resource and Ecological Assessment of Carigara Bay, Leyte, Philippines. Volume 2. Fisheries Resources and Exploitation - Final Report. Silliman University Marine Laboratory, Dumaguete City, Philippines.

Chan, E.H. and H.C. Liew. 1986. A study on tropical demersal species (Malaysia). Final report submitted to IDRC. Faculty of Fisheries and Marine Science, Universiti Pertanian Malaysia. 64 p.

Chang, J.H. 1972. The result of experimental trawl fishing in South China Sea, p.585-596. *In* IPFC, Proceedings of the 13th Session, 14-25 October 1968, Brisbane, Queensland, Australia. Section III - Symposium on Demersal Fisheries. IPFC Proc. 13(3), 664 p.

Charnprasertporn, T. 1982. An analysis of demersal fish catches taken from the otterboard trawling survey in the Gulf of Thailand, 1980. Demersal Investigation Section, Marine Fisheries Division, Fisheries Department, Ministry of Agriculture and Cooperative, Bangkok, Thailand. 91 p.

Chowdhury, S.H. 1983. Completion reports of sea exploratory fishing scheme, East Pakistan. *In* Survey report of the fishery resources of Bay of Bengal by the survey research vessels *Chosui Maru, Kagawa Maru, Kinki Maru and Jalwa.* Research survey of marine fisheries under Directorate of Fisheries, Government of Bangladesh 1983. Mar. Fish. Bull. 1, 24 p.

Chowdhury, W.N., G. Khan, S. Myklevoll and R. Saetre. 1979. Preliminary results from a survey on the marine fish resources of Bangladesh, November-December 1979. Reports on surveys with *R.V. Dr. Fridtjof Nansen*, May 1980. Institute of Marine Research, Bergen, Norway.

Chowdhury, W.N., S.A. Iversen, G. Khan and R. Saetre. 1980. Preliminary results from a survey on the marine fish resources of Bangladesh, May 1980. Reports on surveys with *R.V. Dr. Fridtjof Nansen*. Institute of Marine Research, Bergen, Norway.

Cinco, E. and L. Perez. 1996. Results of the Sorsogon Bay trawl survey. *In* E. Cinco, G. Trono and D.J. Mendoza (eds.) Resource and Ecological Assessment of Sorsogon Bay, Vol. III. Final Reports. UB Tech, Inc. and FSP-Department of Agriculture.

Cinco, E., J. Diaz, R. Gatchalian and G.T. Silvestre. 1995. Results of the San Miguel Bay trawl survey. *In* G. Silvestre, C. Luna and J. Padilla (eds.) Multidisciplinary assessment of the fisheries in San Miguel Bay, Philippines (1992-1993). ICLARM Tech. Rep. 47.

CMFRI (Central Marine Fisheries Research Institute), Cochin, India. Waltair Research Center. 1980. Industrial fisheries off Visakhapatnam coast based on exploratory surveys during 1972-1978. Mar. Fish. Inf. Serv. Tech. Ext. Ser. 15: 1-15.

Compact Cambridge. 1991. ASFA (Aquatic Sciences and Fisheries Abstract) CD-ROM Reference Manual. Cambridge Scientific Abstracts, Bethesda, MD, USA.

Cushing, D.H. 1971. Survey of resources in the Indian and Indonesian area. Dev. Rep. Indian Ocean Programme, 2. 123 p.

de Jesus, A.S. 1976. A report on the design, construction and operation of German (Herman-Engel type) trawl net in Manila Bay. Philipp. J. Fish. 117-145 p.

de Jesus, A.S. and A.R. Maniulit. 1973. Preliminary report on the trial of bottom trawl rigged with plastic roller bobbins and oval board. Philipp. J. Fish. 11 (1/2): 36-50.

Demidenko, U. 1972. Information about the results of the joint Soviet-Lankian fishery investigations carried out in waters adjacent to Ceylon Island. Manuscript report Fisheries Research Station, Colombo.

DOF (Department of Fisheries) 1968. Report on the 1968 trawling survey. Fisheries Department, Ministry of Development, Brunei Darussalam (Unpublished).

Druzhinin, A.D. and U. Phone Hliang. 1972. Observations on the trawl fishery of Southern Burma, p. 151-209. *In* IPFC, Proceedings of the 13th Session, 14-25 October 1968, Brisbane, Queensland, Australia. Section III - Symposium on Demersal Fisheries. IPFC Proc. 13(3), 664 p.

Dwiponggo, A. and M. Badrudin. 1978a. Demersal resources survey in coastal areas of the Java Sea, 1977. Contribution of the Demersal Fisheries Project, No. 5. Marine Fisheries Research Institute, Indonesia. 14 p.

Dwiponggo, A. and M. Badrudin. 1978b. *R.V. Mutiara IV* survey data in 1977. Mar. Fish. Res. Rep. (Spec. Rep.)/Contrib. Demersal Fish. Proj., Jakarta. 5A. 133 p.

Dwiponggo, A. and M. Badrudin. 1979a. Variation in catch rates and species composition of trawl surveys in Java Sea subareas, 1978. Contribution of the Demersal Fisheries Project, No. 6. Marine Fisheries Research Institute, Indonesia. 19 p.

Dwiponggo, A. and M. Badrudin. 1979b. Data of trawl survey by *R.V. Mutiara IV* in the Java Sea subareas, 1978. Contribution of the Demersal Fisheries Project, No. 6A. Marine Fisheries Research Institute, Indonesia. 128 p.

Dwiponggo, A. and M. Badrudin. 1980a. Results of the Java Sea inshore monitoring survey, 1979. Contribution of the Demersal Fisheries Project, No. 7. Marine Fisheries Research Institute, Indonesia.

Dwiponggo, A. and M. Badrudin. 1980b. Data of the Java Sea inshore monitoring survey by *R.V. Mutiara IV*, in 1979. Contribution of the Demersal Fisheries Project, No. 7A. Marine Fisheries Research Institute, Indonesia.

Eiamsa-ard, M. 1981. An analysis of demersal fish catches taken from otterboard trawling survey in the Gulf of Thailand, 1979. Demersal Fish. Rep. No. 1, Mar. Fish. Div., Dept. of Fish., Bangkok. (in Thai).

Eiamsa-ard, M. and D. Dhamniyom. 1979. An analysis of demersal fish catches from the otterboard trawling survey in the Gulf of Thailand, 1977. Demersal Fish. Rep. No. 1, Mar. Fish. Div., Dept. of Fish., Bangkok. (in Thai).

Eiamsa-ard, M. and D. Dhamniyom. 1980. An analysis of demersal fish catches from the otterboard trawling survey in the Gulf of Thailand, 1978. Demersal Fish. Rep. No. 2, Mar. Fish. Div., Dept. of Fish., Bangkok. (in Thai).

Eiamsa-ard, M. and D. Dhamniyom. 1981. An analysis of demersal fish catches from the otterboard trawling survey in the Gulf of Thailand, 1979. Demersal Investigation Section, Marine Fish. Div., Fisheries Department, Ministry of Agriculture and Cooperative, Bangkok Thailand. 105 p.

Eiamsa-ard, M., D. Dhamniyom and S. Pramokchutima. 1979. An analysis of demersal fish catches taken from trawling survey in the Gulf of Thailand, 1977. Department of Fisheries. Demersal Fish. Rep. 1. 21 p.

Encina, V.B. 1972. A report on two-boat bottom trawl fishing in the Philippines, p. 653-664. *In* IPFC, Proceedings of the 13th Session, 14-25 October 1968, Brisbane, Queensland, Australia. Section III - Symposium on Demersal Fisheries. IPFC Proc. 13(3), 664 p.

Froese, R. and D. Pauly. 1997. FishBase 97: Concepts, design and data sources. ICLARM, Manila.

Gambang, A.C. 1981. Demersal fish resources in Malaysian waters. Sixth trawl survey off the coast of Sarawak (26 June-8 August, 1980) (Unpublished)

Gambang, A.C. 1986a. Demersal fish resources in Malaysian waters. Seventh trawl survey off the coast of Sarawak (16 May-6 July 1981). Bul. Perikanan (Malays.) Fish. Bull. No. 36, 41p. Department of Fisheries. Ministry of Agriculture, Malaysia.

Gambang, A.C.1986b. Demersal fish resources in Malaysian waters. Eight trawl survey off the coast of Sarawak (20 April - 10 June 1982). Bul. Perikanan (Malays.) Fish. Bull. No. 37, 43 p. Department of Fisheries. Ministry of Agriculture, Malaysia.

Gambang, A.C.1987. Demersal fish resources in Malaysian waters. Tenth trawl survey off the coast of Sarawak (12 May - 17 June 1987). Bul. Perikanan (Malays.) Fish. Bull. No. 51, 30 p. Department of Fisheries. Ministry of Agriculture, Malaysia

Gjøesaeter, J. 1984. Mesopelagic fish, a large potential resource in the Arabian Sea, p.1019-1035. In M.V. Angel (ed.) Marine science of the northwest Indian Ocean and adjacent waters. Deep-Sea Res. 31(6-8A).

Gloerfelt-Tarp, T. and P.J. Kailola (eds.) 1984. Trawled fishes of southern Indonesia and northwestern Australia. NP-Australia-Australian Development Assistance Bureau. 406 p.

Godo, O.R. 1979. Fish resources off Vietnam, Part I. Report on the Investigation by R.V. Bien Dong in Vinh Bac Bo (Gulf of Tonkin), June 1977-June 1978. Marine Fisheries Research Institute, Haipong Vietnam and Institute of Marine Research, Bergen Norway. September 1979.

Godo, O.R. 1980. Fish resources off Vietnam, Part II. Report on the Investigation by R.V. Bien Dong off the Southeastern part of Vietnam, July 1978-December 1979. Marine Fisheries Research Institute, Haipong Vietnam and Institute of Marine Research, Bergen, Norway. August 1980.

Gonzales, C.C., J.C. Muñoz, A.R. Tanedo and F.N. Lavapie. 1981. Report on the first finding survey of the Visayan Sea. Fish. Res. Div. Tech. Rep.

Grosslein, M.D. and A. Laurec. 1982. Bottom trawl surveys: design, operation and analysis. Interregional Fisheries Development and Management Programme - INT/79/019, CECAF/ECAF Ser. 81/22, 25 p.

Halidi, H.A.M.S. 1987. A preliminary analysis of catch-per-effort data from demersal trawl surveys off Negara Brunei Darussalam (1979-1986). Report of a training stage at ICLARM, Mar-Apr 1987. Manila, Philippines.

Hayase, S. 1983. Preliminary assessment of the demersal stocks along the Indian Ocean coast of Thailand. Cur. Tech. Pap. Train. Dep. Southeast Asian Fish. Dev. Cent. Pharpradaeng Thailand-SEAFDEC. 18, 32 p.

Henriksson, T.G. 1980. Survey of demersal fish resource off the northwest, north and northeast coasts of Sri Lanka. Field Document 7. FAO, February 1980. FI:DP/SRL/72/051.

Hida, T.S. and W.T. Pereyra. 1966. Results of bottom trawling in Indian Seas by R.V. Anton Bruun in 1963. Proc. Indo-Pac. Fish. Counc. 11:156-171.

Humayon, M., G. Mustafa, M. Begum, S.C. Paul, N. Sada and G. Khan. 1983. Results from the 15th cruise (Dec. '83) of the R.V. Anusandhani to the demersal fishing grounds of the northern Bay of Bengal. Marine Fisheries Research, Management and Development, Chittagong, Bangladesh.

Hussain, A.G., M.A. Burney, S.Q. Mohiuddin and S. Zupanovic. 1972. Analysis of demersal catches taken from exploratory fishing off the coast of West Pakistan, p.61-84. In IPFC, Proceedings of the 13th Session, 14-25 October 1968, Brisbane, Queensland, Australia. Section III - Symposium on Demersal Fisheries. IPFC Proc. 13(3), 664 p.

IMR (Institute of Marine Research, Bergen, Norway). 1978. Survey results of R.V. Dr. Fridtjof Nansen Jan-Jun 1977. Joint NORAD/Pakistan project fish assessment survey, Pakistan waters. Bergen Norway Inst. Mar. Res. 39 p.

IMR (Institute of Marine Research, Bergen, Norway). 1986. R.V. Dr. Fridtjof Nansen surveys of Pakistan fishery resources, September 1983 to June 1984. Summary of findings. UNDP/FAO Programme-GLO/82/001. 52 p.

Isarankura, A.P. 1971. Assessment of stocks of demersal fish off the west coasts of Thailand and Malaysia. FAO/UNDP Dev. Rep. Indian Ocean Programme.

Jean, Y. 1957. Summary of operations of Canadian and North Star, April 1954-March 1957, Colombo Plan (Manuscript).

Joseph, K.M. 1974. Demersal fisheries resources off the northwest coast of India. Bull. Expl. Fish. Proj., India No. 1, 47 p.

Joseph, K.M. 1980. Comparative study of the demersal fishery resources of the Indian waters as assessed by the 17.5 m trawlers. Bull. Expl. Fish. Proj. 10, 46 p.

Joseph, K.M., N. Radhakrishnan and K.P. Philip. 1976a. Demersal fisheries resources off the west coast of India. Bull. Expl. Fish. Proj., India, No. 3, 56 p.

Joseph, K.M., N. Radhakrishnan, A. Joseph and K.P. Philip. 1976b. Results of demersal fisheries resources survey along the east coast of India, 1959-1974. Bull. Expl. Fish. Proj., India, No. 5, 53 p.

Jothy, A.A., G. Rauck. S.A.L. Mohammed Shaari, K.S. Ong, P.T. Liong and J.L. Carvalho. 1975. Demersal fish resources in Malaysian waters. Second trawl survey of the coastal waters off the east coast of Peninsular Malaysia (March-May, 1971). Fish. Bull. (Malaysia) No. 4. 36 p.

Kesteven, G.L., O. Nakken, T. Strømme (eds.). 1981 The small pelagic and demersal fish resources of the northwest Arabian Sea, further analysis of the results of the IMR (Institute of Marine Research, Bergen, Norway). 1978. Survey results of R.V. Dr. Fridtjof Nansen Jan-Jun 1977. Joint NORAD/Pakistan project fish assessment survey Pakistan waters. Bergen, Norway Inst. Mar. Res. 39 p.

Khan, G., G. Mustafa, N.U. Sada and Z.A. Chowdhury. 1989a. Bangladesh: Offshore marine fishery resources studies with special reference to the Penaeid shrimp stocks, 1988-89. A report based on R.V. Anusandhani shrimp trawling survey results (Cruise# GOB 49 to 54). Marine Fisheries Survey Management and Development Project, Department of Fisheries.

Khan, G., G. Mustafa, N.U. Sada and Z.A. Chowdhury. 1989b. Bangladesh: Inshore water fishery resources survey, demersal trawling with R.V. Machhranga, 1988-1989. A report based on R.V. Machhranga (Cruise # 8801/06-8903/11). Marine Fisheries Survey Management and Development Project, Department of Fisheries.

Khan, M.G. 1983. Results of the 13th cruise (July 1983) with the R.V. Anusandhani to the demersal fish and shrimp grounds of the Bay of Bengal, Bangladesh. A cruise report submitted to the Project Director Marine Fisheries Research Management and Development Project, Chittagong, 11 p. (Mimeo).

Khan, M.G., M. Humayon, M.G. Mustafa, B. Mansura, S.C. Paul, and M.N. Sada. 1983. Results from the 15th cruise (December 1983) of the R.V. Anusandhani to the demersal fishing grounds of the northern Bay of Bengal (Bangladesh). Report submitted to the Project Director Marine Fisheries Research Management and Development Project, prepared by M. Humayon, M. Golam, M. Begum, C.P. Swapan and M.N. Sada. 8 p. (Mimeo).

Krishnamoorthi, B. 1976. An assessment of the demersal fishery resources off the Andhra-Orissa coast based on exploratory trawling. Indian J. Fish. 21(2): 557-565.

Lam, W.C., S.A.L. Mohammed Shaari, A.K. Lee and W. Weber. 1975a. Demersal fish resources in Malaysian waters-5. Second west coast trawl survey off the west coast of Peninsular Malaysia. Fish. Bull. 7. 29 p. Ministry of Agriculture and Rural Development, Kuala Lumpur, Malaysia.

Lam, W.C., W. Weber, A.K. Lee, K.S. Ong and P.C. Liong. 1975b. Demersal fish resources in Malaysian waters-7. 3rd East coast trawl survey off the east coast of Peninsular Malaysia, 14 August - 20 September 1972. Fish. Bull. 9. 18

p. Ministry of Agriculture and Rural Development, Kuala Lumpur, Malaysia.

Lamboeuf, M. 1987. Bangladesh demersal fish resources of the continental shelf, *R.V. Anusandhani* trawling survey results September 1984-June 1986. A report prepared for the FAO/UNDP Project, Strengthening of the National Programme for Marine Fisheries Resources Management, Research and Development.

Lamp, F. and S.A.L. Mohammed Shaari. 1976. Demersal fish resources in Malaysian waters-10. Fourth trawl survey of the coastal waters off the east coast of the Peninsular Malaysia, 13 July-12 August 1974. Fish. Bull. 12, 25 p. Ministry of Agriculture and Rural Development, Kuala Lumpur, Malaysia.

Legasto, R.M., C.M. del Mundo and K.E. Carpenter. 1975a. On the hydrobiological and socio-economic surveys of San Miguel Bay for the proposed fish nurseries/reservations. Philipp. J. Fish. 13(2): 205-246.

Legasto, R.M., C.M. del Mundo and K.E. Carpenter. 1975b. On the socio-economic and hydrobiological surveys of Maqueda Bay, Villareal Bay and part of Zumarraga Channel for the proposed nurseries/reservations. Philipp. J. Fish. 13(1): 102-146.

Liu, H.C., H.L. Lai and S.Y. Yeh. 1978. General review of demersal fish resources in the Sunda Shelf and the Australian waters. Acta Oceanogr. Taiwan 8: 109-140.

Lohmeyer, U. 1996. Narrative and major results of the Indonesian-German Module (II) of the JETINDOFISH Project, August 1979-July 1981, p. 77-90. *In* D. Pauly and P. Martosubroto (eds.) Baseline studies of biodiversity: the fish resources of Western Indonesia. ICLARM Stud. Rev. 23, 312 p.

Losse, G.F. and A. Dwiponggo. 1977. Report of the Java Sea southeast monsoon trawl survey, June-December 1976. Contribution of the Demersal Fisheries Project, No. 3. Marine Fisheries Research Institute, Indonesia. 119 p.

Lui, Y.P. 1981. Demersal fish resources in Malaysian waters-14. Fourth trawl survey off the west coast of Peninsular Malaysia, 17 October-18 November 1978. Fish. Bull. 26. 19 p. Ministry of Agriculture and Rural Development, Kuala Lumpur, Malaysia.

Lui, Y.P., S.A.L. Mohammed Shaari, M. Ismail Taufid and F. Lamp. 1976. Demersal fish resources in Malaysian waters-12. Third trawl survey off the coast of Sarawak (16 July-8 August 1975). Fish. Bull. 14. 35 pp. Ministry of Agriculture and Rural Development, Kuala Lumpur, Malaysia.

MADECOR (Mandala Agricultural Development Corporation) and National Museum. 1994. FSP-Resource and Ecological Assessment in Manila Bay. Vol. 2. Fishery Stock and Habitat Assessment Studies. Final Report. MADECOR and FSP, BFAR, Quezon City.

Majid, A. and M. Arshad. 1985. Survey of demersal fish resources along Pakistan coast by *R.V. Dr. Fridtjof Nansen* in autumn 1983. Fish. Newsl. Mar. Fish. Dep. Pak. 3(1-2): 33-41.

Maliel, M.M. 1979. Exploratory fishing in the inshore waters of Arabian Sea of Mangalore. Mysore J. Agric. Sci. 13(1): 122 (Summary only).

Manacop, P.R. and S.V. Laron. 1956. The 'dug-out' trawl fishing of Manila Bay. Philipp. J. Fish. 4(2): 103-113.

Marine Fisheries Dep., Karachi (Pakistan). 1985. Activities of the two departmental research vessels Machhera and Tehkik during January-March, 1985. Fish. Newsl. Mar. Fish. Dep. Pak. 3(1-2): 23-31.

Marine Fisheries Laboratory, Ministry of Agriculture, Thailand and the Fisheries Research Institute, Ministry of Agriculture and Cooperatives, Malaysia. 1967. Results of the joint Thai-Malaysia-German trawling survey off the coast of Malay Peninsula, 1967. Ministry of Agriculture, Thailand and the Ministry of Agriculture and Cooperatives, Malaysia. 64 p.

Martosubroto, P. 1996. Structure and dynamics of the demersal resources of the Java Sea, 1975-1979, p. 62-76. *In* D. Pauly and P. Martosubroto (eds.) Baseline studies of biodiversity: the fish resources of Western Indonesia. ICLARM Stud. Rev. 23, 312 p.

Martosubroto, P., T. Sujastani and D. Pauly. 1996. The mid-1970s demersal resources in the Indonesian side of the Malacca Strait, p. 40-46. *In* D. Pauly and P. Martosubroto (eds.) Baseline studies of biodiversity: the fish resources of Western Indonesia. ICLARM Stud. Rev. 23, 312 p.

Megia, T.G., I.A. Ronquillo and R.R. Medina. 1953. Notes and observation on the Danish deep-sea expedition of the *Galathea* to the Mindanao Deep. Philipp. J. Fish. 2(1/2): 197-204.

Meiyappan, M.M. and K.P. Nair. 1990. Preliminary studies on the cephalopods collected from the deep scattering layers of the Indian Exclusive Economic Zone and adjacent seas, p. 397-401. *In* K.J. Mathew (ed.) Proceedings of the First Workshop on Scientific Results of *F.O.R.V. Sagar Sampada*, 5-7 June 1989. Central Marine Fisheries Research Institute, Cochin, India.

Menavesta, D. 1970. Potential demersal fish resources of the Sunda Shelf, p. 525-559. *In* J. Marr (ed.) The Kuroshio, a symposium on the Japan current. Honolulu, East-West Center Press.

Menon, N.G. 1990. Preliminary investigation on the fish biomass in the deep scattering layers of the EEZ of India, p.273-280. *In* K.J. Mathew (ed.) Proceedings of the First Workshop on Scientific Results of *F.O.R.V. Sagar Sampada*, 5-7 June 1989. Central Marine Fisheries Research Institute, Cochin, India.

Mines, A.N. (ed.) 1984. Assessment of the trawlable areas in the Philippines II: Ragay Gulf, Burias Pass, Ticao Pass and waters north of Samar Sea. Tech. Rep. Dept. Mar. Fish. 5, 66 p.

Mines, A.N. 1986. An assessment of the fisheries of Lingayen Gulf. PCARRD/NSTA-UPV. 26 p.

Mohamed, M.I.H., G. Kawamura and Z. Haron. 1986. Trawl catch composition and fish abundance of *Matahari Expedition '85*, p. 147-163. *In* A study on the offshore waters of the Malaysian EEZ. Occ. Pap. 3. Faculty of Fisheries and Marine Science, Universiti Pertanian Malaysia. 233 p.

Mohammed Shaari, S.A.L. and H.L. Chai. 1976. Demersal fish resources in Malaysian waters-11. Third trawl survey off the west coast of Peninsular Malaysia (16 November-11 December 1974). Fish. Bull. 13. 17 p. Ministry of Agriculture and Rural Development, Kuala Lumpur, Malaysia.

Mohammed Shaari, S.A.L., G. Rauck, K.S. Ong and S.P. Tan. 1974. Demersal fish resources in Malaysian waters-2. Trawl survey of the coastal waters off the west coast of Peninsular Malaysia (12 December 1970-22 January 1971) Fish. Bull. 3. 41 p. Ministry of Agriculture and Rural Development, Kuala Lumpur, Malaysia.

Mohammed Shaari, S.A.L., W. Weber and P.C. Liong. 1976a. Demersal fish resources in Malaysian waters-8. Trawl survey off the west coast of Peninsular Malaysia (Southern part of the Malacca Straits). Fish. Bull. 10. 21 p. Ministry of Agriculture and Rural Development, Kuala Lumpur, Malaysia.

Mohammed Shaari, S.A.L., W. Weber, A.K. Lee and W.C. Lam. 1976b. Demersal fish resources in Malaysian waters. First trawl survey off the coasts of Sarawak, Brunei and the west coast of Sabah, 27 March - May 1972. Fish. Bull. 8. 64 p. Ministry of Agriculture and Rural Development, Kuala Lumpur, Malaysia.

Mohammed Shaari, S.A.L., W. Weber and A.K. Lee. 1976c. Demersal fish resources in Malaysian waters-9. Second trawl survey off the coast of Sarawak, 17 August-18 September 1973. Fish. Bull. 11. 28 p. Ministry of Agriculture and Rural Development, Kuala Lumpur, Malaysia.

Mohan, P.C. and G.V. Rayudu. 1986. Squid and cuttlefish in nearshore bottom waters off Visakhapatnam, India, p.373-378. *In* M.F. Thompson, R. Sarojini and R. Nagabhushanam (eds.) Biology of benthic marine organisms: techniques and methods as applied to the Indian Ocean. Indian Ed. Ser. 12.

MSU (Mindanao State University at Naawan). 1993. Resource and ecological assessment of Panguil Bay. Terminal report. MSU and FSP, BFAR, Quezon City.

Mustafa, M.G. and M.G. Khan. 1993. The bottom trawl fishery, p.89-106. *In* Studies of interactive marine Fisheries of Bangladesh. Bay of Bengal Programme for Fisheries Development, Madras, India. BOBP/WP/89.

Mustafa, M.G., M.G. Khan and M. Humayon. 1987. Bangladesh, Bay of Bengal penaeid shrimp trawl survey result, *R.V. Anusandhani*, November 1985-January 1987. Marine Fisheries Research Management and Development Project.

Naik, S.S., A.K. Malik, A. Tiburtias, K.N.V. Nair and S.G. Patwari. 1993. Results of demersal resources survey along Karnataka-Goa-South Maharashtra coast between latitudes 11°N and 18°N. Seafood Export J. 25(7): 39-43.

Nair, K.P. 1976. Exploratory trawl fishing in Bombay-Saurashtra waters during 1968-70. Indian J. Fish. 21(2): 406-426.

Nair, K.P., K.S. Rao, R. Sarvesan, M.M. Meiyappan, G.S. Rao, M. Joseph and D. Nagaraja. 1990. Oceanic squid resources of the EEZ of India, p. 403-407. *In* K.J. Mathew (ed.) Proceedings of the First Workshop on Scientific Results of *F.O.R.V. Sagar Sampada*, 5-7 June 1989. Central Marine Fisheries Research Institute, Cochin, India.

Nakken, O. 1983. Cruise report *Dr. Fridtjof Nansen*. Fisheries resources survey. Pakistan. 5-16 September 1983. NORAD/UNDP/FAO-programme-GLO/82/001. 28 p.

Narayanappa, G., D.A. Narasimha Raju and A.V.V. Satyanarayana. 1972. Certain observations on the otter trawl operations carried out in the inshore and deep-waters off Kakinada, p. 450-455. *In* IPFC, Proceedings of the 13th Session, 14-25 October 1968, Brisbane, Queensland, Australia. Section III - Symposium on Demersal Fisheries. IPFC Proc. 13(3): 664 p.

Nasir, M.T.M., and I. Johari. 1990. The second and third prawn trawling surveys off the west coast of Peninsular Malaysia. Fish. Bull. Dep. Fish. Malays. Bul. Perikanan Jabatan Perikanan Malays. 61. 81 p.

Ochavillo, D., H. Hernandez, S. Resma and G.T. Silvestre. 1989. Preliminary results of a study of the commercial trawl fisheries in Lingayen Gulf, Philippines, p.31-42. *In* G.T. Silvestre, E. Miclat and T.E. Chua (eds.) Towards sustainable development of the coastal resources of Lingayen Gulf, Philippines. ICLARM Conf. Proc. 17, 200 p.

Okera, W. and J.S. Gunn. 1986. Exploratory trawl surveys by *F.R.V. Soela* in the Australian fishing zone sector of the Timor-Arafura seas and in the Gulf of Carpentaria, 1980-81. CSIRO Marine Laboratory Report 150, Commonwealth Scientific and Industrial Research Organization, Marine Research Laboratory, Australia. 104 p.

Ommaney, F.D. 1961. Malayan offshore trawling grounds. The experimental and exploratory fishing cruises of the *F.R.V. Manihine* in Malayan and Borneo waters, 1955-56 with a note on temperatures and salinities in the Singapore Strait. Colonial Office Fisheries Publications No. 18. H.M.S.O., London.

Ong, K.S. and W. Weber. 1977. First prawn trawling survey off the west coast of Peninsular Malaysia. Fish. Bull. 18. 28 p. Ministry of Agriculture, Kuala Lumpur, Malaysia.

Ordoñez, J.A. 1985. A study of the trash fish caught by the otter trawl in the Visayan Sea. Philipp. J. Fish. 18(1/2): 1-76.

Ordoñez, J.A., F.M. Arce, R.A. Ganaden and N.N. Metrillo, Jr. 1975. On the hydro-biological and fisheries survey of Sorsogon Bay, Luzon Island. Philipp. J. Fish. 13(2): 178-204.

Ordoñez, J.A., J.S. Ginon and A.M. Maala. 1977. Report on the fishery-oceanographic observations in Tayabas Bay and adjacent waters. Philipp. J. Fish. 15(1): 41-78.

Pathansali, D., G. Rauck, A.A. Jothy, S.A.L. Mohammed Shaari and T.B. Curtin. 1974. Demersal fish resources in Malaysian waters. Min. Agric. Fish. Malaysia Fish. Bull. 1, 46 p.

Pathansali, D., K.S. Ong, S.S. Latiff and J. Carvalho. 1967. Preliminary results of trawling investigations off Penang. Proc. Indo-Pac. Fish. Coun. 12(11): 181-201.

Pauly, D. 1982. History and status of the San Miguel Bay Fisheries, p. 95-124. *In* D. Pauly and A.N. Mines (eds.) Small-scale fisheries of San Miguel Bay, Philippines: biology and stock assessment. University of the Philippines, Quezon City Philippines. ICLARM Tech. Rep. 7, 124 p.

Pauly, D. 1995. Anecdotes and the shifting baseline syndrome of fisheries. Trends Ecol. Evol. 10(10): 430.

Pauly, D. 1996. Biodiversity and the retrospective analysis of demersal trawl surveys: A programmatic approach, p. 1-6. *In* D. Pauly and P. Martosubroto (eds.) Baseline studies of biodiversity: the fish resources of Western Indonesia. ICLARM Stud. Rev. 23, 312 p.

Pauly, D., P. Martosubroto and J. Saeger. 1996. The *Mutiara 4* surveys in the Java and southern South China Seas, November 1974 to July 1976, p.47-54. *In* D. Pauly and P. Martosubroto (eds.) Baseline studies of biodiversity: the fish resources of Western Indonesia. ICLARM Stud. Rev. 23, 312 p.

Pearsons, J. and A.H. Malpas. 1926. A preliminary report on the possibilities of commercial trawling in the sea around Ceylon. Ceylon J. Sci. 2: 1-12.

Penn, J.W. 1983. Bangladesh, an assessment of potential yields from offshore demersal shrimp and fish stocks in Bangladesh waters (including comments on the trawl fishery 1981-82). A report prepared for the Fisheries Advisory Service (Phase II) Project. Rome, FAO, FI:DP/BGD/81/034. Field Document 4, 22 p.

Philip, K.P. and D.M. Ali. 1989. Population dynamics and stock assessment of the cuttle fish, *Sepia pharaonis* (Ehrenberg) in Wadge Bank, p. 66-75. *In* Studies of fish stock assessment in Indian waters. Spec. Publ. Fish. Surv. India, 2.

Pillai, P.K.M., N. Jayabalan, M. Srinath and S. Subramani. 1983. The catch trend of the commercial trawl fisheries off Rameswaram. Mar. Fish. Inf. Serv. Tech. Ext. Ser. 48: 17-19.

Pura, L.R., E.A. Cinco, Q.P. Sia III, F.C. Alducente, M.J.N. Ayon, F.G. Balleta, A.R. Enerlan, B.E. Llevado and F.V. Lumagod. 1996a. Chapter 1, Assessment of the fishery resources of Carigara Bay. *In* G.T. Silvestre, L.R. Garces and A.C. Trinidad (eds.) Resource and ecological assessment of Carigara Bay, Philippines: results of the monitoring activities (1995-1996).

Pura, L.R., E.A. Cinco, Q.P. Sia III, F.L Gonzales, A.P. Luistro and L.M. Rueca. 1996b. Chapter 1, Assessment of the fishery resources of Manila Bay. *In* G.T. Silvestre, L.R. Garces and A.C. Trinidad (eds.) Resource and ecological assessment of Manila Bay, Philippines: results of the monitoring activities (1995-1996).

Qureshi, M.R. 1955. Marine fishes of Karachi and the coasts of Sind and Makram. Government of Pakistan, 80 p.

Qureshi, M.R. and M.A. Burney. 1952. A preliminary report on trawling in Pakistan. Invest. Rep. 1: 1-53.

Rachid, M.H. 1983. Mitsui-Taiyo survey 1976-77 by the survey research vessels *M.V. Santamonica* and *M.V. Orion-8* in the marine waters of Bangladesh. Research/survey of Marine Fisheries under the Directorate of Fisheries, Government of Bangladesh. Mar. Fish. Bull. 2.

Rao, T.A. and B. Krishnamoorthi. 1982. Diurnal variation in the catches of demersal fishes in the north west region of Bay of Bengal during 1959-60. Indian J. Fish. 29(1-2): 134-143.

194

Rao, K.K. and K.K. Kunjipalu. 1989. Estimation of fishery resources based on trawl surveys off Veraval, Gujarat, India, p. 63-68. *In* S.C. Venema and N.P. van Zalinge (eds.) Contributions to tropical fish stock assessment in India. Papers prepared by the participants at the FAO-DANIDA-ICAR National Follow-up Training Course on Fish Stock Assessment, Cochin, India, 2-28 November 1987. GCP/INT/392/DEN/1.

Reuben, S., G.S. Rao, G. Luther, T.A. Rao, K. Radhakrishnan, Y.A. Sastry and G. Radhakrishnan. 1988. An assessment of the demersal fishery resources of the northeast coast of India. CMFRI Spec. Publ. 40: 15 (Summary only).

Rijavec, L. and Htun Htein. 1984. Burma demersal fish resources of continental shelf and slope (results of trawling surveys, 1981-1983). Marine Resources Survey and Exploratory Fishing Project, Rome, FAO FI:DP/BUR/77/003. Field Document 3, 73 p. and 24 figures.

Ritragsa, S. and S. Pramokchutima. 1970. The analysis of demersal fish catches taken from the otterboard trawling survey in the Gulf of Thailand, 1968. Div. Res. Inv. Thailand Contrib. 16: 1-61.

Ritragsa, S., D. Dhamniyom and S. Sittichaikasem. 1968. An analysis of demersal fish catches taken from the otterboard trawling survey in the Gulf of Thailand, 1966. Div. Res. Inv. Thailand Contrib. 11, 104 p. (In Thai)

Ritragsa, S., D. Dhamniyom and S. Sittichaikasem. 1969. An analysis of demersal fish catches taken from the otterboard trawling survey in the Gulf of Thailand, 1968. Contrib. Mar. Fish. Lab., Bangkok. 16. 61 p. (In Thai)

Ronquillo, I.A. 1959. Oceanographic research in the Philippines. Philipp. J. Fish. 7(2): 87-96.

Ronquillo, I.A., P. Caces-Borja and A.N. Mines. 1960. Preliminary observations on the otter trawl fishery of Manila Bay. Philipp. J. Fish. 8(1): 47-56.

Saetersdal, G.S. and G.H.P. de Bruin (eds.) 1979. Report on the survey of the coastal fish resources of Sri Lanka, August - September 1978. Reports on Surveys with the *R.V. Dr. Fridtjof Nansen*. Fisheries Research Station, Colombo and Institute of Marine Research, Bergen. 36 p.

Saetre, R. 1981. Surveys on the marine fish resources of Bangladesh. Nov.-Dec. 1979 and May 1980. Reports on surveys with the *R.V. Dr. Fridtjof Nansen*. Institute of Marine Research, Bergen.

SCS. 1978. Report of the workshop on the demersal resources of the Sunda Shelf, 31 October-6 November 1977, Penang, Malaysia. Part II. SCS/GEN/77/13. 120 p. South China Sea Fisheries Development and Coordinating Programme, Manila.

SEAFDEC (Southeast Asian Fisheries Development Center). 1985. Report of the Thai-SEAFDEC joint fishery oceanographic survey in the central Gulf of Thailand, (16 May-9 June 1984). Res. Pap. Ser. No. 4, SEAFDEC, Training Department.

Sebastian, A.R. 1951. Oceanographic research in the Philippines. Philipp. J. Fish. 1: 147-153.

Shamsudeen, A.R., and K.G. Raj. 1993. Results of demersal resources survey along north west coast between latitudes 18°N and 23°N by Matsya Nireekshani. Seafood Export J. 25(9): 39-43.

Silas, E.G., K. Vidyasagar, K.P. Nair and B.N. Rao. 1985. Cephalopod resources revealed by exploratory surveys in Indian seas, p. 129-136. *In* E.G. Silas (ed.) Cephalopod-bionomics, fisheries and resources of the Exclusive Economic Zone of India. CMFRI Bull. 37. Central Marine Fisheries Research Institute, Cochin, India.

Silvestre, G.T. and H.J.H. Matdanan. 1992. Brunei Darussalam capture fisheries: A review of resources, exploitation and management, p.1-38. *In* G. Silvestre, H.J.H. Matdanan, P.H.Y. Sharifuddin, M.W.R.N. De Silva and T.-E. Chua (eds.) The coastal resources of Brunei Darussalam: status, utilization and management. ICLARM Conf. Proc. 34, 214 p.

Silvestre, G.T., R.B. Regalado and D. Pauly. 1986. Status of Philippine demersal stocks-inferences from underutilized catch rate data, p.47-96. *In* D. Pauly, J. Saeger and G. Silvestre (eds.) Resources, management and socio-economics of Philippine marine fisheries. Tech. Rep. Dep. Mar. Fish. Tech. Rep. 10 , 217 p.

Sivasubramaniam, K. and R. Maldeniya. 1985. The demersal fisheries of Sri Lanka. Work. Pap. Bay Bengal Programme, No. 41, 41 p.

Sparre, P., E. Ursin and S.C. Venema. 1989. Introduction to tropical fish stock assessment, Part 1-Manual. FAO Fish. Tech. Pap. 306/1, 337 p.

Sreenivasan, P.V. and R. Sarvesan. 1989. On the cephalopods collected during the exploratory survey by *F.O.R.V. Sagar Sampada* in the Andaman-Nicobar Seas, p.409-413. *In* K.J. Mathew (ed.) Proceedings of the First Workshop on Scientific Results of *F.O.R.V. Sagar Sampada*, 5-7 June 1989. Central Marine Fisheries Research Institute, Cochin, India.

Strømme, T., O. Nakken, Sann Aung and G. Saetersdal (eds.) 1981. Surveys on the marine fish resources of Burma, September to November 1979 and March-April 1980. Reports on surveys with the *R.V. Dr. Fridtjof Nansen*. Institute of Marine Research. Bergen, Norway.

Sudarsan, D. and P.J. Joseph. 1975. On the location of a potential prawn fishing ground off the west Bengal coast. J. Mar. Biol. Assoc. India. 17(2): 247-249.

Sudarsan, D. M.E. John and A. Joseph. 1988. An assessment of demersal stocks in the southwest coast of India with particular reference to the exploitable resources in outer continental shelf and slope. CMFRI Spec. Publ. 40: 102-103.

Sudradjat, A. and U. Beck. 1978. Report on southern South China Sea demersal trawl survey, June and July 1978, p. 81-140. *In* Contribution of the Demersal Fisheries Project, No. 4. Marine Fisheries Research Institute, Indonesia.

Suppachai-Ananpongsuk. 1989. Report on some offshore demersal resources of the Andaman Sea. Res. Pap. Ser. Train. Dep. Southeast Asian Fish. Dev. Cent. Samutparakarn, Thailand, SEAFDEC Training Department, No. 20, 30 p. SEAFDEC TD/RES/20.

Survey and Research Division, Directorate of Fisheries, Government of Gujarat, India. 1972. Exploratory operations off Gujarat coast. Proc. Indo-Pac. Fish. Counc. 13(3): 456-567.

Suseelan, C., G. Nandakumar and K.N. Rajan. 1990. Results of bottom trawling by *F.O.R.V. Sagar Sampada* with special reference to catch and abundance of edible crustaceans, p. 337-346. *In* K.J. Mathew (ed.) Proceedings of the First Workshop on Scientific Results of *F.O.R.V. Sagar-Sampada*, 5-7 June, 1989, Cochin. Cochin India Central Marine Fisheries Research Institute.

Tiews, K. and P. Caces-Borja. 1959. On the availability of fish of the family Leiognathidae Lacepède in Manila Bay and San Miguel Bay and on their accessibility to controversial fishing gears. Philipp. J. Fish. 7(1): 59-86

Vacarella, R., A.M. Pastorelli and G. Marano. 1985. Observations on the trawl fisheries of the Bari oceanic fleet carried out along the Somalian coasts (Indian Ocean). Oebalia 11(2): 633-653.

Vakily, J.M. 1982. Catch and effort in the trawl fishery, p. 65-94. *In* D. Pauly and A.N. Mines (eds.) Small Scale Fisheries of San Miguel Bay, Philippines: Biology and Stock Assessment. University of the Philippines, Quezon City, Philippines. ICLARM Tech. Rep. 7.

Venema, S.C. (ed.). 1983. Fishery resources in the North Arabian Sea and adjacent waters, p.1001-1018. *In* M.V. Angel Marine science of the Northwest Indian Ocean and adjacent waters. Deep-Sea Res. 31(6-8A).

Virabhadra Rao K., K. Dorairaj and P.V. Kagwade. 1972. Results of the exploratory fishing operations of the Government of India vessels at Bombay base for the period 1961-1967. Proc. Indo-Pac. Fish. Counc. 13(3): 402-430.

Vivekanandan, E.V. and B. Krishnamoorthi. 1985. Estimated resources of demersal fisheries off North Tamil Nadu-South Andhra coasts based on exploratory surveys, p.64-76. *In* Harvest and Postharvest Technology of Fish. Society of Fisheries Technologists, Cochin, India.

Warfel, H.E. and P.R. Manacop. 1950. Otter trawl explorations in Philippines waters. Research Report 25, Fish and Wildlife Service, U.S. Department. of the Interior, Washington, D.C.

West, W.Q.B. 1973. Fishery resources of the Upper Bay of Bengal. Indian Ocean Programme, Indian Ocean Fisheries Commission, Rome, FAO IOFC/DEV/73/28, 44 p.

White, T.F. 1985a. Marine fisheries resource survey: demersal trawling. Survey cruise report, No.4, November 9-20, 1984. BGD/80/025/CR4, 72 p. Marine Fisheries Research, Management and Development Project, UNDP/FAO, Chittagong, Bangladesh.

White, T.F. 1985b. Marine fisheries resource survey: demersal trawling. Survey cruise report, No. 5, November 27-December 6, 1984. BGD/80/025/CR5, 65 p. Marine Fisheries Research, Management and Development Project, UNDP/FAO, Chittagong, Bangladesh.

White, T.F. 1985c. Marine fisheries resource survey: demersal trawling. Survey cruise report, No.6, December 13-21, 1984. BGD/80/025/CR6, 68 p. Marine Fisheries Research, Management and Development Project, UNDP/FAO, Chittagong, Bangladesh.

White, T.F. 1985d. Marine fisheries resource survey: demersal trawling. Survey cruise report, No.7, January 6-16, 1984. BGD/80/025/CR7, 78 p. Marine Fisheries Research, Management and Development Project, UNDP/FAO, Chittagong, Bangladesh.

White, T.F. 1985e. Marine fisheries resource survey: demersal trawling. Survey cruise report, No.8, January 31-February 11, 1984. BGD/80/025/CR8, 71 p. Marine Fisheries Research, Management and Development Project, UNDP/FAO, Chittagong, Bangladesh.

White, T. F. 1985f. Marine fisheries resource survey: demersal trawling. Survey Cruise Report, No.9, February 17-24, 1985. BGD/80/025/CR9, 48 p. Marine Fisheries Research, Management and Development Project, UNDP/FAO, Chittagong, Bangladesh.

White, T. F. 1985g. Marine fisheries resource survey: demersal trawling. Survey Cruise Report, No.11, May 19-24, 1985 and July 12-17, 1985. BGD/80/025/CR11, 35 p. Marine Fisheries Research, Management and Development Project, UNDP/FAO, Chittagong, Bangladesh.

White, T.F. and M.G. Khan. 1985a. Marine fisheries resource survey: demersal trawling. Survey cruise report, No.1, September 15-25, 1984. BGD/80/025/CR1, 73 p. Marine Fisheries Research, Management and Development Project, UNDP/FAO, Chittagong, Bangladesh.

White, T.F. and M.G. Khan. 1985b. Marine fisheries resource survey: demersal trawling. Survey cruise report, No.2, October 3-13, 1984. BGD/80/025/CR2, 81 p. Marine Fisheries Research, Management and Development Project, UNDP/FAO, Chittagong, Bangladesh.

White, T.F. and M.G. Khan. 1985c. Marine fisheries resource survey: demersal trawling. Survey cruise report, No. 3, October 20-31, 1984. BGD/80/025/CR3, 62 p. Marine Fisheries Research, Management and Development Project, UNDP/FAO, Chittagong, Bangladesh.

Yeh, S.Y. 1981. Dynamics of the demersal fish resources in the Sunda Shelf area of the South China Sea., University of Washington. Ph.D. dissertation.

Yeh, S.Y., L.L. Loow and H.-C. Liu. 1981. Assessment on groundfish resources in the Sunda Shelf area of the South China Sea. Acta Oceanogr. Taiwan. 12: 175-189.

Appendix IV

Author Index

200

Geographic Index

202

Species Index